NATIONAL PARKS AND PROTECTED AREAS

National Parks and Protected Areas: Their Role in Environmental Protection

Edited by R. Gerald Wright
Professor, Department of Wildlife Resources, and
Unit Leader, National Biological Service,
Cooperative Park Studies Unit,
University of Idaho
Moscow, Idaho

Consulting Editor:
John Lemons
Department of Life Sciences
University of New England
Biddeford, Maine

Blackwell
Science

Blackwell Science

Editorial offices:
238 Main Street, Cambridge, Massachusetts 02142, USA
Osney Mead, Oxford OX2 0El, England
25 John Street, London WC1N 2BL, England
23 Ainslie Place, Edinburgh EH3 6AJ, Scotland
54 University Street, Carlton, Victoria 3053, Australia

Other Editorial offices:
Arnette Blackwell SA, 224, Boulevard Saint Germain, 75007 Paris, France
Blackwell Wissenschafts-Verlag GmbH Kurfürstendamm 57, 10707 Berlin, Germany
Zehetnergasse 6, A-1140 Vienna, Austria

Distributors:
USA
Blackwell Science, Inc.
238 Main Street
Cambridge, Massachusetts 02142
(Telephone orders: 800-215-1000 or 617-876-7000; Fax orders: 617-492-5263)
Canada
Copp Clark, Ltd.
2775 Matheson Blvd. East
Mississauga, Ontario
Canada, L4W 4P7
(Telephone orders: 800-263-4374 or 905-238-6074)
Australia
Blackwell Science Pty., Ltd.
54 University Street
Carlton, Victoria 3053
(Telephone orders: 03-9347-0300; fax orders 03-9349-3016)
Outside North America and Australia
Blackwell Science, Ltd.
c/o Marston Book Services, Ltd.
P.O. Box 269
Abingdon
Oxon OX14 4YN
England
(Telephone orders: 44-01235-465500; fax orders 44-01235-465555)

Acquisitions: Jane Humphreys
Development: Kathleen Broderick
Production: Irene Herlihy
Manufacturing: Lisa Flanagan
Typeset by BookMasters, Inc.
Printed and bound by Capital City Press

Printed in the United States of America
96 97 98 99 5 4 3 2 1

Library of Congress Cataloging-in-Publication Data

National parks and protected areas : their role in environmental protection / [edited by] R. Gerald Wright.
p. cm.
sqcludes bibliographical references and index.
ISBN 0-86542-496-9
1. National parks and reserves—United States—Management. 2. National parks and reserves Canada—Management. 3. Ecosystem management—United States 4. Ecosystem management—Canada
I. Wright, R. Gerald.
SB482.A4N3742 1996
333.78′0973—dc20 96-8572
CIP

1

The Origin and Purpose of National Parks and Protected Areas

R. Gerald Wright
David J. Mattson

As of this writing, there are more than 2700 national parks and protected areas in over 120 different countries, and the number increases yearly. These areas are as diverse as the physical settings and cultural patterns of the nations that have established them, yet they all have one thing in common—they serve as special places where people go for spiritual, cultural, and physical renewal. Parks reflect a nation's desire to preserve for generations unborn its floral, faunal, and landscape diversity, as well as elements of its national and cultural heritage. In a world of rapid environmental change, parks and protected areas ideally represent islands of stability—places where environmental changes are dictated by the rhythms of nature rather than by human demography and economic demands.

M. A. Badshah, a wildlife officer for the State of Madras, India, presented this vision of a national park in his introductory remarks at the First World Conference on National Parks in Seattle, Washington, in 1962:

> So far as I can visualize, the nearest approach to a paradise on earth—short of a fabled land with milk and honey flowing and angels and fairies in attendance—is a national park, a combination of some or all of the ingredients: forested hills and valleys, sparkling multicolored lakes, crystal clear streams, rippling brooks, placid rivers, silvery beaches, scented flowers, luscious fruits, sweet berries, snow-clad mountains, and wildlife. In the parks, coexistence,

tolerance, goodness, love and attention prevail, and both man and beast can go each his own way and wander freely.

Unfortunately, because of unchecked human population growth, social instability, and growing environmental degradation throughout the world, this vision is becoming ever more illusory. At the same time, governments are reacting to increasing societal demands to enhance the economic well-being of their citizens, often by fostering short-term resource consumption at the expense of long-term environmental protection. One outgrowth of these actions is that more and more countries are viewing landscapes set aside as national parks and protected areas as being economic and social luxuries they can ill-afford; i.e., parks and protected areas are perceived as a "lock-up" of lands that could better be devoted to uses serving immediate human needs.

Concerns such as these have led organizations like the International Union for the Conservation of Nature and Natural Resources (IUCN) and the World National Park Congress (McNeely & Miller, 1984) to recommend diverse management approaches for protected areas that seek to accommodate the economic and social requirements of society while still protecting the entire range of a nation's resources. These different management approaches range from strictly protected nature reserves, designed to maximize the protection of biological diversity, to multiple-use management areas that provide for sustainable human development (IUCN, 1984) (Table 1.1).

The first four of these management categories—category I: scientific reserve/strict nature reserve; category II: national park; category III: natural monument/natural landmark; and category IV: nature conservation reserve/managed nature reserve/wildlife sanctuary—are representative of the types of areas discussed in this book. All four categories possess some or all of the following characteristics:

1. Outstanding ecosystems, features, and/or species of flora and fauna of national scientific importance;
2. Fragile ecosystems or life forms, areas of important biological or geological diversity, or areas of particular importance to the conservation of genetic resources; and/or
3. Representative samples of major natural regions, features, or scenery, where plant and animal species, geomorphological sites, and habitats are of special scientific, educational, and recreational interest.

The one key concept that distinguishes areas in these four categories from other landscapes is that they are places where species assemblages and natural

Table 1.1 Categories of protected areas by the International Union for the Conservation of Nature and Natural Resources

I. Scientific reserve/strict nature reserve: These areas possess some outstanding ecosystems, features, and/or species of flora and fauna of national scientific importance; they often contain fragile ecosystems or areas of particular importance to the conservation of genetic resources.

II. National park: Relatively large areas that contain representative samples of major natural regions, features, or scenery, where plant and animal species, geomorphological sites, and habitats are of special scientific, educational, and recreational interest.

III. Natural monument/natural landmark: One or more specific natural features of outstanding national significance, which, because of uniqueness or rarity, should be protected; these features are not the size or of the diversity of features that would justify their inclusion as a national park.

IV. Nature conservation reserve/managed nature reserve/wildlife sanctuary: These are specific sites or habitats whose protection is essential to the continued well-being of resident fauna or migratory fauna of national or global significance.

V. Protected landscape or seascape: Natural or scenic areas found along coastlines and lake and river shores, sometimes adjacent to visitor-use areas or population centers, with potential to be developed for a variety of recreational uses.

VI. Resource reserve/interim conservation unit: These areas normally comprise an extensive and relatively isolated and uninhabited area having difficult access, or include regions that are lightly populated yet may be under considerable pressure for colonization or resource development.

VII. Natural biotic area/anthropological reserve: Natural areas where the influence or technology of modern humans has not significantly interfered with or been absorbed by the traditional ways of life of the inhabitants.

VIII. Multiple-use management area/managed resource area: Areas designed to provide for sustained production of or access to water, timber, fauna, pasture, marine products, and outdoor recreation.

IX. Biosphere reserve: Areas containing unique communities with unusual natural features, examples of harmonious landscapes resulting from traditional patterns of land use, and examples of modified or degraded ecosystems that are capable of being restored.

X. World heritage site: Areas that are of true international significance.

Source: From International Union for Conservation of Nature and Natural Resources. Categories, Objectives, and Criteria for Protected Areas. In JA McNeely, KR Miller (eds), *National Parks Conservation and Development.* Washington, DC: Smithsonian Institution Press, 1984:47–53. (Reprinted by permission of the publisher. Copyright 1984.)

processes, even those inimical to modern society, are allowed to exist, and where direct human interference with ecological processes is prohibited or restricted in scope.

Even so, humans must define in very pragmatic terms how protected areas should be managed, where they should be located, and how many there should be. Because human influences are prevalent, this process becomes a matter of determining not whether natural processes will be modified but to what extent, and not whether all species will be saved but how many and which ones. In part,

Preface

The desire to gain greater knowledge of the resources of national parks and equivalent areas—primarily in the United States—has been my professional goal for the past 25 years. I have been guided by the conviction and hope that greater scientific knowledge of parks would in turn lead to better and more enlightened management of their resources and a clearer understanding of the important role that such areas play in protecting biodiversity and ecological processes.

Throughout my career, I have been privileged to view firsthand the tremendous diversity of natural, cultural, and historic resources contained within national park systems in the United States and Canada and to participate in a wide variety of research, planning, and resource management actions, from Alaska to Maine to coastal Texas. Many insights have emerged.

I have gained a tremendous appreciation for the great talent and dedication of the scientists, historians, educators, and administrators who are working to solve the myriad problems confronting parks and protected areas. I am proud to call many of these people my friends, and I feel fortunate that some of these individuals agreed to write chapters in this book.

I have come to realize that parks cannot be viewed as isolated entities but must be understood and managed in the context of their surrounding ecological and cultural landscapes. This process requires not only the efforts of multidisciplinary professional teams but a better ecological understanding by the public and political leaders. Without the latter, the efforts of the former will likely be in vain.

Finally, I have learned that the parks themselves are tremendous reservoirs of knowledge about their resources. Most of this information is unappreciated and much is undocumented and thus limited in its accessibility. I am pleased to have been able to direct programs designed to locate and catalog such information. If, as so often has been posited, parks are to serve as baselines from which future environmental change will be monitored and evaluated, there needs to be a clear record of the ecological and cultural changes that have occurred within them. In many cases, I have found, such information exists if we dig deep enough.

It is my hope that the chapters that follow will provide the reader with some revelations into these and other aspects of parks and protected areas and will provide some insights into why such areas are indeed unique and deserving of continued protection. An ecologically enlightened public may in the long run be the only salvation of our society.

R.G.W.

Acknowledgments

First and foremost I thank the 28 authors who contributed to the individual chapters in this book. They have endured unreasonably short time frames for writing and editing, and the drastic rewrites, changes in content, length, orientation, and interpretation that I sometimes have made to the individual chapters. Their forbearance has been greatly appreciated, and their timely submissions and understanding of the demands of the publishing process speak volumes.

My colleague David Mattson deserves particular recognition. In addition to his own chapter, he assisted in editing selected chapters and, more importantly, provided encouragement and advice when my enthusiasm for the effort waned. Special thanks also go to Debra Lance, my Development Editor at Blackwell Science, for her diligence in keeping all individuals on schedule and her understanding when deadlines were not met. I also acknowledge the support and encouragement I have received through the Cooperative Research Unit program of the National Biological Service.

Contributing Authors

James K. Agee
Professor, Division of Ecosystem Science and Conservation, College of Forest Resources, University of Washington, Seattle, Washington

Stephen C. Bunting
Professor, Department of Range Resources, University of Idaho, Moscow, Idaho

John C. Carnes
Research Associate, Department of Fish and Wildlife Resources, College of Forestry, Wildlife, and Range Resources, University of Idaho, Moscow, Idaho

Joseph C. Dunstan
Landscape Architect, National Park Service, Pacific Northwest Regional Office, Seattle, Washington

William L. Halvorson
Unit Leader and Research Biologist, National Biological Service, Cooperative Park Studies Unit, University of Arizona, Tucson, Arizona

Larry D. Harris
Professor, Department of Wildlife Ecology and Conservation, University of Florida, Gainesville, Florida

Catherine Hawkins Hoffman
Chief, Division of Natural Resource Management, Olympic National Park, Port Angeles, Washington

Stephen Herrero
Professor, Faculty of Environmental Design, The University of Calgary, Calgary, Alberta, Canada

Thomas Hoctor
Program in Landscape Ecology, Department of Wildlife Ecology and Conservation, University of Florida, Gainesville, Florida

Katherine L. Jope
Chief, Division of Natural Resources, National Park Service, Pacific Northwest Regional Office, Seattle, Washington

Kirk Junker
Office of Chief Counsel, Department of Environmental Resources, Commonwealth of Pennsylvania Southwest Regional Office, Pittsburgh, Pennsylvania

Robert B. Keiter
James I. Farr Professor of Law, Director, Wallace Stegner Center for Land, Resources and the Environment, University of Utah, College of Law, Salt Lake City, Utah

Gary L. Larson
Research Scientist, National Biological Service, Department of Forest Resources, College of Forestry, Oregon State University, Corvallis, Oregon

John Lemons
Professor of Biology, Department of Life Sciences, University of New England, Biddeford, Maine

James G. MacCracken
Wildlife Biologist, Longview Fiber Company, Timber Division, Longview, Washington

Dave Maehr
Department of Wildlife Ecology and Conservation, University of Florida, Gainesville, Florida

David J. Mattson
Research Scientist, National Biological Service, Cooperative Park Studies Unit, University of Idaho, Moscow, Idaho

L. Scott Mills
Associate Professor, Wildlife Biology Program, School of Forestry, University of Montana, Missoula, Montana

Reed F. Noss
Director of the Wildlands Project, Editor Conservation Biology, *Faculty member Department of Fisheries and Wildlife, Oregon State University, Corvallis, Oregon*

Craig M. Pease
Professor, Department of Zoology, University of Texas at Austin, Austin, Texas

James M. Peek
Professor, Department of Fish and Wildlife Resources, College of Forestry, Wildlife, and Range Resources, University of Idaho, Moscow, Idaho

David L. Peterson
Research Scientist, Cooperative Park Studies Unit, University of Washington, Seattle, Washington

William F. Porter
Professor of Wildlife Ecology, State University of New York, College of Environmental Science and Forestry, Syracuse, New York

Jim Sanderson
Program in Landscape Ecology, Department of Wildlife Ecology and Conservation, University of Florida, Gainesville, Florida

J. Michael Scott
Unit Leader, Cooperative Fish and Wildlife Research Unit, University of Idaho, Moscow, Idaho

D. Scott Slocombe
Associate Professor, Department of Geology and Cold Regions Research Centre, Wilfred Laurier University, Waterloo, Ontario, Canada

Stephanie S. Toothman
Chief of Cultural Resources, National Park Service, Pacific Northwest Regional Office, Seattle, Washington

Brian D. Winter
Elwha Project Coordinator, Olympic National Park, Port Angeles, Washington

R. Gerald Wright
Professor, Department of Wildlife Resources, and Unit Leader, National Biological Service, Cooperative Park Studies Unit, University of Idaho, Moscow, Idaho

Introduction

The roles of protected areas need to be viewed in the context of evolving physical, social, and legal environments. This book originated with my concern that this message was being lost or at least not fully appreciated. I thus set out to develop a book that would attempt to explore and elucidate the roles that protected areas in general and national parks in particular play in protecting and understanding biodiversity and related ecosystem processes. The authors contacted to write individual chapters enthusiastically agreed with this premise. All of the authors have had a long association, in either a research, management, or policy role, with national parks in the United States and Canada and have deep feeling for and commitment to these institutions. For many authors, writing their chapters provided a welcomed and often long-sought opportunity to synthesize many years of data and research results and to reflect on the insights this information revealed. Each chapter is valuable in its own right, but I believe that together the whole is greater than the sum of the individual contributions. It will be up to the reader to judge how successful I have been in this endeavour.

Layout of the Book

Today, there is increasing awareness that parks and protected areas cannot be managed in isolation but instead must be viewed as integral parts of the structure and ecological processes of the landscapes in which they exist. This concept and its associated principles have been incorporated into the language of contemporary land stewardship by the term *ecosystem management.* I have used the concepts of ecosystem management as espoused by Grumbine (What is ecosystem management? *Cons Biol* 1994; 8:27–38.) and others to organize this book. The 21 chapters are arranged into five major sections. Each has a separate introduction. The first section outlines the reasons why ecosystem management can serve as a functional paradigm for parks and protected areas and presents the historical, political, ecological, and legal bases for this viewpoint.

The guiding principles of ecosystem management have rarely been employed in the planning and design of protected areas. The second section illustrates the role that these principles can play in protecting landscapes at various scales, ranging from a complete system of protected areas to an individual park. Providing for the continued existence of natural ecological processes in the landscape or the reestablishment of processes truncated by human alterations is a critical element of ecosystem management. In the third section, Chapters 12–16 provide background and understanding, with case examples, of the often contentious role natural processes play in park ecosystems.

Ecosystem management also recognizes that humans cannot be divorced from the system but, rather, are an integral part of it. The fourth section evaluates the role of humans in ecosystem in contemporary society, in their demands for the protection of culture and history, and as functional elements in the wilderness areas of North America.

Finally, it is recognized that ecologically sound management of protected areas requires constant surveillance and "monitoring" of existing conditions in order to track current conditions and identify changes and trends. Virtually all of the chapters discuss this need: however, the final two chapters relate it in a formal sense and thus clearly identify the real values of maintaining pristine protected areas in the world as we know it.

I

Protected Areas in Context

A major premise of this book is that the roles of national parks and protected areas must be examined in the context of their surrounding landscapes. These areas cannot function as isolated entities surrounded by, in most instances, increasingly fragmented landscapes. Instead, protected areas and the surrounding lands must be managed as an integrated whole. This concept, contemporarily referred to as *ecosystem management,* is examined in Chapters 2–5 from historical, political, ecological, and legal perspectives, respectively. It is recognized that even though the concept of ecosystem management has been proclaimed during the last decade as an important method to integrate parks and protected areas into their surrounding landscapes, putting the concept into practice has not been easy. In fact, there are few existing successful working applications of ecosystem management.

The discussions in the different chapters serve to point out some significant differences of opinions as to what is needed to achieve protection of biodiversity at the ecosystem level. The concept of *protected landscapes* as opposed to the more traditional sacrosanct parks advocated in Chapter 1, while workable in some societies, certainly is not universally acclaimed as the best method to protect our diminishing biodiversity.

There is no clear agreement, in either principal or practice, as to whether the concept of ecosystem management is appropriate for parks and protected areas. Although Chapter 3 suggests that it is appropriate, the author reminds us that ecosystem management means unique things to each participant or land manager involved in the process, and this can certainly influence how it is put into practice.

Ecosystem management likewise implies that management is concerned with both the structure and function of ecosystems. However, the traditional emphasis by the National Park Service and most other protected area land managers has typically focused only on the structure of the system, i.e., the elements of the biota and landscape features, while largely ignoring ecosystem processes. The implications of ecosystem management in terms of the ecological processes that link parks and adjacent landscapes are discussed in Chapter 4.

Finally, it is well recognized that park and protected area managers, for many reasons, have traditionally considered that their responsibilities end at the boundaries of their particular units. The legal and political constraints that underlie this reasoning, as well as the legal responsibilities managers have to view their areas in an ecosystem context, have important repercussions in the effort to protect biodiversity and are thoroughly reviewed in Chapter 5.

control. They were thus a long way from being "parks" in the sense the term is used in this book.

It is logical, considering the social climate of the times, that in both North America and Europe the earliest protected areas were products of the upper class and were managed exclusively for their use and enjoyment. The idea of large urban parks was virtually unknown in the United States prior to the mid-1800s. In fact, when Frederick Law Olmsted designed New York's Central Park in 1850, he was criticized for laying out the park on what was then the periphery of the city, beyond the reach of most residents. The average person of that era had little or no leisure time, limited education, and virtually no ability to travel. It took the development of the middle class and the advent of the automobile for this situation to change.

We do not mean to suggest that the idea of large natural area parks, open to all citizens, was completely foreign to the human psyche. One of the earliest suggestions for a great natural area park that would serve more than the needs of the social elite was made in the United States in 1832 by a young artist named George Catlin, while gazing at the wilderness expanse surrounding the confluence of the Yellowstone and Missouri Rivers in Montana. He wrote in his journal that this place ought to be set aside as a "nation's park, containing man and beast, in all the freshness of their nature's beauty" (Catlin, 1841). Catlin was at least 50 years ahead of his time; however, his dream for the area never materialized.

Yellowstone as Precursor to the National Park System and Protected Natural Areas

It is generally agreed that the modern concept of a national park being a large protected natural area originated with the establishment of Yellowstone National Park in 1872. Yellowstone, in turn, has served as a model for many of the parks subsequently established in the United States and throughout the world. It is thus informative to trace briefly the history of this park's establishment.

The first descriptions of the Yellowstone region of northwest Wyoming reached the settled portion of the United States in 1807 through the accounts of John Colter, probably the first English-speaking person to visit the area. His descriptions, reinforced and embellished by the tales of mountain man Jim Bridger, told of an area of such geologic wonder that few believed them to be true. The Yellowstone region remained little more than a curiosity for most of the busiest years of the westward movement and was largely avoided because of difficult access. Between 1804 and 1870 there were 110 scientific explorations west of the Mississippi River, but not until 1859 was one assigned to the Yellowstone region and it was halted by the outbreak of the Civil War (Zaslowsky & Watkins, 1994). Accurate and verifiable descriptions of the Yellowstone region thus awaited expeditions in 1870 by a group of explorers from the Montana Territory and in 1871 by the Geological and Geographical Survey of the Territories, under the direction of Ferdinand Hayden. The latter party included artist Thomas Moran and photographer William Henry Jackson (Haines, 1977).

The combined evidence, both written and visual, produced by these parties, describing the wonders of the region, was compelling and resulted in a rush of support for congressional legislation to set aside and "protect" the area. This support quickly culminated in the Yellowstone Act, passed on March 1, 1872, which created Yellowstone National Park as "a public park or pleasuring ground for the benefit and enjoyment of the people." However, beyond the relatively simple notion that the geologic wonders of Yellowstone be maintained in public ownership for all to experience (and not be overwhelmed with private entrepreneurs like Niagara Falls), the Yellowstone legislation provided little guidance as to what the area should be and how it should be managed.

In 1872, of course, there was no model for what a large natural area national park should be. Yellowstone was established at a time when it was believed that natural resources were inexhaustible and should be utilized for the maximum benefit to society. There was little public sentiment—and relatively little scientific evidence available—to support preservation of the seemingly unlimited biological resources found on the North American continent, particularly birds, mammals, and fish, as well as reptiles and amphibians. This was likely the main reason that concern over protection of the fauna was not raised in debates about the establishment of the park and the reason that the act itself is almost silent on the protection of biological resources (Wright, 1992). The Yellowstone Act prohibited only the wanton destruction of fish and game for profit. Hunting, trapping, and fishing were otherwise permitted for recreation or to supply food for visitors or park residents. This relative indifference was a sign of the times and did not anticipate the rapid depletion of game that was occurring nor the awareness of which would soon overtake the national consciousness.

Growth of the National Park System

For 18 years following the establishment of Yellowstone National Park, there was little interest in creating additional large protected natural areas in the

United States. This was a time of tremendous westward expansion, and efforts to settle and subdue rather than preserve the wilderness dominated the American agenda. The next new parks were not established until 1890 when Yosemite, Sequoia, and General Grant National Parks were created in California. During the next 25 years, the number of national parks in the United States grew slowly. Each proposal was met with varying degrees of opposition because lands that could be used for consumptive economic purposes were being withheld from development. Parks added to the system at this time were primarily the result of local support, often driven by only a few influential individuals with close ties to elected representatives, and generally included lands that most concluded were not useful for other purposes.

The passage of the Antiquities Act of 1906 produced another venue for the establishment of protected areas by allowing the president of the United States, through executive order, to preserve unilaterally the resources of a given geographic area. The primary intent of this law was, however, to preserve small sites containing important archaeological and scientific features. Through use of the Antiquities Act and congressional legislation, the number of park areas continued to increase, and by 1916 the Department of the Interior had overseen the creation of 14 national parks, 21 national monuments, and 2 reservations. Other monuments and military parks were administered by the Departments of Agriculture and War (Mackintosh, 1991).

Among the different park units, there was little uniformity in the management of resources. Some of the more remote areas received limited public and congressional support and therefore had little ability to undertake any level of resource management or protection. Other units were managed primarily for timber harvest and domestic livestock grazing. Virtually all areas suffered from poaching, market hunting, theft of resources, and vandalism—situations that, because of limited funds and personnel, national parks were ill-suited to remedy.

Frustrated with the inability of parks to control such problems, in 1886 Secretary of the Interior Lewis Lamar requested that army troops be stationed in larger parks like Yellowstone and Sequoia to protect the resources and administer the parks. The United States Army remained in control of Yellowstone until 1918. (See Hampton [1971] and Haines [1977] for a fascinating account of this era.) Suffice it to say, the military had the regimentation and personnel needed to bring order to the chaos that characterized many park administrations in the early twentieth century. In general, the military superintendents had a strong commitment to resource protection, and their actions set the groundwork for future resource management policies (Wright, 1992).

Creation of the National Park Service

The lack of uniform management policies for different park areas and the different needs of the various units were the primary motivators behind efforts to create a central bureau to manage all park units. These efforts came to fruition in 1916 when Congress passed the National Park Service Act, creating the U.S. National Park Service (NPS), and making the service responsible for the 35 national parks and monuments that were then under the control of the Department of the Interior.

The 1916 act was a milestone on the long road toward the creation of a system of protected areas in the United States; however, it was not without its faults. For example, it provided only vague guidance for the mission of the new agency and how it was to manage the existing parks. It stipulated that the role of the NPS was "to conserve the scenery and the natural and historic objects and wildlife therein and to provide for the enjoyment of the same in such manner and by such means as will leave them unimpaired for the enjoyment of future generations." Although apparently unappreciated by the legislators who passed this act, the contradictory mandate advocating both preservation and use of parks has haunted the NPS since its inception, and it has been a major source of debate, confusion, and frustration within and without the agency. Elements of this debate are described in Chapter 19.

The act also provided no guidance as to the kind of areas considered suitable for national park status. It is probably safe to assume that most individuals then in a position to influence park establishment did not view parks as a mechanism to protect areas containing unique biological resources. Rather, they likely viewed parks primarily as a means of preserving segments of the nation's most exquisite landforms. This reasoning is seen most clearly in a May 13, 1918, letter from Secretary of the Interior Franklin Lane to NPS Director Stephen Mather that stated:

> In studying new park projects, you should seek to find scenery of supreme and distinctive quality or some natural feature so extraordinary or unique as to be of national interest and importance. You should seek distinguished examples of typical forms of world architecture; such, for instance as the Grand Canyon, as exemplifying the highest accomplishment of stream erosion, and the high, rugged portion of Mount Desert Island as exemplifying the oldest rock forms in America and the luxuriance of deciduous forests. (quoted in Mackintosh, 1991)

Judging by the character of many of the parks added during this era (e.g., Mount Rainier, Mount Lassen, and Petrified Forest), this advice was clearly heeded.

The first two decades of the NPS were characterized by persistent efforts of the limited NPS staff to keep the system intact and to fend off continual efforts by special interests to eliminate or exploit certain areas. In retrospect, the survival of the NPS during the threatening financial and political climate of its early years is a testament to the political skills, strategic compromises, and adherence to principle displayed by its first director and assistant director, Stephen Mather and Horace Albright. It is not surprising, therefore, that in the heat of political and fiscal crises the NPS administration had little opportunity to reflect on the broader concept of what a national park "system" should be. Not until 1972 did the NPS undertake a formal effort to classify the natural regions of the United States and to identify those zones that were not represented by a unit of the national park system (NPS, 1972). This study showed that, at the time, approximately 50 percent of the defined physiographic provinces in the United States were not represented adequately in the national park system (Table 1.2). However, far from being a blueprint to guide new park establishment, the recommendations of this report were larely ignored, and the publication received limited acknowledgment from policymakers.

The lack of a system-wide plan for park establishment has meant from the beginning that new park establishment typically is driven not by a need to protect unique ecological resources but by factors such as opportunity, aesthetics, local economic impacts, surrounding land uses, and political support (Pressey, 1994; Wright et al., 1994) (see Chap. 7). The result of the present selection process is a park system with areas as diverse as Steamtown National Historic Site and Lowell National Historic Park, which celebrate elements of the industrial revolution; numerous Indian, Revolutionary, and Civil War battlefields; areas that preserve the culture of Native Americans of the Southwest, like Aztec Ruins and Bandelier National Monuments; areas managed primarily for intensive recreation, like Coulee Dam and Glen Canyon National Recreation Areas, which are monuments to the United States' ability to harness, manipulate, and alter nature; and, of course, the "crown jewels" of the system—the scenic large natural area parks, such as Great Smokies and Grand Canyon National Parks. Conversely, relatively few park areas—most notably Olympic, Everglades, and Denali National Parks—were selected primarily for their biological resources.

Some Lessons for Protected Area Management

The history of the U.S. National Park Service reveals a dialectic that has shaped the roles of and visions for protected areas worldwide. This dialectic is characterized by an initial loss of natural values that stimulates concern among citizens

Table 1.2 Adequacy of natural region representation in the U.S. National Park System

Natural region	Representation (%)
North Pacific Border	70
South Pacific Border	25
Cascade Range	95
Sierra Nevada	85
Columbia Plateau	15
Great Basin	10
Mohave-Sonoran Desert	85
Chihuahuan Desert-Mexican Highland	90
Colorado Plateau	75
Northern Rocky Mountains	85
Middle Rocky Mountains	85
Wyoming Basin	0
Southern Rocky Mountains	70
Great Plains	35
Central Lowlands	15
Superior Uplands	70
Interior Highlands	45
Appalachian Plateaus	0
Appalachian Ranges	55
Piedmont	0
New England-Adirondacks	30
Atlantic Coastal Plain	20
Gulf Coastal Plain	15
Florida Peninsula	65
Island of Hawaii	55
Maui Island Group	55
Oahu	0
Kauai, Niihau	0
Pacific Mountain System, Alaska	55
Interior and Western Alaska	0
Brooks Range	0
Arctic Lowland	0
Virgin Islands	100
Puerto Rico	0
Guam	0
Samoa	0
Mariana Islands	0
Caroline and Marshall Islands	0

Source: From National Park Service. *Part Two of the National Park System Plan: Natural History.* Washington, DC: U.S. Government Printing Office, 1972.

with power and influence—and in democratic societies concern among the electorate—and efforts toward protection. Huntable game populations usually have been the first valued natural resources to suffer from increased human populations, and such loss has led to the establishment of game reserves. Natural wonders follow suit, being marred primarily by commercialization and visual impairment, which has led to protection of natural areas with aesthetic and cultural values free from incompatible human activities. Most recently, humans are recognizing the loss of other intangibles and more obscure natural values such as wildness, biodiversity, and ecological services. This awareness is fomenting perhaps the most radical redefinition of protected areas and often is at the center of conflicts involving larger cultural change.

The roles of protected areas thus need to be viewed in the context of evolving physical, social, and legal environments. One of the more important features of this dynamic milieu is the often prolonged time lag between changes in the landscape and changes in management practice. Changes in the physical environment inevitably precipitate changes in society's norms, values, and expectations. These, in turn, induce changes in the political process by which protected areas are created and their management determined. Finally, these changes in the social and political context promote and allow for changes in management needed to salvage or preserve threatened natural values. Often, park managers and scientists are among those in the forefront, attempting to increase awareness of the public and policymakers. Even so, however, systemic time lags frequently and often critically impede society's response to accelerating environmental deterioration.

Park managers, policymakers, and scientists, preoccupied with ever-increasing numbers of visitors, a declining physical infrastructure, budgetary constraints, changes in the political agenda, and an array of adverse land uses bordering almost every park, often overlook or at least trivialize the role that parks areas can and should play in protection of the nation's biodiversity. This is unfortunate because, as stated earlier, parks—even more than designated wilderness areas—are the only areas in the United States with a legal mandate that permits species assemblages and natural processes to exist free from human interference (see Chap. 5).

Literature Cited

Badshah MA. Parks: Their Principles and Purposes. In Adams A (ed), *First World Conference on National Parks.* Washington, DC: Smithsonian Institution Press, 1962:24–33.

Catlin G. *Letters and Notes on the Manners, Customs, and Conditions of the North American Indians,* Vol. 1. New York, 1841.

Haines AL. *The Yellowstone Story.* Boulder, CO: Colorado Associated University Press, 1977.

Hampton HD. *How the U.S. Calvary Saved our National Parks.* Bloomington, IN: Indiana University Press, 1971.

International Union for Conservation of Nature and Natural Resources. Categories, Objectives and Criteria for Protected Areas. In McNeely JA, Miller KR (eds), *National Parks Conservation and Development.* Washington, DC: Smithsonian Institution Press, 1984:47–53.

Mackintosh B. *The National Parks: Shaping the System.* Washington, DC: National Park Service, 1991.

McNeely JA, Miller KR (eds). *National Parks, Conservation, and Development.* Washington, DC: Smithsonian Institution Press, 1984.

National Park Service. *Part Two of the National Park System Plan: Natural History.* Washington, DC: U.S. Government Printing Office, 1972.

Page TE (ed). *Plato: The Republic.* English translation by P. Shorey. Cambridge, MA: Harvard University Press, 1935.

Palmer TS. National reservations for the protection of wildlife. U.S. Department of Agriculture, Circular No. 87, 1912.

Pressey RL. Ad hoc reservations: Forward or backward steps in developing representative reserve systems. *Cons Biol* 1994;8:662–668.

Talbot LM. The Role of Protected Areas in the Implementation of the World Conservation Strategy. In McNeely JA, Miller KR (eds), *National Parks, Conservation, and Development.* Washington, DC: Smithsonian Institution Press, 1984:15–17.

Wright RG. *Wildlife Research and Management in the National Parks.* Urbana, IL: University of Illinois Press, 1992.

Wright RG, MacCracken JG, Hall J. An ecological evaluation of proposed new conservation areas in Idaho: Evaluating proposed Idaho National Parks. *Cons Biol* 1994;8:207–216.

Zaslowsky D, Watkins TH. *These American Lands.* Washington, DC: Island Press, 1994.

2

Changes in Landscape Values and Expectations: What Do We Want and How Do We Measure It?

William L. Halvorson

Conservation and Protection: Changing Attitudes Toward Natural Resources

The establishment and maintenance of a system of protected natural areas mark the culmination of a long history of changing attitudes toward natural resources. Human attitudes toward natural resources have evolved in response to changes in spiritual understanding and technological knowledge. For most of recorded history, the earth's resources were believed to exist solely for the well-being of humans. Depending on the technological capacity of the time, rampant resource exploitation was the norm, with minimal accounting of costs to others and what seemed minimal depletion of natural capital. In keeping with this viewpoint, landscapes were valued primarily in terms of their potential for human exploitation. Deserts, mountains, and swamps, for example, were considered to have little value because they could be exploited only with great difficulty.

According to Simmons (1993), the major chronological developments that occurred in Western society were:

1. The change from a hunting-gathering to an early agricultural society. This change was facilitated by knowledge of how to domesticate plants and animals.
2. The development of riverine civilizations. These were the great irrigation-based economies, such as those of the Nile and Mesopotamia. Lasting approximately from 6000 BC until 2000 BC, these civilizations used technology to store and move water and therefore freed themselves from some of the constraints of an arid environment. However, most of the land was still used by hunter-gatherers.
3. The creation of agricultural empires. From 1500 BC until the eve of the full industrial revolution (circa 1800) a number of city-dominated core areas developed throughout the world. These city-states were political as well as commercial empires. Many adopted technology to overcome environmental barriers to greater production; for example, water storage, terracing, draining of swamps, and selective breeding were practiced.

After 1850, there was movement toward development of commodities, such as increased cattle and sheep herds, and large-scale farming. Development of more and larger cities was possible because of better water storage, transfer, and treatment systems; improved waste handling capacities; and better transportation systems to bring goods to the city.

Landscapes came to be valued for the commodities they could produce, e.g., grazing, timber, minerals, and irrigation water. As transportation systems improved, more and more land became economically exploitable. Large cities continued to develop, resulting in further encroachment on undeveloped lands. Landscapes were affected dramatically by poor agricultural and grazing practices that reduced productivity and dramatically altered the biotic composition of entire ecosystems. Wilderness areas were reduced to fragmented remnants. In arid regions, lowered water tables caused loss of species and subsidence of the land.

Improvement in these harmful practices came slowly. As discussed in Chapter 1, there has been a growing recognition of the need for natural area protection throughout the twentieth century. However, many early supporters of conservation viewed land protection only from a "park" perspective; that is, they believed conservation could be achieved simply by setting aside parcels of land, thus theoretically saving them from exploitation. Leaders of the conservation movement soon recognized that conservation goals could be achieved only by viewing protected areas in the context of their surrounding landscapes. Core protected areas demonstrate this holistic view. Core protected areas are surrounded by a buffer zone in which sustainable human use is permitted; the human-managed landscapes of agriculture and silviculture and housing developments occur outside the buffer zone. In a pluralistic democratic society such as that of the United States, this type of conservation action requires that all interests in a landscape (farmers, ranchers, developers, businesses, government agencies) be involved in planning.

Humans have achieved mastery over most of the earth's surface. With this mastery has come an acute awareness of the fact that the land will not sustain humans unless humans sustain the land. Instead of an exploitive attitude, a more cooperative and appreciative approach toward others and toward the earth itself is necessary.

The diminishment of earth's natural resources, the absence of new frontiers to conquer, and increased awareness of the "wholeness" and interconnectedness of earth's systems has led to greater appreciation and understanding of "wilderness" and "nature" (Silver & DeFries, 1990). This new attitude incorporates a wide spectrum of emotional, spiritual, philosophical, and scientific values that are manifested in Thoreau's belief that "[i]n wilderness is the preservation of the world" (Shanley, 1971).

Thoreau's view resonates with Jackson's definition of *landscape:* a composition of human-made or human-modified spaces to serve as infrastructure or background for our collective existence (1984). In this framework, the entire earth becomes a series of landscapes (Dubos, 1980; Nisbet, 1991), and national parks, preserves, and wilderness areas large-scale counterparts to the city park—they are human-managed spaces serving our collective existence (Runte, 1984).

Today, it is clearly understood that protection of biodiversity requires a commitment to whole systems, and that within the landscape, context is just as important as content. Despite vast expenditures on charismatic megafauna, the landscape continues to be fragmented through destruction of natural areas. Small protected areas surrounded by development have much less ability to protect biodiversity than do larger ones or protected areas that are surrounded by less-developed lands. Protection of biodiversity must be a primary goal of conservation plans at both the landscape and regional level, thereby facilitating the protection of unfragmented communities over a broad spectrum of environments, such as elevation or other environmental gradients (Noss & Cooperrider, 1994).

Current Concepts of Biodiversity

Biodiversity is more than a list of species. Conserving or preserving biodiversity does not mean simply maintaining a given number of species on a particular plot of ground; rather, it is ensuring that each of these species continues to play its unique role in the ecosystem. Despite the numerous federal and state laws and regulations emphasizing the preservation of species, the real issue is not that we keep another plant or animal alive—the real issue is that we maintain an important thread in the web of life. When we rely on zoos and botanical gardens to maintain a species outside of its natural habitat, we maintain the genetics of only a few individuals. Lost is the unique role that the particular species plays within the ecosystem. The maintenance of ecologically significant biodiversity requires that we safeguard ecosystem health and integrity (Woodley et al., 1993).

Because of the way knowledge of our world has been organized, it often is difficult to appreciate the complex connections that exist in an ecosystem. Managing the transition from individuals to populations to species at the taxonomic level is difficult for many biologists. Making the transition from individuals to populations to communities to ecosystems, while at the same time incorporating the concepts of energy flow and nutrient cycling (see Jones & Lawton, 1995), is far more difficult. For example, it is difficult to understand how clear-cutting and conversion of native vegetation to cropland can be long-term detriments to the livelihood of a local population; more challenging to understand is how trees in the Amazon Forest affect the climate in Arizona. Understanding these difficult transitions is important when we discuss conservation, preservation, and the value of biodiversity. To achieve a change in attitude, from exploiter to caretaker, requires that we see and understand all of the energetic linkages occurring between the natural world and society.

In order to understand how natural area conservation can protect biodiversity, it is important to understand the levels at which biodiversity can be viewed. All are important, potentially and essentially, to conservation and protection of natural resources:

1. *Genetic diversity* has to do with genetic variability within a taxon. Species with little genetic diversity are at greater risk for extirpation because they have less ability to respond to changing conditions. More widespread species usually have greater genetic diversity.
2. *Species diversity* has to do with numbers of species per unit area. Species diversity is usually described in terms of communities or ecosystems. In general, areas with less stressful environments have greater species diversity; that is, there is a reduction in species diversity from tropical environments to polar or arid environments.
3. *Community diversity* has to do with the number of ecological communities within a given geographic area. Community diversity generally responds to topographic variability; regions that have high topographic

variation (e.g., Arizona and California) have higher community diversity than areas with little topographic variation (e.g., Kansas). This also holds true for smaller geographic areas like the Channel Islands off the coast of southern California.

4. *Trophic diversity*, the complexity of system organization, may be relevant to monitoring and managing ecosystems (Primack, 1993). It is important to know how many producers, primary consumers, and secondary consumers the system has when trying to understand and protect system function.
5. So-called *keystone* or *system-directing species* have disproportionately large effects on overall system diversity. If removed, these species would have dramatic effects on the system because a number of other species are substantially affected by their presence or absence (Orians & Kunin, 1990). Therefore, from an ecological perspective, it is more important to protect system-directing species than those that do not fill such an important role.

Even though there has been a considerable amount of research into the effects of adding or removing species from an ecosystem, ecologists still are unable to predict confidently the outcome of such events. Systems can be predicted to change, but it is difficult to identify what will change specifically and to what extent.

Spatial Scale

Any discussion of the role of parks and preserves in the protection of biodiversity necessarily involves issues of scale. Focusing a discussion on the role of parks means that the scale of the discussion has already been defined at a smaller ecological level. Scale can be viewed from many spatial perspectives (e.g., the earth, continent, ecoregion, watershed, or habitat type). The biodiversity of the earth could be preserved theoretically by saving examples of each species without preserving their respective habitats. However, few people would consider this acceptable. At the other extreme, one could propose that in order to protect the biodiversity of the Colorado Plateau or the Great Plains, the entire region should be preserved from human development. This, too, would be unacceptable to many. Somewhere between these extremes is a hard-to-identify, "acceptable" mix of preserved land, multiple-use land, and developed land.

Landscape Management: Changing Concepts and Expectations

Humans require places to live and land dedicated to the production of commodities. Human livelihood therefore needs to be balanced with conservation. Conservationists need to effectively communicate the fact that human life depends on natural systems. This is a difficult task, and, for the most part, current efforts have been ineffective. It seems that unless natural resources generate some economic benefit, many believe their protection not worthwhile. Developing effective solutions to this problem is likely the single biggest challenge facing the conservation community. The answer involves managing landscapes on a sustained basis and requires the participation of all interests. Tough decisions must be made concerning the harvest of fish, wildlife, and vegetation, and the number of people a given landscape can support for the long-term. A balance must be struck between private rights and public responsibilities.

Most conservationists have concluded, though the debate rages on (Grumbine, 1990; Brussard, 1991; Noss & Cooperrider, 1994), that managing the landscape for the greatest good of the greatest number of people requires preserved areas (e.g., national parks, wilderness areas, other reserves) and more ecologically sensitive management of the remaining landscape. Landscapes run the gamut, from vacant lots in inner cities to the most protected natural areas. Both extremes are ill-suited to serve the daily needs of the human population. Somewhere between these extremes, the human population can live in sustainable harmony (Leopold, 1949). This balance is found in the concept of "protected landscapes."

Protected Landscapes

The protected landscape concept reinforces the positive aspects of relationships between humans and nature and ameliorates negative influences that may damage or destroy the harmony between them. At the same time, the goal of protected landscapes is to provide opportunities for the public to visit and experience natural areas in ways that enhance the local economy but do not harm the landscape's natural, cultural, and social values (Lucas, 1992).

The protected landscape differs significantly from some of the other categories of protected areas recognized by the International Union for Conservation of Nature and Natural Resources (IUCN) (see Table 1.1), specifically because most of the land in protected landscapes is owned privately or communally or is occupied with resident populations.

It is the combination of an outstanding landscape and a resident population in harmonious interaction that distinguishes the protected landscape from the national park, which remains the best known of the categories of protected areas. National parks and protected landscapes are described by Phillips (1988) as follows: A *national park* comprises an extensive natural area that is protected from exploitation, protected from occupation, is the responsibility of the national government, and is owned publicly. A *protected landscape* comprises an extensive and outstanding seminatural area that is in productive use, inhabited, is primarily the responsibility of local government, and is, for the most part, owned privately.

The importance of the protected landscape in safeguarding the biodiversity and well-being of the human population was the focus of an international symposium, the results of which are detailed by Foster (1988). From the symposium came the Lake District Declaration, an affirmation of the value of protected landscapes. The declaration began with the following set of beliefs established by symposium participants:

1. People, in harmonious interactions with nature, have in many parts of the world fashioned landscapes of outstanding value, beauty, and interest.
2. These landscapes, although often much changed from their natural state, make their own special contribution to the conservation of nature and of biological diversity, for many of the ecosystems they contain have evolved and continue to survive because of human intervention. As large areas of undisturbed land become scarcer because of rapidly rising human populations and intensified land use, these landscapes will greatly increase in importance as repositories of biological richness. Moreover, they can serve as vital buffer zones around more strictly protected areas.
3. They preserve the evidence of human history in monuments, buildings, and the traces of past land-use practices. Their continuing use to provide living space and livelihood for indigenous populations allows traditional ways of life and traditional values to endure and to evolve in harmony with the environment.
4. They make an important contribution to the physical and mental health of people subject to the stresses of present-day life, and they offer beauty, pleasure, and recreation to many. They give inspiration to writers and artists. They provide young and old with opportunities to learn about their surroundings and comprehend the cultural diversity of the world.
5. These landscapes are living models of the sustainable use of the land and natural resources on which the future of this planet and its people depend. They demonstrate that it is possible to design durable systems of use that

provide economic livelihoods, are socially and spiritually satisfying, are in harmony with nature, are aesthetically pleasing and preserve the cultural identity of communities.

As a result of these beliefs, the symposium participants declared that:

1. The protection of such landscapes is vital both for their present value and for the contribution that they will make to spreading the philosophy and practices of sustainable development over much larger areas of the world.
2. There should therefore be universal recognition for this concept of landscape protection, much greater priority should be given to it, and there should be an active exchange of experience between nations.
3. These inhabited landscapes are in delicate and dynamic equilibrium; they cannot be allowed to stagnate or fossilize. But change must be guided so that it does not destroy but will indeed increase their inherent values. This means for each protected area a clear definition of objectives, to which land-use policies within it should conform as needed. It means also a style of management that is sensitive to ecological and social conditions. This will be possible by building on spiritual and emotional links to the land and by the operation of flexible systems of graded incentives and controls.
4. The protection of these landscapes depends on maintaining within them a vigorous economy and social structure, and a population that is sympathetic to the objectives of conservation.

The symposium concluded that:

1. Governments, international organizations, development agencies, and nongovernmental organizations should recognize the crucial role that such landscapes can play in sustainable development and in the conservation of the cultural and natural heritage of nations and should develop programs accordingly.
2. Governments should adopt the protection of these landscapes as a part of their public policies for the use of natural resources and provide sufficient funds to make this effective; and they should use these protected areas as models for the sustainable management of the wider countryside.
3. Governments and development agencies should direct funds destined for the support of agriculture or other economic objectives in these areas toward kinds of development that favor conservation.

4. National and international organizations should promote a worldwide exchange of information and experience on the management of such landscapes and should encourage and extend training in this field.

The protected landscape provides an alternative to outright preservation in a national park–like management unit in areas where the presence and impacts of resident populations and private ownership rule out such a designation; it also provides a vehicle for conservation in which the very harmony of people and nature creates an environment of quality and distinctiveness. The concept could be enlarged to encompass broader areas and to include management units set aside for preservation and other government-managed entities, such as state and national forests, game refuges, and state natural-area parks. These ideas parallel the concept of biosphere reserve.

Biosphere Reserves

The biosphere reserve concept was developed by the United Nations' Man in the Biosphere Program. Its aim was to protect natural biodiversity while protecting the economic well-being of local human populations. At its most basic level, each protected area (biosphere reserve) has a core area or core areas of maximum protection (wilderness or natural area), surrounded by a zone of minimal use, and an outer zone, or buffer, of moderate use (forestry, range, campgrounds, mineral and energy development). The entire reserve resides in a matrix of land used for agriculture or urban development. It is similar to the multiple-use module of Noss (1987).

Inventory and Monitoring: Changing Concepts and Expectations

National parks and wilderness areas are threatened by habitat fragmentation, invasions of alien species, development on their boundaries, and illegal use or collection of resources. In the past, beliefs such as those about fire, predators, and the static nature of a "climax" ecosystem sufficed to direct conservation strategies. Today, managers need information about park ecosystems and threats to their resources that is scientifically reliable. Without reliable knowledge, managers cannot protect and maintain resources effectively or restore them when protection fails (Halvorson & Davis, 1996).

Three factors have helped bring about changes in concepts and expectations about natural areas. First, monitoring and research have produced long-term views of ecosystems that yield dramatically different understandings than short-term studies (Parsons & van Wagtendonk, 1996; Wright, 1996). Management of natural areas needs to be viewed as an experimental, iterative process that is carried out as a cooperative effort and is monitored. The knowledge gained from sustained ecosystem-level research may complicate management options, but it also allows for treatment of causes—not just symptoms—such as trying to repair system dysfunction one species at a time. Differences among short-term managerial views, often based on beliefs in static, isolated landforms, and long-term scientific knowledge of dynamic, interconnected ecosystems are frequently in conflict. Nevertheless, the understanding that comes from sustained monitoring and research provides hope for undertaking such daunting tasks as ecological restoration and sustainable use, and it helps build public consensus. Monitoring and long-term research have shown that natural areas are dynamic, not static, and need to be managed accordingly.

Second, no park is an island (Shaw, 1996; Stone & Loope, 1996). Transboundary forces influence natural-area ecosystems, and they must be identified and addressed to protect natural resources adequately. The myth of isolated natural areas and wilderness, separate and apart from the rest of the world, has been deposed by research, and the message is quickly reaching the public at large.

Third, and finally, experience shows that scientific processes help to better balance resource protection and visitor-use dilemmas than belief-based decisions. Monitoring programs show how scientifically derived knowledge resolves issues such as how much use is possible without impairment, and how large parks need to be to assure protection of system function and avoid losses from habitat fragmentation.

Restoration

Restoration is an increasingly important part of ecosystem and natural-area management, partly because conservation efforts have brought degraded lands into protected status and partly because already "protected" natural areas have been invaded by nonnative species. In order to maintain natural systems, most land management agencies must now develop strategies to restore naturalness.

Landscape Management Programs

Ideally, landscape management must start with an inventory and a map. These tools help to identify areas of high or significant biodiversity and facilitate association of these sites with land ownership and development patterns. In many areas of the United States, new partnerships that allow inventory data commonly scattered among many different offices to be centralized at one site are being created between the government and private groups. For example, in Arizona, state agencies are joining with a number of federal agencies to develop one database from information of the state heritage program, The Nature Conservancy, the Gap Analysis Program (see Chap. 7), and other programs or agencies that wish to participate. This kind of program emphasizes provision of the best scientific information possible to aid in policy development and implementation.

The next step in landscape management brings together as much geographic information as possible into one database so that realistic maps can be produced. Based on mapped information for a reasonably sized area, land-use decisions may then be made that incorporate the full range of available information.

Resource Management with Knowledge: What Does It Take?

Although some management units have organized inventories of macroflora and macrofauna, managers still lack complete, exhaustive knowledge of all organisms found on any reasonably sized plot of ground anywhere in the world. That is to say, all the elements that compose any ecosystem are yet to be known, much less the health of the system's populations or the system itself. Thus, managers are forced to manage simply the major species in a natural area and frequently discover the importance of minor components to the system as a whole only when problems arise.

As stated above, natural area management needs to be an iterative process. This process needs to involve inventory, monitoring, research, and management action on a continuing basis (Fig. 2.1). By invoking management actions and monitoring the effects of these actions, understanding of the system and its dynamics is increased. As understanding increases, we hone our abilities to adjust management strategies and manage more effectively. Research is needed to fine-tune understanding. Generally, monitoring does not provide information on cause and effect; rather it gives information on trends and changes. It often is necessary to supplement monitoring with research into the "why" of the change (Davis and Halvorson, 1988). Also, periodic inventories need to be conducted because changes take place continually. For example, a seabird inventory conducted at Channel Islands National Park in 1992 found that tufted puffins were breeding in the park for the first time in decades. This was a new finding despite the fact that the park had a seabird monitoring program in place since

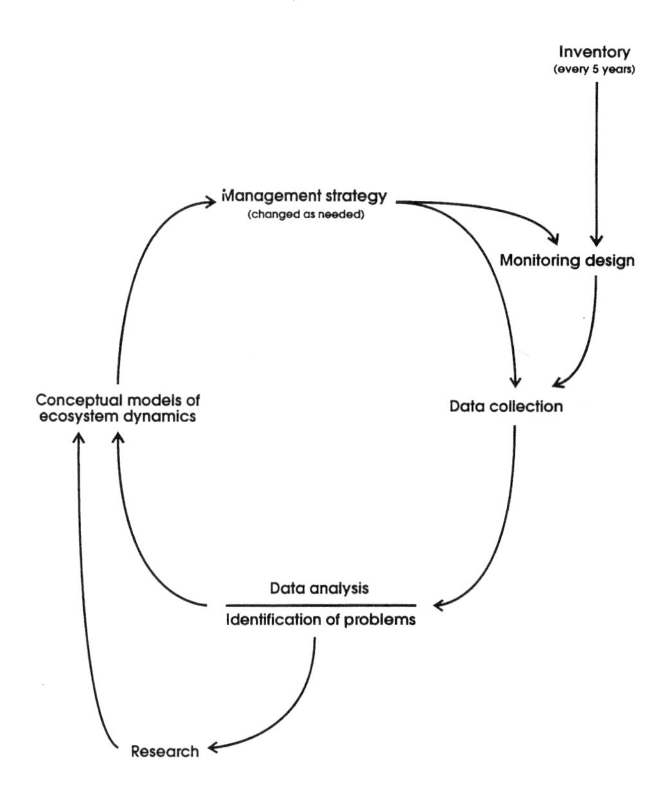

Figure 2.1 A model of knowledge-based iterative resource management.

1985. When considering natural biodiversity, these periodic surveys are important to understand species distributions and movements.

Systematic Protocols and Criteria

Although the need for systematic and comprehensive monitoring in natural areas has existed for decades, there are very few examples of such efforts. There is more experience with monitoring individual species and individual system characteristics. Administrators of natural areas in the United States usually have better information on how park visitation has changed during the past five years than they do on changes in plant or animal species numbers.

No "cookbook" has been written on how to monitor ecosystems. Any manager who begins to develop a program must understand that each area is unique and needs to have a program designed individually. There are now available many resources that can aid and assist the development of such a program (Davis & Halvorson, 1988; Davis, 1989; Davis et al., 1994; Noss & Cooperrider, 1994). The most important step to ensure that a monitoring program will meet the needs of management is careful program design. For a long-term ecological monitoring program to provide useful information, managers and scientists together must make a commitment to long-term continuity and select ecological components, locations, and methods for monitoring. Monitoring that is subject to inconsistent or inadequate funding leads to inconsistent data collection, analysis, and reporting. This kind of program usually results in a few bits of poorly stored, never-used data. Lack of administrative commitment, in these cases, creates an inadequate program.

Once a commitment is made and adequate and consistent funding is appropriated, the science part of a credible monitoring program needs to be ensured. Ecological characteristics to be monitored should be selected through a scoping process, with knowledgeable experts from a variety of agencies and academic institutions participating. Being too narrowly focused at this stage can lead to programs that ask the wrong questions or leave out important aspects that should have been considered.

Because it is impossible to monitor everything, the initial selection process should be based on a set of selection criteria with which all participants agree. Once the specific ecological components are selected, procedures must be established and institutionalized to allow regular, reliable sampling, data analysis, and reporting over the long term.

An ecological monitoring program for any natural area should include the following steps:

1. Scoping, i.e., determine what to monitor and what questions the monitoring program will answer; at this stage, it also is a good idea to determine some of the questions that monitoring will *not* answer. Selection criteria might include:
   a. Taxa from a broad array of ecological roles.
   b. Taxa with special legal status.
   c. Endemic taxa.
   d. Alien taxa, especially those that are most invasive.
   e. Keystone taxa, those that dominate or control.
   f. Taxa that are exceptionally common.
   g. Heroic taxa, those that have general public support.
   h. Driving functions: e.g., rainfall, soil moisture, temperature.
2. Determine how and when to monitor through research.
3. Establish data management procedures.
4. Establish reporting procedures.
5. Document the monitoring protocols; these protocols should be permanently established in written documents.

a. How, when, and where to monitor.
b. How to manage the data, including data analysis.
c. How to summarize, synthesize, and report the data and results.

It is important in any monitoring program to institutionalize the data collection, data management, and information reporting. Too often program administrators are content with simply collecting data for years, hoping that at some point in the future there will be time and money available to analyze and synthesize the data and complete a report on the status of resources. This thinking is faulty. The belief that you are not done monitoring until the paperwork is done needs to be infused into everyone associated with an effective program. Nothing actually happens until a report on status and trends has been completed.

Conclusion

On examining the role of parks and preserves in the protection of natural biodiversity, we recognize major changes occurring in the basis for management and the value and expectations of landscapes. We are moving quickly from management by belief-based directives of the few to management by scientific understanding and broad consensus. With this change, the notion that natural areas can be protected simply by putting a fence around them is giving way to the realization that natural areas can be protected only in the context of a protected landscape. That is, the core natural area must be surrounded by landscape units that have minimal use and buffer the natural area from damage due to inappropriate or illegal uses.

Management of landscapes that will provide for the long-term well-being of both natural populations and human populations requires the cooperative efforts of all who live in and manage that landscape. It also requires that decisions be based on scientific knowledge about the health and dynamics of that landscape, including all of the existing systems, both natural and human-made.

In order to arrive at this scientific knowledge, extensive ecological monitoring programs must be developed so that information on changes to the systems will be available. Research is necessary when there is a need to know why changes are happening. The management of the information that is being generated for landscape-scale areas requires a jump in the size and complexity of computer systems to handle the data. In the coming years, as the demand for data increases, geographic information systems and computer networks will become standard equipment.

The continual increase in the earth's population is one reason we must manage natural areas on a landscape scale. We must learn to live within our means, actively conserving available resources on a long-term basis. Failure to develop protected landscapes that safeguard natural biodiversity and provide for conservation of natural resources, while at the same time encouraging harmony between humans and nature, may allow development and resource exploitation to overrun all natural areas. Without careful planning and stewardship, it is possible that our landscapes will become barren eroded lands. If such were to occur, there would be few left to consider the state of biodiversity.

Literature Cited

Brussard PF. The role of ecology in biological conservation. *Ecol Appl* 1991;1:6–12.

Davis GE. Design of a long-term ecological monitoring program for Channel Islands National Park, California. *Nat Areas J* 1989;9:80–89.

Davis GE, Halvorson WL. *Inventory and Monitoring of Natural Resources in Channel Islands National Park.* Ventura, CA: Channel Islands National Park, 1988.

Davis GE, Faulkner KR, Halvorson WL. Ecological Monitoring in Channel Islands National Park, California. In Halvorson WL, Maender GJ (eds), *The Fourth California Islands Symposium: Update on the Status of Resources.* Santa Barbara, CA: Santa Barbara Museum of Natural History, 1994;465–482.

Davis GE, Halvorson WL. Channel Islands assess ecosystem health: 1987 highlights of natural resources management. *Natural Resources Report.* Washington, DC: National Park Service, 1988;20–23.

Dubos R. *The Wooing of Earth.* New York: Charles Scribner's Sons, 1980.

Foster J (ed). Protected Landscapes: Summary Proceedings of an International Symposium, Lake District, United Kingdom. Gland, Switzerland: International Union for Conservation of Nature and Natural Resources, 1988.

Grumbine RE. Viable populations, reserve size, and federal lands management: A critique. *Cons Biol* 1990;4:127–134.

Halvorson WL, Davis GE (eds). *Ecosystem Management in the National Parks.* Tucson, AZ: University of Arizona Press, 1996.

Jackson JB. *Discovering the Vernacular Landscape.* New Haven, CT: Yale University Press, 1984.

Jones CG, Lawton JH. *Linking Species and Ecosystems.* New York: Chapman and Hall, 1995.

Leopold A. *A Sand County Almanac.* London: Oxford University Press, 1949.

Lucas PHC (ed). *Protected Landscapes: A Guide for Policymakers and Planners.* New York: Chapman and Hall, 1992.

Nisbet EG. *Leaving Eden: To Protect and Manage the Earth.* New York: Cambridge University Press, 1991.

Noss RF. Protecting natural areas in fragmented landscapes. *Nat Areas J* 1987;7:2–13.

Noss RF, Cooperrider AY. *Saving Nature's Legacy: Protecting and Restoring Biodiversity.* Washington, DC: Island Press, 1994.

Orians GH, Kunin WE. Ecological Uniqueness and Loss of Species. In Orians GH, Brown GM, Kunin WE, Swierzbinski JE (eds), *The Preservation and Valuation of Biological Resources.* Seattle: University of Washington Press, 1990;146–184.

Parsons DJ, van Wagtendonk JW. Fire Research and Management in the Sierra Nevada National Parks. In Halvorson WL, Davis GE (eds), *Ecosystem Management in the National Parks.* Tucson, AZ: University of Arizona Press, 1996:25–47.

Phillips A. Landscape conservation: British experience and world conservation needs. *Papers for the British Association for the Advancement of Science,* London, 1988.

Primack RB. *Essentials of Conservation Biology.* Sunderland, MA: Sinauer Associates, 1993.

Runte A. The Origins of the Park Idea in the United States. In Stilgoe JR, Nash R, Runte A (eds), *Indiana Historical Society Lectures, 1983: Perceptions of the Landscape and Its Preservation.* Indianapolis, IN: Indiana Historical Society, 1984:53–74.

Shanley HP (ed). *Walden.* By HD Thoreau. Princeton: Princeton University Press, 1971.

Shaw WW. Urban Encroachment at Saguaro National Monument. In Halvorson WL, Davis GE (eds), *Ecosystem Management in the National Parks.* Tucson, AZ: University of Arizona Press, 1996:184–200.

Silver CS, DeFries RS. *One Earth, One Future: Our Changing Global Environment.* Washington, DC: National Academy Press, 1990.

Simmons IG. *Environmental History: A Concise History.* Cambridge, MA: Blackwell Scientific Publishers, 1993.

Stone CP, Loope LL. Alien Species in Hawaiian National Parks. In Halvorson WL, Davis GE (eds), *Ecosystem Management in the National Parks.* Tucson, AZ: University of Arizona Press, 1996:132–158.

Woodley S, Kay J, Francis G. *Ecological Integrity and the Management of Ecosystems.* Delray Beach, FL: St. Lucie Press, 1993.

Wright RG. Moose and Wolf Populations in Isle Royal National Park. In Halvorson WL, Davis GE (eds), *Ecosystem Management in the National Parks.* Tucson, AZ: University of Arizona Press, 1996:74–95.

3

Ecosystem Management: An Appropriate Concept for Parks?

James K. Agee

Ecosystem management is a concept that managers of nature preserves, national forests, and industrial forests have adopted over the past decade as a paradigm for sustainability. These representatives of seemingly disparate interests have been able to adopt a common approach because each defines ecosystem management in a different way. There is no unique definition of ecosystem management. A semantic battlefield has resulted, with each group espousing adherence to a very broad and fuzzily-defined concept. The multiple definitions of ecosystem management mirror similar semantic battles over concepts such as carrying capacity, stability, and diversity. Some of the confusion over carrying capacity for wildlife was resolved with more specific definition over time; that is, economic carrying capacity was differentiated from ecological carrying capacity (Caughley, 1979). Ecosystem stability was a vaguely defined notion related to resistance to or resilience from disturbance. It was embraced as a good thing for ecosystems to have and associated with late-successional systems (Odum, 1969), an attribute that, as a universal truth, was later rejected (Christensen, 1988). Ecosystem management, as a relatively new concept, will inevitably undergo refinement and redefinition, but for now the term remains somewhat vague. However, there is value in vagueness, as it may provide an umbrella wide enough for representatives of many interests to sit together and discuss resource issues that encompass broad areas of landscapes that have multiple, often overlapping objectives. Competing interests may be drawn to a table large enough to accommodate, at least initially, the interests of all, with inherent flexibility in defining the goals and processes to be adopted by the group.

A Goal- or Process-Oriented Approach?

The concept of ecosystem management can be traced back to the 1930s (Grumbine, 1994). The Ecological Society of America and the National Park Service independently proposed that protection of biodiversity would require core areas ("sanctuaries" or national parks) and attention to surrounding areas. However, the concept lay largely dormant for nearly five decades. A review of literature on ecosystem management (Grumbine, 1994) showed that 29 of the 33 most detailed analyses were published or in press from 1987 to 1994. This review, appropriately titled "What is ecosystem management?," lists ten dominant themes of ecosystem management (Table 3.1). Only two of the papers reviewed included consideration of all ten themes and subthemes, although the majority addressed most of the themes.

The first book-length treatment of ecosystem management by Agee and Johnson (1988) defines it this way: "Ecosystem management involves regulating internal ecosystem structure and function, plus inputs and outputs, to achieve socially desirable conditions. It includes, within a chosen and not always static geographic setting, the usual array of planning and management activities but conceptualized in a systems framework . . ." This definition is inherently value neutral and avoids a goal-oriented focus. The review by Grumbine (1994) attempts an updated definition: "Ecosystem management integrates scientific knowledge of ecological relationships within a complex sociopolitical and values framework toward the general goal of protecting native ecosystem integrity over the long term."

The change that has occurred in the definition of ecosystem management in less than a decade reflects an evolution toward a more preservation-oriented focus. Agee and Johnson's definition is broad and process-oriented, although throughout their book it is applied to the specific case of ecosystem management for parks and wilderness. Grumbine's definition more clearly articulates the goal of "protecting native ecosystem integrity." In both cases, the concept is clearly applicable to national parks and protected areas. Within an ecosystem management process, the challenge is to maintain active involvement of others who may have goals other than the integrity of native ecosystems.

Goal-oriented definitions of ecosystem management tend to polarize the populations of interest groups that should be interacting. For example, a private landowner residing next to a nature reserve may have little problem with a

Table 3.1 Dominant themes of ecosystem management

Theme	Description
1. Hierarchical context	Use a "systems" perspective, and avoid focusing on only one level of organization (e.g., genus, species, landscapes, etc.); seek connections between levels
2. Ecological boundaries	Work across administrative boundaries; use ecological boundaries
3. Ecological integrity	Protect native diversity: viable populations, species reintroductions, and ecosystem patterns and processes
4. Data collection	Increase research and inventory databases; better manage existing data
5. Monitoring	Track actions for results; monitoring provides a feedback loop to management
6. Adaptive management	Assume scientific knowledge is provisional and focus on management as an experiment
7. Interagency cooperation	Using ecological boundaries requires cooperation between adjacent landowners and managers
8. Organizational change	Implementing ecosystem management may require minor to major organizational change
9. Humans embedded in nature	Humans are part of the ecosystem, influencing ecological processes and being affected by them
10. Values	Values play an important role in defining ecosystem management goals

Source: Adapted from RE Grumbine. What is ecosystem management? *Cons Biol* 1994;8:27–38.

park-oriented goal as long as the landowner's goals are also recognized at the beginning of the process. If protecting native ecosystem integrity is the only stated goal within the definition of ecosystem management, then it may be more difficult to bring all parties into the discussion. A process-oriented definition of ecosystem management avoids the initial polarization but implies that the goals must be part of the initial process. For parks and protected areas, Grumbine's (1994) goals of ecosystem management appear quite appropriate:

1. Maintain viable populations of all native species in situ.
2. Represent, within protected areas, all native ecosystem types across their natural range of variation.
3. Maintain evolutionary and ecological processes (i.e., disturbance regimes, hydrological cycles, nutrient cycles).
4. Manage over periods of time long enough to maintain the evolutionary potential of species and ecosystems.
5. Accommodate human use and occupancy within these constraints.

With a process-oriented approach, a park or reserve manager would bring these goals to the table, where other goals—some complementary and some competing—would also be presented. Obviously, it would be a rare event for all goals to be complementary. Competing goals have to be either spatially separated or dealt with through other mechanisms (usually political or legal). Frequently, the competing goals deal with commodity-versus-preservation issues, and ecosystem management is not a panacea for their resolution. It can, however, define appropriate ecological, spatial, and temporal components that will improve the decision-making process.

Ecosystem Management as a Process

Characteristics of Natural Systems

The first step in ecosystem management is to define the characteristics of ecosystems. The simplest definition of an ecosystem is any part of the universe chosen as an area of interest, with the line around that partial universe the ecosystem boundary, and anything crossing that boundary being input or output (Agee & Johnson, 1988). Several elements of that definition have significant implications for ecosystem management.

First, ecosystems are spatially variable: problems in one region are unlikely to have the same spatial boundaries as in another. Elk (*Cervus elaphus*) ranges in the vicinity of reserves, for example, typically overlap reserve boundaries, with the summer range often in the reserve and the winter range at least partially outside. This problem will not be the same in Yellowstone National Park as it is in Olympic or Redwood National Park. The migration patterns may range from large to nonexistent, and all to very little of the winter range may be on unprotected lands. Many animal migration patterns will be site specific, so that local knowledge will be needed to define the scope of the issue. Similarly, the meaning and function of an old-growth Douglas fir (*Pseudotsuga menziesii*) forest is quite different on the Olympic Peninsula of Washington State, with a natural fire-return interval of 500 years, than it is in the Kalmiopsis Wilderness of southern Oregon, where the natural fire-return interval is 30 to 50 years. The disturbance regime responsible for seemingly similar forest structures created these structures in ways that are quite different, and the challenges to restoring these regimes are also quite different.

Second, ecosystems are temporally variable, such that a given problem may have one set of boundaries today and yet another in a decade or two. We have a growing awareness that many plant species have migrated across the continent over the past millennia, and that what we see on the landscape today may not represent highly coevolved plant communities (Brubaker, 1988; Davis, 1976). Ecosystems can change rapidly, too. The forests of Yellowstone National Park, which burned in the massive, intense fires of 1988, are quite different today than they were in 1987 and provide a different wildlife habitat that favors some species over others. Our planning time frames usually are so short that we do not incorporate longer-term change into management plans. Unless such change is very abrupt, such as the Yellowstone fires, we may not even be aware that it is happening. We may see change, such as recent declines in amphibian populations (Barinaga, 1990; Blaustein et al., 1994), but we may not be able to differentiate between natural fluctuations and those due to some level of environmental degradation.

Third, administrative boundaries are usually filters rather than barriers. Most ownership boundaries are permeable membranes, and across these boundaries—political lines on a map—pass energy, organisms, people, and values: essentially much of the substance of resources management problems and solutions. Seasonal wildlife migrations, the movements of park visitors or wildlife poachers, and the invasions of alien species such as knapweeds (*Centaurea* spp.) or feral animals are all examples of the permeable-membrane concept.

When different ecosystem components, such as the distribution of plant communities or wildlife ranges, are spatially defined, a complex set of overlapping regions emerge, and if we follow them over time, we find they weave, merge, and diverge. It is also clear that politically defined boundaries frequently do not contain all the ingredients (whether resources or people) necessary to resolve resources management issues. Political boundaries are clearly the most controllable boundaries under current local, state, and federal law, but it is equally clear that ecosystem management problems are neither definable nor solvable within these boundaries.

Implications of the Ecosystem Concept

One of the most important implications of the ecosystem concept is that ecosystems are not static. The physical, biological, and social components of these systems are continually changing, sometimes in a cyclic manner and sometimes chaotically. This dynamic variability will affect composition, structure, and function of the system. The result may occasionally be a cyclic stability, as evidenced by the frequent, low intensity natural fires in mixed conifer forests of the Sierra Nevadas of California (van Wagtendonk, 1985). Yet, fire has not always acted as an agent of ecosystem stability (Agee, 1993). In the Boundary Waters Canoe Area of Minnesota, no spatial scale exhibited a temporally stable patch mosaic, because the fire regime itself was variable over time (Baker, 1989). Alien species may alter successional dynamics to such an extent that the patterns we may have seen in the past will not be seen in the future. Social components of the system are changing, too. While population pressures have mounted in the twentieth century, the structure of the population has not been constant, and "waves" such as the post–World War II "baby boom" will affect park, wilderness, and preserve visitation, and demand for access well into the twenty-first century.

There may be substantial spatial heterogeneity between actions and impacts. Nature reserve management is often centered on the "core" preserve, with the effect of outside activities generally assumed to be diluted with distance from the preserve. This may not always be the case. Air pollution effects and their ecological consequences may be more significant many miles from the source than nearer the source. Plant damage can occur in nature reserves from activities that might seem to be too far away to be significant. The Sierra Nevada national parks of California experience ozone damage from the San Francisco Bay Area more than 200 air kilometers away, and Mount Rainier National Park in Washington receives ozone from the upwind Seattle-Tacoma area, which has much lower levels near the source (Basabe, 1992). These "soups" of hydrocarbons "simmer" photochemically as they move downwind, and ozone increases with distance from the source, reaching levels injurious to plants many kilometers from the source. Nature reserves in the Pacific Northwest have become core areas for preservation of the northern spotted owl (*Strix occidentalis* var. *caurina*) through a process known as Federal Ecosystem Management Assessment Team (FEMAT) and its chosen Option 9, known as President Clinton's Forest Plan (Thomas, 1994). Due to harvesting of much of the old growth timber in the Pacific Northwest, preserve areas with high proportions of old growth will serve as old growth reserves, constraining any plans for allowing natural fires to burn. Timber harvesting well beyond the borders of nature preserves sets this plan into motion.

Ecosystems may exhibit several levels of stable behavior. When goats were removed from Hawaiian parks, alien trees and grasses increased, while native trees were unable to compete in the previously grazed areas (Mueller-Dombois & Spatz, 1975). Succession theory, particularly in grasslands, has previously assumed a linear response, with community replacement considered a deterministic event in terms of changes with grazing and recovery after overgrazing (Joyce, 1993). Much of the sage steppe of eastern Washington and Oregon and southern Idaho was heavily grazed in the late nineteenth century, and cheatgrass (*Bromus tectorum*) has invaded these lands. The native perennial grasses are unable to compete against this aggressive annual. Fire has been removed from these systems as well, allowing native western juniper (*Juniperus occiden-*

talis) to expand and effectively compete against the grasses (Young & Evans, 1981; Harris, 1991). These can be relatively stable ecosystem states, and there is no known successional pathway back to the "natural" state.

Although ecosystems are so complex that we cannot know everything about them, management can and must proceed with the limited knowledge at hand. Integrating knowledge gained from other ecosystems can be an efficient way to amass a critical level of information, if this information is interpreted with sufficient caution; it requires testing and adaptation in the system of concern (Gordon, 1993). Certain links in the system have to be ignored if not currently seen to be related to the management objective. While there is an organized connection between parts, everything is not necessarily connected to everything else (Holling, 1978).

Characteristics of Successful Ecosystem Management

Successful ecosystem management results in a mutually acceptable agreement or decision reached by affected interests through a negotiation process that reconciles and integrates the legitimate interests of all parties (Washington Sea Grant Program, 1989). For nature reserves, given a goal orientation, this means successfully protecting native ecosystem integrity. If consensus does not occur, parties to the negotiation are free to fall back on other alternatives (legislative, judicial), but the intent of successful ecosystem management is to avoid these avenues by building consensus and establishing strong foundations for continued cooperation.

The first objective is to have the affected parties sit down together. Diverse agencies and neighbors have differing mandates and goals and usually are encouraged to cooperate when they are dissatisfied with the way past decisions have been made. To some extent, the diverse definitions of ecosystem management encourage such initial meetings, as they allow philosophical room for different parties with disparate hopes to communicate effectively.

In ecosystem management, success is measured by results, not by the amount of coordination occurring during the process. Two principles flow from this notion. The first is that time is important but progress cannot be rushed; this seems to be in conflict with the need for results, but really it is not. One of the major barriers to progress is lack of trust between parties. It takes some time to develop trust between people representing diverse views, and emphasis on quick action can be frustrating and, in the long run, ineffective. The second principle is that results will be best obtained by clearly defining the problems and goals identified by the parties. Clearly defined problems and goals have the highest probability of successful resolution. Goals must be defined in terms of ecosystem condition (viable populations of wildlife, forest health) as well as ecosystem output (a more traditional view of defining resource management goals).

High-quality information is essential to define effective ecosystem management strategies. Rarely is all the needed information available when the process begins. Some can be obtained from existing information, some can be collected during the process, and some will never be available (e.g., population numbers for most species of North American wildlife before European intervention). Many types of data on biodiversity are now being collected at myriad scales, ranging from less than 1 square meter to the 635 square kilometer hexagons of the Environmental Protection Agency's Environmental Monitoring and Assessment Program (EMAP). The emergence of geographic information systems and microcomputers powerful enough to easily manipulate such databases has provided the potential for easy access to high-quality information (Machlis et al., 1994).

Information is best collected if driven by clear problem definition. However, over time, new problems will arise, and new uses will be made for existing data. It is therefore important to define overall strategies for data collection and management as being tied to, but not totally dependent on, problem-driven data needs. Data must be compatible with regional database needs as well as local site needs, and this implies that data-sharing must occur much more widely than it has in the past.

An example of an excellent framework for inventory and monitoring is one identified by several levels of scale and several indicators or components at each scale level. For example, Noss (1990) defines four levels of scale: regional landscape, community-ecosystem, population-species, and genetic, and three indicator levels: composition, structure, and function. He also defines a number of inventory and monitoring tools appropriate to each level. A nature reserve manager might categorize existing inventory in such a matrix and identify major gaps, filling them not only with problem-oriented data but fitting the collection of such data into a larger framework of biodiversity protection for the nature reserve. Considerable attention has been given various biodiversity indices (listed in Chapter 2) as appropriate measures of ecosystem health. It is tempting to adopt such simplified measures as adequate indicators of ecosystem "health," but they may measure only a small part of the real diversity of the ecosystem, which we are poorly equipped to measure adequately. Not all elements of diversity require equal attention and many mathematical indices of biological diversity may be biased towards the "more is better" concept (Franklin, 1993a).

One of the central paradigms of ecosystem management is the assumption that all management is a long-term experiment, with decisions made with information that is less than complete. Even if we have all the information we think we need, there are still major limitations to accurate prediction of the effect of management actions on the ecosystem. These limitations can occur at many scales. At the scale of the tree, when prescribed fires were reintroduced in ponderosa pine forests of Crater Lake National Park, prescriptions were defined to limit the flame lengths within the historically low, benign "natural" levels. However, many decades of fuel buildup allowed these fires to smolder in thick forest floors for hours, increasing the duration of heat far above "natural" levels, with the unanticipated consequence of killing old-growth pines that had high crowns and thick bark (Swezy & Agee, 1991). Monitoring of the fires uncovered this effect. At the scale of the landscape, the Yellowstone fires of 1988 were not predictable from the 15 years of experience with prescribed natural fires in the park, although localized research in the park and subsequent expansion of that work uncovered fires of close to the same extent several hundred years previous to 1988 (Romme & Despain, 1989).

Because we cannot fully predict outcomes of management, management must be treated as an experiment, with well-defined, explicit hypotheses about system structures and processes, clear statements of goals, and a set of targeted actions (Walters, 1986; Walters & Holling, 1990). This process is now commonly called *adaptive management* and is assumed to be part of ecosystem management (U.S. Forest Service, 1994). By recognizing uncertainty in decision making, feedback loops are built into the process at all levels (Fig. 3.1). Assessments are made, followed by decisions and implementation. Natural and social components and processes are recognized throughout the process (Slocombe, 1993) (Table 3.2); attention to both is critical to the success of ecosystem management. The effectiveness of the implementation is then evaluated and may show need for new assessments, new decisions, or new implementation procedures.

An Appropriate Concept for Parks?

The short answer is "yes, of course it is," and for many reasons. First, ecosystem management is an appropriate way to frame park, wilderness, and other protected area goals and to articulate them to neighbors. There are many reasons why neighbors may be willing to cooperate, if only for the practical reason that the protected area may serve to meet some, if not many, of the large reserve needs for the area, so that they need not be met on adjacent areas. We know that the SLOSS (single large or several small) reserve argument is not quite that simple: certain conservation objectives can be met with a few large reserves, but not

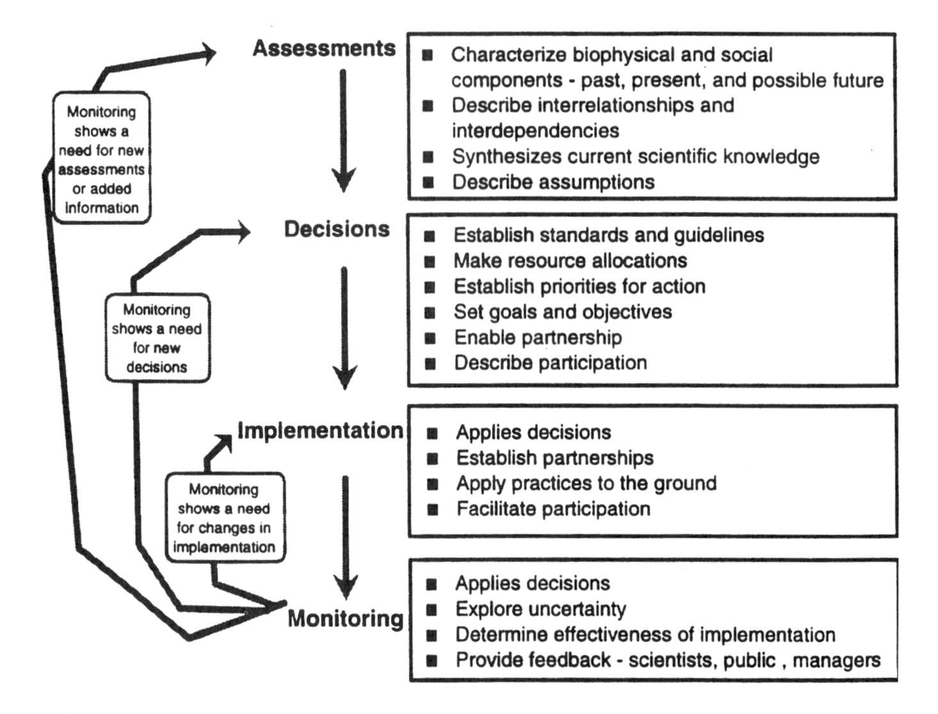

Figure 3.1 The iterative process of adaptive management. (From U.S. Forest Service, Scientific framework for ecosystem management in the interior Columbia River basin. Walla Walla, WA: Eastside Ecosystem Management Project, 1994.)

Table 3.2 Methods used in ecosystem management*

Substantive methods	Process methods
Multidisciplinary studies with integrative modeling and GIS	Facilitated, representative scoping workshops and ongoing consultation
Comprehensive studies using theory and detailed knowledge	Incentives and methods for institutional cooperation
Innovative approaches to evaluation and definition of criteria	Consensus goal-definition and related planning for their achievement
Ongoing, multilevel monitoring	Newsletters and consultation to disseminate information
Use of expert and public knowledge to develop hypotheses and models	Testing and revising results and processes
Using scenarios and working backward from desired future scenarios	Developing visions of desired futures and scenario-development exercises

*Substantive methods focus on information and analysis; process methods focus on wider processes of human involvement.
Source: From DS Slocombe, Implementing ecosystem-based management. *BioScience* 1993;43:612–622.

all regional conservation objectives can be so met. Medium-to-small reserves and attention to the matrix (lands not so reserved) are also important (Franklin, 1993a). Conversely, many neighbors of protected areas are also adopting ecosystem management, albeit with a range of objectives, e.g., the U.S. Forest Service (Overbay, 1992) and commercial forest products companies (Messinger, 1994). The key is to recognize that ecosystem management means unique things to each participant, but the common phrase should bring all to the table.

Some might say that ecosystem management is just an old idea dressed up in new semantics. In part, this is correct, because natural area managers have been dealing with these issues for a long time. Certain of the charismatic megafauna, for example, have been missing from parks and wilderness areas in the United States for decades; the only way to restore these components, such as the wolf (*Canis lupus*) and the grizzly bear (*Ursus horribilis*) was a regional approach to biodiversity. In the United States, neither the organic legislation for parks (1916) nor the Wilderness Act (1964) has focused much agency attention on ecosystem management. Instead, actions on commodity lands have driven the application of ecosystem management, although nature reserves (such as Yellowstone National Park) have been geographically at the center of ecosystem management evolution. Legislation that has obligated land managers to broaden their management perspectives include the National Environmental Policy Act (1969), Endangered Species Act (1973), National Forest Management Act (1976), and Federal Land Policy and Management Act (1976) (Keiter, 1994). It took about 15 to 20 years for the policy implications of these laws to become realized on the ground, and it is ecosystem management which has encompassed the new directions of the federal land management agencies.

An adaptive management approach for parks and protected areas must begin with a definition of goals. Goals for natural areas have historically been framed in very vague terms. The changing physical, biological, and cultural environments in and around our nature reserves will force us more specifically to define the values we wish to sustain or we are likely to see them erode. These goals need not be restricted within the boundaries of the reserve, but it should be recognized that the larger the area considered, the more likely that the goals of neighbors will influence the final set. Among the appropriate goals will be those focusing on composition, structure, and function at hierarchical scales (the community ecosystem, the landscape, etc.). Goals dealing with maintaining viable populations of native species and preserving functions such as natural disturbance patterns may be included in the set. Conservation strategies can include both coarse-filter (ecosystem-level) and fine-filter (species-level) approaches (Hunter, 1990). When cooperative strategies are developed and implemented, they may range from reserve-specific to regional in nature (e.g., the President's Forest Plan for the Pacific Northwest), and they will commit management and monitoring efforts into the future.

Ecosystem management approaches can be either formal, informal, or a combination of both (Agee & Johnson, 1988). In some cases, formally structured groups may be mandated by law, and, as discussed in Chapter 5, existing law does not preclude, and in part demands, such approaches for federal land managers (Keiter, 1994) of park and wilderness areas. In other cases, more informal, case-by-case approaches with fluid membership may be appropriate and might be a good way to start when there is little history of cooperative management.

Ecosystem management is not a panacea for all that ails resource management. It can only be successful as a cooperative management approach if the various interests believe their views are incorporated into the process. Power, politics, and perception do not disappear when ecosystem management begins. The "greater ecosystem" approach is inherent in the adoption of ecosystem management (U.S. Forest Service, 1994), but this is a controversial concept, as neighbors may interpret their importance as only serving as a buffer to a core nature reserve rather than as an integral part of the landscape (see Chap. 2). Boundary mentalities are still a dominant theme of land managers; interagency training can help to break down these barriers, particularly if they include developing appropriate skills such as effective communication, effective negotiation, and constructive challenging (the ability to reconcile discrepancies in information and perception without provoking adversarial reactions) (Washington Sea Grant Program, n.d.).

Not everything can be maximized, and some issues may be so value-based that compromise is impossible. The upfront recognition that outcomes have uncertainty associated with them is bound to make some cooperators, if not all, nervous. This is why the cooperative approach is iterative, so that each of the players has checkpoints in order to evaluate progress toward the objectives. A principal constraint to successful ecosystem management is the cost of adequate research and monitoring. Nature reserves have historically been underfunded for management, but much better funded for management than for inventory or monitoring. Data on biological diversity in national parks, and specifically spatially explicit data, are very limited (Stohlgren et al., 1995). As managers of nature reserves begin to negotiate with neighbors on cooperative management strategies, commitment far beyond the efforts of the past will be necessary for managers to hold up their end of the bargain. Financial limits will probably be much more constraining than the technical issues associated with inventory and monitoring.

As this new era of ecosystem management begins in North America, it is important to remember that it is a paradigm under development. While still in its infancy, ecosystem management has yet to prove itself. It appears to hold promise for many of the emerging landscape problems faced by natural area managers, but it has a short history of development and application. Ecosystem management will likely have a new face a decade from now as theory develops and application proceeds. Such change is inevitable and hopefully will be reflected in more effective stewardship of our nation's national parks, protected areas, and the lands that link them together.

Literature Cited

Agee JK. *Fire Ecology of Pacific Northwest Forests.* Washington, DC: Island Press, 1993.

Agee JK, Johnson DR (eds). *Ecosystem Management for Parks and Wilderness.* Seattle: University of Washington Press, 1988.

Baker WL. Landscape ecology and nature reserve design in the Boundary Waters Canoe Area, Minnesota. *Ecol* 1989;70:23–35.

Barinaga M. Where have all the froggies gone? *Science* 1990;247:1033–1034.

Basabe FA III. Ozone in western Washington forests and Douglas-fir seedling response. Ph.D. Diss., University of Washington, 1992.

Blaustein AR, Wake DB, Sousa WP. Amphibian declines: Judging stability, persistence, and susceptibility of populations to local and global extinction. *Cons Biol* 1994;8:60–71.

Brubaker LB. Vegetation History and Anticipating Future Vegetation Change. In Agee JK, Johnson DR (eds), *Ecosystem Management for Parks and Wilderness.* Seattle: University of Washington Press, 1988:41–61.

Caughley G. What Is This Thing They Call Carrying Capacity? In Boyce MS, Hayden-Wing LD (eds), *North American Elk: Ecology, Behavior, and Management.* Laramie WY: University of Wyoming Press, 1979:2–8.

Christensen NL. Succession and Natural Disturbance: Paradigms, Problems, and Preservation of Natural Systems. In Agee JK, Johnson DR (eds), *Ecosystem Management for Parks and Wilderness.* Seattle: University of Washington Press, 1988:62–86.

Davis MB. Pleistocene biogeography of temperate deciduous forests. *Geoscience and Man* 1976;13:13–26.

Franklin JF. The Fundamentals of Ecosystem Management with Applications in the Pacific Northwest. In Aplet GH, Johnson N, Olson JT, Sample VA (eds), *Defining Sustainable Forestry.* Washington, DC: Island Press, 1993a:127–144.

Franklin JF. Preserving biodiversity: Species, ecosystems, or landscapes? *Ecol Appl* 1993b;3:202–205.

Gordon JC. Ecosystem Management: An Idiosyncratic Overview. In Aplet GH, Johnson N, Olson JT, Sample VA (eds), *Defining Sustainable Forestry.* Washington, DC: Island Press, 1993:240–244.

Grumbine RE. What is ecosystem management? *Cons Biol* 1994;8:27–38.

Harris GA. Grazing lands of Washington state. *Rangelands* 1991;13:222–227.

Holling CS (ed). *Adaptive Environmental Assessment and Management.* London: John Wiley and Sons, 1978.

Hunter ML. *Wildlife Forests, and Forestry: Principles of Managing Forests for Biological Diversity.* Englewood Cliffs, NJ: Regents/Prentice-Hall, 1990.

Joyce, LA. The life cycle of the range condition concept. *J Range Manage* 1993;46:132–138.

Keiter RB. Beyond the boundary line: Constructing a law of ecosystem management. *Univ Colorado Law Rev* 1994;65:294–333.

Machlis GE, Forester DJ, McKendry JE. Biodiversity gap analysis: Critical challenges and solutions. Moscow, ID: Idaho Forest, Wildlife, and Range Experiment Station Contribution 736, 1994.

Messinger B. Ecosystem management at Boise Cascade Corporation. La Grande, OR: *Blue Mountains Nat Resour News* 1994;4(4):5–11.

Mueller-Dombois D, Spatz G. The influence of feral goats on the lowland vegetation in Hawaii Volcanoes National Park. *Phytocoenologia* 1975;3:1–29.

Noss RF. Indicators for monitoring biodiversity: A hierarchical approach. *Conserva Biol* 1990;4:355–364.

Odum EP. The strategy of ecosystem development. *Science* 1969;164:262–270.

Overbay JC. Ecosystem Management. In Proceedings of the national workshop: Taking an ecological approach to management. *USDA For Serv Rep* WO-WSA-3, 1992:3–15.

Romme WH, Despain D. Historical perspective on the Yellowstone fires of 1988. *BioScience* 1989;39:695–699.

Slocombe DS. Implementing ecosystem-based management. *BioScience* 1993;43:612–622.

Stohlgren TJ, Quinn JF, Ruggiero M, Waggoner GS. Status of biotic inventories in US national parks. *Cons Biol* 1995;7:97–106.

Swezy DM, Agee JK. Prescribed fire effects on fine root and tree mortality in old growth ponderosa pine. *Can J For Res* 1991;21:626–634.

Thomas JW. Forest ecosystem management assessment team: Objectives, processes, and options. *J For* 1994;92:12–19.

U.S. Forest Service. *Scientific Framework for Ecosystem Management in the Interior Columbia River Basin.* Walla Walla, WA: Eastside Ecosystem Management Project, 1994.

van Wagtendonk JW. Fire Suppression Effects on Fuels and Succession in Short-Fire Interval Wilderness Ecosystems. In Lotan JE (ed), Proceedings, Symposium and workshop on wilderness fire. *USDA For Serv Gen Tech Rep INT-182* 1985:119–126.

Walters CJ. *Adaptive Management of Renewable Natural Resources.* New York: McGraw-Hill, 1986.

Walters CJ, Holling CS. Large-scale management experiments and learning by doing. *Ecol* 1990;71:2060–2068.

Washington Sea Grant Program. How agencies can promote cooperative management. University of Washington, Seattle, WA: Washington Sea Grant Program, no date.

Young JA, Evans RA. Demography and fire history of a western juniper stand. *J Range Manage* 1981;34:501–505.

4

Ecosystem-Based Management: Natural Processes and Systems Theory

Katherine L. Jope
Joseph C. Dunstan

The great ecological issues of our time have to do in one way or another with our failure to see things in their entirety.

—David Orr

There is something magical about entering a place like Yosemite Valley. A child visiting for the first time is struck with awe at the towering granite cliffs, the roaring waterfalls, the smell of pine needles and campfires. In a child who is open to it all, the place evokes a profound sense of wonder.

When a person enters a place such as this—or even an everyday landscape—what is it that he or she is experiencing? It is not the individual elements of the landscape, but rather the whole that brings it such beauty and is the essence of experience. Experiencing a place such as Yosemite Valley is at once both emotional and intuitive, and at the same time rational: an experience unique to each individual yet sharing similarities with the experience of others.

David Orr (1993) suggests that we experience nature as a medley of sensations that play on us in complex ways—as sights, sounds, smells, touches, and tastes. We know at a subliminal level that it is the whole that is important. Yet, in seeking to learn about nature, we divide landscapes into discrete compartments of soil, water, vegetation, geology, and slope, and then attempt to reconstruct the ecosystem based on what we have learned about each piece.

Since the time of Descartes, we have accepted quantitative, reductionist science as a filter for our perception of reality—how we define questions, the approach that we take in addressing them, and the alternative solutions and explanations that we see as possible. This approach has led us to dissect and compartmentalize the world around us in the ways that we relate to it and attempt to better understand it (Capra, 1982). Ironically, in doing so, we may be building more barriers than bridges to our understanding of the world.

There is no question that Cartesian reductionist science has resulted in spectacular progress in certain areas and continues to produce exciting results. For example, tremendous progress has been made in understanding the structures and functions of many of an ecosystem's subunits.

There are other problems for which Cartesian reductionism is inappropriate, however, and, as a consequence, vast areas of inquiry—such as the integrative processes of living systems—have been left unstudied. Systemic properties—the interrelationships, patterns, and dynamics—are destroyed, lost, or ignored when a system is dissected, either physically or theoretically, into isolated elements (Capra, 1982). A reliance on this approach has hindered our understanding of ecosystems and therefore obstructs our ability to conserve them. As Ehrenfeld (1993) noted, ". . . the methods of reductionist science skillfully applied over many years have not solved the problem of how animals find their way home. Is it permissible to wonder whether the reductionist approach *can* answer the important questions that life poses?"

Professional land managers and scientists would be wise to approach their relationship to nature as would a child. While an expert presumes to know what the answers should be and is oblivious to what they might be, a child is unconstrained by what "should" be and is open to all possibilities. Perhaps a child understands the question better than the expert does (Carson, 1956; Maser, 1994).

In this chapter we explore the application of systems theory to the concept of ecosystems. We examine ecosystem processes with a systems perspective and discuss their implications to ecosystem-based management of human activities across broad landscapes that include national parks.

Ecosystem Defined

The term *ecosystem* is not easily defined. O'Neill et al. (1988) recognize two distinct views of ecosystems that have been used: (1) a population-community view espoused by Clements (1916, 1936), Braun (1950), Holling (1986), and others; and (2) a view based on flows of energy, nutrients, and other materials (Tansley, 1935; Lindeman, 1942; Odum, 1993).

An ecosystem, in fact, has characteristics that fit both of these definitions. A definition incorporating both views is offered by Likens (1992), who suggested that an ecosystem may be defined as "the processes influencing the distribution and abundance of organisms, the interactions among organisms, and the interactions between organisms and the transformation and flux of energy and matter."

The term *ecosystem* has often been used loosely to refer to the next higher scale of a grouping of organisms above the community or watershed level. However, an ecosystem—and ecological system—is not a specific area on the ground (O'Neill et al., 1988; Levin, 1992). It is a given, therefore, that ecosystems can exist at any scale, and they are linked with others across space and time (Holling, 1986; O'Neill, 1988; King, 1993). At each scale in a hierarchy, the system is the "whole" at that level and, at the same time, a part of the larger-scale system above it. Higher levels of the hierarchy are, to various extents, the environment of lower levels. While they do not determine the internally driven behavior of systems at lower levels, higher levels do give that behavior a context and constrain it within certain side-boards (Allen & Starr, 1982; Collins & Glenn, 1991).

Ecosystem Processes

Ecosystem processes are associated with all components of the ecosystem. They may involve biotic or abiotic components of the ecosystem, and they occur in living systems at all scales. Processes also involve interactions among ecosystem components. These interactions are often nonlinear and can be characterized by sharp thresholds, time lags, and complex feedback loops (Bormann & Likens, 1979; Likens, 1992; Costanza et al., 1993). We use the term *ecosystem process* to refer to the operation of the ecosystem, *not* its role or job. This distinction is analogous to the difference between the operating processes of an automobile and its function as a means of transportation (King, 1993).

Ecosystem processes can change in an ecosystem over time. This change may consist of qualitative change, quantitative change, or spatial change—change in geographic location or extent—over time. Landres (1992) and Dolan et al. (1978) describe three general types of change in ecosystems:

1. Regular cycles, such as changes or movements that occur over a daily or annual cycle.
2. Directional trends, such as steady erosion of soil, growth of a plant, or warming of a lake.

3. Apparently chaotic or erratic, often catastrophic, changes, such as landslides or hurricanes.

Considering that even the most impoverished ecosystem harbors thousands of species, the variety of processes are virtually infinite. For example, ecosystem processes that occur at the suborganismic level include metabolism, catabolism, respiration, and photosynthesis.

At the level of the organism, biotic processes include reproduction, birth, growth, death, and decay. Abiotic processes at this level occur in three-dimensional space. They include solar radiation, wind, variation in temperature and humidity, precipitation, and nutrient transport (Anderson, 1973).

Ecosystem processes also include interactions between biotic components of the system. They involve interactions among organisms, such as social interactions, as well as herbivory, predation, avoidance, and decomposition. In addition, ecosystem processes include the vast diversity of ways in which organisms interact with their abiotic environment. Plants take up water and nutrients from the soil and give off water vapor and oxygen. The roots of plants work their way through the soil and crack and break apart rocks; they also produce carbon dioxide, which forms carbonic acid and hastens the chemical breakdown of rock.

An example of more complex ecosystem processes involves mycorrhizal fungi, which absorb and translocate water to the host, absorb nutrients from soil, and produce enzymes that increase the availability of nutrients to the higher plants with which the fungi are associated (Trappe & Fogel, 1977; Trappe & Maser, 1977; Amaranthus et al., 1989; Perry & Amaranthus, 1990; Trappe & Luoma, 1992).

At global scales, physical, chemical, and biological processes are responsible for coupling major reservoirs, such as the atmosphere, oceans, and terrestrial ecosystems, to form the cycles of elements. Pronounced global cycles are formed by coupling of terrestrial and oceanic reservoirs via the atmosphere (Emanuel et al., 1987).

These many examples point out the impossibility of understanding ecosystems through our traditional approach. Examining every ecosystem process individually will tell us little about how an entire ecosystem works. We can only hope to understand ecosystems—and conserve them—by adopting an ecological world view based on a systems perspective.

Systems

This brings us to the concept of *systems*. Three general types of systems have been recognized:

1. There are *small-number systems,* consisting of only two components, whose "organized simplicity" readily yields to mathematical analysis. Some schools of ecology artificially isolate ecological entities as though they were small-number systems and use classical Newtonian mathematics to describe and explain their behavior. Formal theories of population and community dynamics, for example, are based on this approach.
2. A second type of system involves large numbers of essentially identical components. Study of these *large-number systems* is typified by the statistical mechanics approach to gases, where the individual behavior of 10^{23} molecules is unknowable and does not need to be known. In the "disorganized complexity" of large-number systems, the interactions of large numbers of nearly identical components are random, and overall system averages are easily computed.
3. Ecosystems are what Weinberg (1975) calls *medium-number systems.* Medium-number systems are made up of many dissimilar components with structured interrelationships. They are characterized by "organized complexity," which has been described as the fine edge between order and chaos (Lewin, 1992; Waldrop, 1992; Costanza et al., 1993). Neither a mechanical nor a statistical approach will suffice for medium-number systems. There are too few parts to average their behavior reliably and too many parts to manage each separately with its own equation (Allen & Starr, 1982; O'Neill et al., 1988).

Emergent Properties

A system is not simply an aggregation of its component parts. The key to a system is that its components are linked. In fact, the essence of an ecosystem is not its components or its flows and processes, but the dynamic interrelationships among them. It is through these linkages and interactions that *emergent properties* appear (Costanza et al., 1993). Emergent properties are properties of a system that are not obvious from its component parts (Allen & Starr, 1982; Lewin, 1992; Waldrop, 1992). They may include:

1. Properties that are unexpected by the observer because of incomplete data related to the phenomenon at hand.
2. Properties that emerge as a coarser-grained level of resolution is used by the observer.
3. Properties which cannot be derived a priori from the behavior of the parts.

Ecosystem function may be seen as an emergent property. Wetlands modify fluctuations in the flow of adjacent streams, as well as influencing stream water-quality parameters (Johnston et al., 1990). Moderation of water-flow fluctuation and cleansing of water quality are not the result of any single species inhabiting the wetland, but are a function of the system as a whole. Our experience of a landscape also is an emergent property. It is a property that emerges from the whole, which includes the observer.

It follows that the behavior of a system cannot be deduced by simply aggregating information on the system's component parts (Holling, 1992). In any study of an ecosystem, rather than taking a bottom-up view of community structure and trying to assemble an understanding of the ecosystem based on information about its components, it is imperative to define the properties that emerge at higher levels of the system and then work in a top-down fashion (Collins & Glenn, 1991).

Three concepts are critical in gaining a holistic perspective of ecosystems: self-organization, disturbance, and boundaries.

Self-Organization

The first concept is an ecosystem's emergent property of self-organization. Although seemingly contrary to the laws of thermodynamics, in which organized systems tend toward entropy, when a certain array of components are brought together and form linkages among themselves, with certain forces acting on them, the system will organize itself in a certain way (Nicolis & Prigogine, 1977; Naveh, 1982; Allen & Starr, 1982; O'Neill et al., 1988).

Bormann and Likens (1979), Vitousek (1985), and Holling (1987) suggest that the organization of living systems moves through a cycle that consists of four phases. These phases illustrate the property of self-organization in response to a change in the component parts of a system or the forces acting on it.

1. The first is a disturbance phase, or "creative destruction," caused by factors such as fire, storm, or pest infestation. O'Neill et al. (1988) state that when a system becomes unstable following a disturbance, it is the functioning of the unconstrained components that tears the system apart. Eventually, a new set of constraints will form, and a new structure emerges.
2. The second is the mobilization or reorganization phase. During this phase, the system, "digs deeply" into its nutrient capital to effect rapid repair. In essence, the ecosystem draws on a bank account of energy and nutrients built up over a long period of time to solve an immediate crisis (Bormann & Likens, 1979).

3. The exploitation phase follows, characterized by "*r*-strategy," pioneer, opportunist species that can readily move into an available habitat. These species tend to be sensitive and respond rapidly to any changes in their environment, whether positive or negative.
4. Conservation is the final phase, characterized by "*K*-strategy," climax species. Relatively small-scale endogenous disturbances operating over long periods of relative quiescence play a major role in development of the system.

Related to self-organization is the concept of *resilience*—the capability of the system to reestablish itself and recover its previous structure and patterns of behavior, after a perturbation—and cyclic renewal. Timmerman (1986) suggests that the concept of resilience is a myth. It assumes that the system is capable of some sort of adaptive memory. Horn (1976) and Jørgensen (1992) state that a system cannot renew itself and replicate the previous system due to differences in initial conditions, such as differences in seed sources, soil biota, soil nutrients, and even differences in the weather that happen to occur. If a disturbance occurs that completely ruptures the system's constraining linkages and feedbacks, then the original structure is lost, and it is very unlikely that the system will return to its original organization. Even with relatively small disturbances, the system is changed, and recovery to exactly the same organizational state would be a "highly unusual case" (O'Neill et al., 1988).

Disturbance

A second critical concept in a systems perspective of ecosystems is the concept of *disturbance,* which is an intrusive external event out of tune with local frequencies. If the disturbance is small or ephemeral, it passes unnoticed. If it is large and occurs only occasionally, generations of system components have sufficient periods in between to complete their lives without ever experiencing the disturbance. When the disturbance does occur, the system is perturbed by the event, initiating the four-phase process described in the preceding section (Bormann & Likens, 1979; Allen & Starr, 1982; O'Neill et al., 1988).

A disturbance can occur only rarely and only a few times, however, before it is tuned out of existence. If a disturbance occurs frequently or is so long-lasting that system components live out their lives knowing nothing but the disturbance, then the disturbance becomes incorporated as part of the system. In the case of frequent, repeated disturbances, the system reorganizes to incorporate the "disturbance" as an ecosystem process.

A "disturbance" perturbs less as it becomes incorporated. It is no longer external to the system and ceases to be a disturbance at all as it is brought under the feedback control of system. The effect of what was a disturbance has been to change the formerly perturbed system into an entirely new system. What was a disturbance is now an ecosystem process that is linked with other system processes and components. Once the disturbance is incorporated, ironically it is its removal that would be the disturbance (Allen & Starr, 1982; O'Neill et al., 1988).

Herbivory and predation may be viewed as disturbances that have been incorporated into the ecosystem. They alter the behavior of the food species, cause changes in life history strategy and community composition, or induce physical and chemical defenses. These responses reverberate throughout all trophic levels, causing alterations in the ecosystem as the "disturbances" are incorporated as ecosystem processes (Naiman, 1988).

Incorporation of a disturbance into the system is also evident in the relationship between fire regimes and forest structure and processes (O'Neill et al., 1988; Payette, 1992). Fire also illustrates the role of scale in the concept of disturbance. What occurs as a disturbance at one level may be a stabilizing force at another. At the scale of the burning tree, fire is a perturbation, but at the scale of the fire cycle (the scale that integrates multiple fires), fire is essential to maintain the forest diversity that has adapted to it (Allen & Starr, 1982).

Boundaries

A third critical concept in the understanding of systems is the importance of boundaries. *Boundaries* represent a sudden change in the system's organization. A naturally occurring ecotone is an example of a boundary. Ecotones occur when environmental conditions reach a threshold for tolerance and the system changes to a different organization (Von Bertalanffy, 1968; Wiens et al., 1985; Gosz, 1991; Wiens, 1992; Sirois, 1992).

Feedback processes occur at an ecotone that sharpen the contrast on either side of the boundary. For example, a spruce forest may extend across a range of elevations, persevering as the climate gradually changes with increasing elevation. Positive feedbacks within the forest contribute to maintaining conditions conducive to the forest community. With increasing elevation, environmental conditions present ever-greater challenges until, relatively suddenly, a threshold is reached, and the system reorganizes to a subalpine community. The loss of the trees magnifies the change in environmental conditions experienced by other species as this boundary is crossed.

Boundaries can also represent breaks in the linkages that are fundamental to a system. Boundaries have been likened to a semipermeable membrane whose effectiveness in stopping or altering system flows depends, in part, on the contrast between the two sides, the characteristics of the boundary itself, and the responses of different organisms and processes to the boundary. Abiotic processes, such as wind and water, are strongly affected by structural features of the boundary, such as topography, vegetation height, and soil texture. Whether an organism crosses a boundary—and survives there—depends on features on the other side of the boundary, such as habitat structure, resource availability, the organism's susceptibility to heat or water stress, or the presence or absence of competitors or predators (Wiens et al., 1985).

Movements of species and other ecosystem flows for which a boundary is relatively impermeable may be deflected by the boundary. If an organism crosses a boundary unaware of predators, lack of resources, intolerable temperature or moisture conditions, or other hostile factors, the area beyond the boundary may become a sink that drains populations and flows from within the boundary. Many system processes may become increasingly self-contained within the boundary, as outputs and inputs across the boundary are reduced. Imbalance between inputs and outputs across the boundary can destabilize boundary integrity and ecosystem processes within the boundary (Gosz, 1991; Chen et al., 1992; Franklin, 1993; Kalkhoven, 1993).

Most park managers recognize the need to manage not for individual species but in terms of ecosystems. However, an ecosystem-based approach to national park management must go beyond the consideration of multiple species and cooperation with adjacent landowners. It is imperative that an ecosystem-based approach incorporate concepts of systems theory, including an awareness of emergent properties and the implications of self-organization, disturbance, and boundaries. We provide three examples of the application of these concepts in national park management.

Zoning Within Parks

Management of national parks is based on boundaries, even within the parks (National Park Service, 1988). General Management Plans for parks in the U.S. National Park System delineate federally owned land in the parks into three zones: natural zones, cultural zones, and development zones.

The emphasis of management in *natural zones* is to conserve and restore natural ecosystem processes. Ecological processes are relied on to the extent possible to regulate wildlife populations and other aspects of the ecosystem. Where these processes have been disrupted, park managers often make an effort to restore them or to simulate them as closely as possible.

The emphasis in *cultural zones* is to conserve cultural landscapes, historic or prehistoric structures, or other cultural resources. The primary emphasis in *development zones* is to accommodate human use. In both zones, management policies call for natural ecosystem processes to be conserved "where compatible with [other] resource objectives" (National Park Service, 1988).

This type of zoning creates a system of de facto boundaries within a national park. Conditions are often quite different on opposite sides of these boundaries, reflecting different management objectives. Some wildlife species are at risk when they venture across the boundaries from natural zones; developed areas become a population sink for such species. Other species, such as scavengers, that can exploit human culture are at an advantage in developed zones; their concentrated populations can impact other species with which they are linked, inhabiting adjacent areas of the natural zone. Surface or subsurface water sheet flow is often interrupted and channelized into culverts where a road crosses a hillside or at the boundary of a developed area. Streams are constrained within their channels, altering the dynamics of backwater areas, flow and flood regimes, and riparian community dynamics.

Mutually conflicting objectives in different ways within a given national park will inevitably impair the whole. Management objectives with a systems perspective will focus on conserving a thriving ecosystem throughout the park. Fundamental to a systems approach is the need to recognize and respect the tremendous complexity of ecosystems, with their thousands of species, their linkages, and their emergent properties. In parks, we should strive to see people as part of the ecosystem—not separate from it nor at its center but simply as a part of it.

Parks in the Larger Landscape

It is well recognized that park ecosystems cannot be conserved as isolated islands surrounded by inhospitable landscapes. Since the mid-1980s, park managers have made a concerted effort to look beyond park boundaries and work cooperatively with adjacent landowners and stakeholders. Yet, the well-being of park ecosystems continues to erode. Impacts on species, flows, and other ecosystem processes that extend beyond park boundaries inevitably affect the park ecosystem. With pervasive habitat destruction beyond park boundaries, the concepts of metapopulation and minimum viable population have become critical issues for the long-term conservation of species within parks. There is evidence of widespread general declines in songbirds, waterfowl, and amphibians. One can only speculate on the consequences to the park ecosystem of declines in entire classes of species such as these.

Kalkhoven (1993) and Franklin (1993) note that the impacts of a boundary can be reduced by reducing the magnitude of the contrast across the boundary. In doing so, however, there is a tendency for people to focus on visual attributes. "Softening" boundaries also may simply push back by some distance the inevitable impact of a degraded landscape.

The entire ecosystem must be considered and an effort made to reestablish broken linkages and truncated flows. Franklin (1993) rightly asserts that the conservation of biodiversity depends not on reserves and protected areas, but on our activities in the intervening matrix. He states that "human activities can either produce very hostile conditions in the matrix—deep seas full of sharks, barren of food, characterized by lethal temperatures. Or activities can be designed to enhance dispersion and in-place survival of organisms" (Franklin, 1993).

Land uses that tear apart the linkages and self-organization of the ecosystem cannot sustain the productivity of the system for the long-term and will inevitably affect national park ecosystems. A systems perspective involves eliminating the boundaries that break the linkages of an ecosystem. It involves modifying land uses that result in deadly ecological sinks. A systems perspective makes it clear that an ecosystem on one side of a boundary can be conserved only if ecosystems on both sides of the boundary are conserved.

An ecosystem perspective involves envisioning parks as one component in a landscape at all scales, where there are a spectrum of land uses that all respect the "system" properties of the ecosystem. Human uses should fit harmoniously into the ecosystem or, in the context of Halvorson's discussion in Chapter 2, the protected landscape, conserving the components and processes, linkages and interrelationships, organization, function, and well-being of the system.

Ecosystems and People

We have traditionally seen people as separate from nature. This sets up a boundary. Parks, wilderness, and other reserves from which people are excluded or allowed only as visitors inadvertently perpetuate this sense of separateness. Rather than perpetuate this sense of separateness, a systems perspective accepts that people are part of an ecosystem. They influence, and are influenced by, birds, mammals, plants, soil biota, as well as a myriad of ecosystem processes. Cultural systems are indelibly linked with natural systems. The health of people is linked with the well-being of the ecosystem. A degraded environment affects people directly—in their health and their emotional well-being, as well as their access to food, water, and other resources. The condition of natural systems also affects people indirectly. Suppression of spruce budworm populations, eradication of malaria, effective protection of salmon and consequent increases in fishing, and conversion of semi-arid lands to cattle grazing range—each of these resulted in changes to natural systems as well as changes to institutional, social, and economic systems (Holling, 1987; Myers, 1993).

A sense of connection is essential to the conservation of the ecosystems of which national parks are a part. Economic decisions made far away from parks have profound effects on the well-being of park ecosystems. It is vital that people recognize the effects their daily activities have on the ecosystem around them and, in turn, the consequences that well-being of the ecosystem will have on their daily lives. "Environmental" problems are, in fact, social problems that can be solved only when people have a sense of connection with the environment—an understanding that the well-being of the earth is integral to their own.

National parks can foster this sense of connection and might, in fact, be seen as representing an emergent property of the linkage between natural and cultural systems. Through the design of their trails, structures, and other facilities, and through the experiences they provide to visitors, parks can foster a sense that people are linked with the ecosystem. Park management practices can, if carried out with a systems perspective, provide an opportunity for the public to learn ways of living and working as a harmonious part of the ecosystem, or they can, if done in the traditional manner, continue to reinforce the image of people as separate from nature.

Conclusion

Management of national parks and other lands has been based on a Cartesian reductionist approach. By and large, the only things given consideration in science and management have been those that can be measured and quantified. In the face of the challenges being presented by a rapidly changing world, this approach is no longer adequate.

An ecosystem perspective based on systems theory, on the other hand, can profoundly affect the way in which we perceive the world around us, approach the study of ecosystems, define our objectives for national parks, make decisions, and interact with others. Capra (1994) suggests the following changes for science and management, based on a systems approach:

1. *Shift from focusing on parts to a focus on the whole.* The implications of boundaries need to be recognized. Thriving systems cannot be confined be-

hind boundaries, whether the boundaries represent natural zones within parks, the boundaries that surround parks and other protected areas, or the boundaries that separate human systems from natural systems.

2. *Shift from "truth" to approximate descriptions.* The old paradigm was based on the Cartesian belief in the certainty of scientific knowledge. This is an illusion. Because all natural phenomena are ultimately interconnected, in order to explain any one of them we would need to understand all of the others, which is obviously impossible. What makes it possible to turn the systems approach into scientific theory is the fact that there is such a thing as approximate knowledge. Scientists do not deal with truth in the sense of a precise correspondence between the description and the phenomenon described; they only deal with limited and approximate descriptions of reality (Capra, 1994).
3. *Shift from objective to "epistemic" science.* In the old paradigm, scientific descriptions were believed to be objective. Yet "what we observe [through science] is not nature itself, but nature exposed to our method of questioning" (Heisenberg, 1971).

All models and descriptions of ecological systems are inherently representational. Whether based on words or mathematical formulas, they represent what we consider to be salient features and relationships. Subjective values and perceptions are integral to the questions we ask, the approaches we select to study them, our expectations concerning the results, the measures we therefore use, our perception and interpretation of the results, and the alternative explanations that we consider reasonable as we seek to incorporate the results into our framework of perception and "knowledge" based on our past experience. Maser (1994) states that "scientific knowledge is a product of the personal lens through which a scientist peers."

Holling (1987) proposes the *principle of surprise.* Surprises occur when events depart from what was expected. Expectations develop from two interacting sources: (1) from the metaphors and concepts we develop in an effort to provide order and understanding, and (2) from the events we perceive and remember. Our understanding will be improved if we recognize the imprecise, subjective, representational nature of science. With this in mind, in our management, we must allow latitude for our very incomplete knowledge and the eventuality of surprise. In activities that affect ecosystems, we should manage conservatively and with humility.

A systems perspective makes it clear that the belief that ecosystems can be protected in perpetuity through a series of "protected areas" is an illusion. The long-term conservation of ecosystems depends on the restoration of thriving systems across the landscape, from urban areas to national parks and wilderness. Our activities cannot be truly "sustainable" if they sustain only part of a system and sacrifice the rest.

Saving complete ecosystems is what is at stake (Noss, 1983). Degradation of ecosystems leads to altered performance of the systems, changes in ecological interactions, and, finally, changes in ecosystem functions such as nutrient regimes, decomposition, cycles of water and essential nutrients, primary productivity, energetic costs, production, and functional regulation of ecosystem processes (Sheehan, 1984).

In a world where it has been estimated that 40 percent of the primary productivity of terrestrial ecosystems has been diverted to human purposes (Vitousek et al., 1986), the role of national parks as relatively unmanipulated systems will be of ever-increasing value. If we value the functions that are emergent properties of thriving intact ecosystems, we will strive to conserve them—for future generations of children and the others with whom we share the earth.

Literature Cited

Allen TFH, Starr TB. *Hierarchy: Perspectives for Ecological Complexity.* Chicago: University of Chicago Press, 1982.

Amaranthus MP, Trappe JM, Molina RJ. Long-Term Forest Productivity and the Living Soil. In Perry PA (ed), *Maintaining the Long-Term Productivity of Pacific Northwest Forest Ecosystems.* Portland, OR: Timber Press, 1989:36–52.

Anderson, WA (ed). *Soil Ecology.* Englewood Cliffs, NJ: Prentice-Hall, 1973.

Bormann FH, Likens GE. *Pattern and Process in a Forested Ecosystem: Disturbance, Development, and the Steady State.* New York: Springer-Verlag, 1979.

Braun EL. *Deciduous Forests of Eastern North America.* Philadelphia: Blakiston, 1950.

Capra F. Systems Theory and the New Paradigm. In Merchant C (ed), *Key Concepts in Critical Theory: Ecology.* Atlantic Highlands, NJ: Humanities Press, 1994:334–341.

Capra F. *The Turning Point: Science, Society, and the Rising Culture.* New York: Simon and Schuster, 1982.

Carson R. *The Sense of Wonder.* New York: Harper and Row, 1956.

Chen J, Franklin JF, Spies TA. Vegetation responses to edge environments in old-growth Douglas-fir forests. *Ecol Appl* 1992;2:387–396.

Clements FE. Nature and structure of the climax. *J Ecol* 1936;24:252–284.

Clements FE. Plant succession: An analysis of the developments of vegetation. Washington, DC: Carnegie Institute of Washington; Publication 242, 1916.

Collins SL, Glenn SM. Importance of spatial and temporal dynamics in species regional abundance and distribution. *Ecol* 1991;72:654–664.

Costanza R, Wainger L, Folke C, Mäler K. Modeling complex ecological economic systems. *BioScience* 1993;43:545–555.

Dolan R, Hayden BP, Soucie G. Environmental dynamics and resource management in the U.S. national parks. *Environ Manage* 1978;2:249–258.

Ehrenfeld D. Raritan letter: Animal navigation. *Orion* 1993 (Summer):4–5, 63.

Ehrlich PR, Wilson EO. Biodiversity studies: Science and policy. *Science* 1991;253:758–762.

Emanuel WR, Pastor J, O'Neill RV. Maintaining the integrity of global cycles: Requirements for long-term research. In Draggan S, Cohrssen JJ, Morrison RE (eds), *Preserving Ecological Systems: The Agenda for Long-Term Research and Development.* New York: Praeger Publishers, 1987:23–40.

Franklin JF. Preserving biodiversity: Species, ecosystems, or landscapes? *Ecol Appl* 1993;3:202–205.

Gosz JR. Fundamental Ecological Characteristics of Landscape Boundaries. In Holland MM, Risser PG, Naiman RJ (eds), *Ecotones: The Role of Landscape Boundaries in the Management and Restoration of Changing Environments.* New York: Chapman and Hall, 1991:8–30.

Heisenberg W. *Physics and Philosophy.* New York: Harper and Row, 1971.

Holling CS. Cross-scale morphology, geometry and dynamics of ecosystems. *Ecol Monog* 1992;62:447–502.

Holling CS. The Resilience of Terrestrial Ecosystems: Local Surprise and Global Change. In Clark WC, Munn RE (eds), *Sustainable Development of the Biosphere.* New York: Cambridge University Press, 1986:292–317.

Holling CS. Simplifying the complex: The paradigms of ecological function and structure. *Eur J Oper Res* 1987;30:139–146.

Horn HS. Succession. In May RM (ed), *Theoretical Ecology: Principles and Applications.* Boston: Blackwell Scientific Publications, 1976:253–271.

Johnston CA, Detenbeck NA, Niemi GJ. The cumulative effect of wetlands on stream water quality and quantity: A landscape approach. *Biogeochemistry* 1990;10:105–141.

Jørgensen SE. *Integration of Ecosystem Theories: A Pattern.* Norwell, MA: Kluwer Academic Publishers, 1992.

Kalkhoven JTR. Survival of Populations and the Scale of the Fragmented Agricultural Landscape. In Bunce RGH, Ryszkowski L, Paoletti MG (eds), *Landscape Ecology and Agroecosystems.* Boca Raton, FL: Lewis Publishers, 1993:83–90.

King AW. Considerations of Scale and Hierarchy. In Woodley S, Kay J, Francis G (eds), *Ecological Integrity and the Management of Ecosystems.* Delray Beach, FL: St. Lucie Press, 1993:19–45.

Landres PB. Temporal scale perspectives in managing biological diversity. *Trans N Am Wildl Nat Res Conf* 1992;57:292–307.

Leopold A. *A Sand County Almanac.* New York: Oxford University Press, 1949.

Levin SA. The problem of pattern and scale in ecology. *Ecol* 1992;73:1943–1967.

Lewin R. *Complexity: Life at the Edge of Chaos.* New York: Macmillan, 1992.

Likens GE. *The Ecosystem Approach: Its Use and Abuse.* Oldendorf/Luhe, Germany: Ecology Institute, 1992.

Lindeman RL. The trophic-dynamic aspect of ecology. *Ecol* 1942;23:438–450.

Maser C. *Sustainable Forestry: Philosophy, Science, and Economics.* Delray Beach, FL: St. Lucie Press, 1994.

Myers N. The question of linkages in environment and development. *BioScience* 1993;43:302–310.

Naiman RJ. Animal influences on ecosystem dynamics. *BioScience* 1988;38:750–752.
National Park Service. Management policies. Washington, DC: United States Department of the Interior, National Park Service, 1988.
Naveh Z. Landscape ecology as an emerging branch of human ecosystem science. *Adv Ecol Res* 1982;12:189–209.
Nicolis G, Prigogine I. *Self-Organization and Non-Equilibrium Systems.* New York: John Wiley and Sons, 1977.
Noss RF. A regional landscape approach to maintain diversity. *BioScience* 1983;33:700–706.
Odum HT. *Systems Ecology: An Introduction.* New York: Wiley, 1993.
O'Neill RV, Milne BT, Turner MG, Gardner RH. Resource utilization scale and landscape pattern. *Landscape Ecol* 1988;2:63–69.
Orr DW. The problem of disciplines/the discipline of problems. *Cons Biol* 1993;7:10–12.
Payette S. Fire as a Controlling Process in the North American Boreal Forest. In Shugart HH, Leemans R, Bonan GB (eds), *A Systems Analysis of the Global Boreal Forest.* New York: Cambridge University Press, 1992:144–169.
Perry DA, Amaranthus MP. The Plant-Soil Bootstrap: Microorganisms and Reclamation of Degraded Ecosystems. In Berger JJ (ed), *Environmental Restoration: Science and Strategies for Restoring the Earth.* Washington, DC: Island Press, 1990:94–102.
Sheehan PJ, Miller DR, Butler GC, Bourdeau P. *Effects of Pollutants at the Ecosystem Level.* New York: John Wiley and Sons, 1984.
Sirois L. The Transition Between Boreal Forest and Tundra. In Shugart HH, Leemans R, Bonan GB (eds), *A Systems Analysis of the Global Boreal Forest.* New York: Cambridge University Press, 1992:196–215.
Tansley AG. The use and abuse of vegetational concepts and terms. *Ecol* 1935;16:284–307.
Timmerman P. Mythology and Surprise in the Sustainable Development of the Biosphere. In Clark WC, Munn RE (eds), *Sustainable Development of the Biosphere.* New York: Cambridge University Press, 1986:435–453.
Trappe JM, Fogel RD. Ecosystematic Functions of Mycorrhizae. In Hansen DH (ed), *The Belowground Ecosystem: A Synthesis of Plant-Associated Processes.* Fort Collins, CO: Colorado State University, Range Science Department, Science Series No. 26, 1977:205–214.
Trappe JM, Luoma DL. The Ties that Bind: Fungi in Ecosystems. In Carroll GC, Wicklow DT (eds), *The Fungal Community: Its Organization and Role in the Ecosystem.* New York: Marcel Dekker, 1992:17–26.
Trappe JM, Maser C. Ectomycorrhizal Fungi: Interactions of Mushrooms and Truffles with Beasts and Trees. In Walters T (ed), *Mushrooms and Man: An Interdisciplinary Approach to Mycology.* Washington, DC: U.S. Department of Agriculture, Forest Service, 1977:165–179.
Vitousek PM. Community Turnover and Ecosystem Nutrient Dynamics. In Pickett STA, White PS (eds), *The Ecology of Natural Disturbance and Patch Dynamics.* New York: Academic Press, 1985:325–333.
Vitousek P, Ehrlich PR, Ehrlich AH, Matson PA. Human appropriation of the products of photosynthesis. *BioScience* 1986;36:368–373.
Von Bertalanffy L. *General System Theory: Foundations, Development, Applications.* New York: George Braziller, 1968.

Waldrop MM. *Complexity: The Emerging Science at the Edge of Order and Chaos.* New York: Simon and Schuster, 1992.
Weinberg GM. *An Introduction to General Systems Thinking.* New York: Wiley, 1975.
Wiens JA. Ecological Flows Across Landscape Boundaries: A Conceptual Overview. In Hansen AJ, diCastri F (eds), *Landscape Boundaries: Consequences for Biotic Diversity and Ecological Flows.* New York: Springer-Verlag, 1992:217–235.
Wiens JA, Crawford CS, Gosz JR. Boundary dynamics: Conceptual frame-work for studying landscape ecosystems. *Oikos* 1985;45:421–427.

5

Ecosystem Management: Exploring the Legal-Political Framework

Robert B. Keiter

Once on the verge of extinction, Yellowstone National Park's bison populations now exceed historic population levels and are moving out of the park, causing conflicts with adjacent landowners. Some bison carry the *Brucella abortus* organism, a highly contagious livestock disease. Because brucellosis may be transmissible to domestic cattle on nearby ranches, it is strictly regulated by federal and state agriculture officials. Ranchers fear loss of their state's brucellosis-free status, which could increase their production costs and limit their access to the interstate cattle market. As a result, bison management in Yellowstone has been mired in controversy for most of the past decade (Thorne et al., 1991).

The bison controversy has spawned at least ten different lawsuits, congressional legislative activity, and significant changes in state wildlife law (Keiter & Froelicher, 1993). A lengthy interagency environmental impact statement process is not yet complete. Negative national media attention has forced the state of Montana to cancel its widely publicized bison hunt adjacent to the park, leaving managers uncertain over how to control bison movements. What the bison controversy makes irrefutably clear, though, is that the park is not an island; rather, it is part of a larger ecological complex often referred to as the Greater Yellowstone Ecosystem (Reese, 1991; Clark & Zaunbrecher, 1987). The bison, following natural instincts, simply have no regard for the linear—yet ecologically irrelevant—political boundaries that have been superimposed on the park and the surrounding landscape. In fact, the legal framework governing management of Yellowstone and adjacent lands takes little account of the ecological realities that confront park and wildlife managers today (Keiter, 1989).

The bison, however, are just one of many complex problems confronting park and protected areas managers today. National parks throughout North America are increasingly faced with human pressures emanating from surrounding lands. In the case of Yellowstone, intensive development activities on adjacent lands are fragmenting critical wildlife habitats and cumulatively threatening the environmental and aesthetic integrity of the park's ecosystems, putting vital resources like the grizzly bear at risk (Keiter, 1989; Clark & Minta, 1994) (see Chap. 8). At the same time, Yellowstone's commitment to allowing natural processes to function unabated poses risks for its neighbors. Wandering bison threaten to transmit disease, transplanted wolves will occasionally depredate livestock, and wildfires can jump the boundary and destroy nearby homes and timber.

Historically, the national parks were established to preserve unusual aesthetic features and to provide visitors with pleasant experiences in the natural world. In an era when the parks were seen as monuments and the ecological sciences were in their infancy, park boundaries were rarely established to accommodate the needs of wildlife or other natural processes (Runte, 1987). A similar situation occurred during the 1960s when Congress legalized the wilderness concept to protect attractive undeveloped settings for visitors seeking a truly wild experience (Nash, 1982). Today, however, national parks and protected areas play a vital role in protecting important biological resources and provide the settings where natural processes may unfold without human intervention (Grumbine, 1992). In short, parks and wilderness areas are the critical core of larger ecological complexes with manifold interconnections that dictate management at a scale transcending traditional boundary lines (Agee & Johnson, 1988).

Yet, with the growth in human population and increased demands on natural resources, parks and wilderness areas are in danger of becoming merely islands surrounded by development activities and other human pressures. For more than a decade, the National Park Service (NPS) has documented, repeatedly, mounting external threats to park resources (Keiter, 1985; U.S. General Accounting Office, 1994a). Despite a plethora of high-level reports and extensive scientific documentation, neither Congress nor the NPS has taken much effective action to address the problem. Although less well-documented, a similar scenario has evolved with respect to the nation's wilderness areas. As a result, wilderness-dependent species have been placed at risk, and the pace of endangered species listings has accelerated markedly. With the political process in default, the courts have increasingly been called on to intervene in public land management decisions to protect high-risk species like the grizzly bear and northern spotted owl.

The federal land management agencies, however, have now endorsed the concept of ecosystem management to address transboundary resource issues and to ensure the ecological integrity of public domain natural systems (Congressional Research Service, 1994). As described in Chapter 3, ecosystem management aids in the conservation of biological diversity through management at spatial and temporal scales, designed to sustain productive natural systems. Ecosystem management parallels the NPS commitment to natural process management; the goal in each case is to allow nature to function while minimizing adverse human interference. But whether these goals can be achieved under the banner of ecosystem management remains to be seen. These administrative initiatives are constrained by existing law, which does not explicitly provide for ecosystem management on the public domain (Keiter, 1994). Moreover, many people perceive the concept as a threat to traditional economic activities and access rights, both outside and within the parks (Budd, 1991).

This chapter explores the concept of ecosystem management in national parks and wilderness areas from a legal-political perspective. It begins by identifying the contemporary origins of ecosystem management and defining the concept. It then examines the legal underpinnings for federal ecosystem management policy and notes remaining obstacles to broad implementation of ecosystem management approaches. The chapter concludes by analyzing two related and complementary ecosystem management strategies, namely a process-based and a substantive standard-based approach. Concluding that neither the NPS nor other land management agencies can hope to achieve biodiversity or other ecological goals by pursuing only process-based strategies, the chapter advocates adopting a complementary mixture of these two approaches to ensure long-term ecosystem integrity. In short, park and wilderness managers must be prepared to invoke their legal authority to establish the substantive limits of a cooperative relationship—a position that will require political courage and a clear understanding of existing legal obligations and authorities.

Understanding Ecosystem Management

Ecosystem management appears to represent another important stage in the evolutionary development of public land and natural resources management policy. The concept of managing at the ecosystem level can be linked to Aldo Leopold's seminal writings advocating a holistic approach to natural resources management. More recently, the concept has gained currency from mounting public concern over protecting national parks and other pristine areas from transboundary environmental impacts, and from the growing realization that modern society faces a species extinction crisis. Despite these somewhat disparate origins, ecosystem management has been defined with sufficient precision to constitute a viable natural resource management policy. As with any natural resource policy that must address social, environmental, and political complexity, ecosystem management cannot be defined precisely in a few phrases or sentences. Rather, it is best understood as an overarching set of principles designed to accommodate ecological complexity with existing land ownership and management patterns.

Origins of the Doctrine

Current ecosystem management doctrine can be traced directly to Aldo Leopold's holistic view toward nature and natural processes. *A Sand County Almanac* sets forth a "land ethic" that links ecology and philosophy to establish a compelling ecological vision for natural resource management (Leopold, 1949). Leopold's land ethic asserts that "a thing is right when it tends to preserve the integrity, stability, and beauty of the biotic community. It is wrong when it tends otherwise." Leopold's vision was ultimately translated into policy when, in 1963, the NPS adopted the so-called Leopold Report on Wildlife Management in National Parks, which was written in part by one of his sons (Leopold et al., 1963). Urging that national parks be managed to maintain a "vignette of primitive America," the report concludes that the unimpeded operation of natural processes represented a viable natural resource policy in the parks. Despite sometimes harsh criticism (Chase, 1986), the NPS has continued to adhere to this "hands off" management approach in most large natural area parks—a position that most scientists have endorsed (Boyce, 1991).

It has become apparent, however, that national parks do not exist in isolation and face serious environmental threats from sources beyond park boundaries. During the early 1970s, the problem surfaced prominently in the case of Redwood National Park, which confronted the prospect of significant ecological degradation caused by extensive upstream logging just beyond park boundaries. After court decisions ordered the NPS to take action to protect Redwood's resources, Congress was persuaded to adopt legislation expanding the park and clarifying the agency's authority to protect park resources (Hudson, 1979). In 1980, perceiving that the Redwood crisis was symptomatic of similar problems elsewhere, the NPS completed a State of the Parks Report, identifying myriad external activities that threatened park resources, including air pollution, energy exploration, timber harvesting, and subdivision development (National Park Service, 1980). However, despite repeated attempts to secure passage of comprehensive park protection legislation (Keiter, 1985), Congress has refused to act, except to address a few pressing individual park problems. The NPS has been left to rely on its existing authority to address external problems.

During this same period, scientists have become increasingly convinced that the nation faces a species extinction crisis. It has become equally apparent that national parks and other preserved areas play a critical role in providing secure habitats for threatened species (Wright, 1992; Noss & Cooperrider, 1994). With passage of the Endangered Species Act in 1973, Congress created a powerful, judicially enforceable law giving species recovery priority over other considerations. Under this legislation, the national parks, wilderness areas, and other public lands are playing a major role in many species recovery efforts. In the Northern Rockies, for example, the grizzly bear recovery effort has focused on Glacier and Yellowstone National Parks, along with wilderness lands in the surrounding national forests (U.S. Fish and Wildlife Service, 1993). In the Pacific Northwest, the federal recovery effort for the threatened northern spotted owl, which has generated extensive litigation over national forest logging practices, has relied heavily on Olympic National Park and wilderness lands as critical habitat (Forest Ecosystem Management Assessment Team, 1993). In the case of both the grizzly bear and spotted owl, however, these preserved lands do not provide sufficient habitat to ensure survival; additional habitat beyond park and wilderness boundaries is necessary to ensure population recovery. Recognizing this fact, the courts have invoked the Endangered Species Act to impose severe constraints on logging and other economic activities on the multiple-use public lands, bringing the statute under harsh criticism. Meanwhile, Congress has been largely silent: It has neither altered the Act itself, nor responded legislatively to the Northwest's logging crisis, nor given serious consideration to comprehensive legislation addressing national biodiversity conservation concerns.

From these related developments, a potential solution to the national parks' external-threats problem as well as the species extinction crisis has emerged under the banner of ecosystem management. Driven by the compelling scientific logic of biodiversity conservation research as well as court decisions highlighting the shortcomings of conventional managerial approaches, the federal land management agencies have embraced ecosystem management as a means to meet environmental protection requirements and to reduce the level of conflict on the public lands (Congressional Research Service, 1994). With crises simmering in the Pacific Northwest and elsewhere across the public domain, and with Congress providing very little leadership, the agencies have been forced to respond administratively by shifting toward a new ecosystem-based management policy. In short, ecosystem management is properly viewed as an administrative creation linked to park protection, endangered species preservation, and related scientific and legal developments.

Governing Principles

The concept of ecosystem management derives from several widely accepted scientific and related propositions (Keiter, 1994):

1. Ecosystems are complex, dynamic, and inherently unstable, which means that boundaries cannot be easily defined or maintained in conventional jurisdictional terms.
2. All ecological components and species are interrelated and merit consideration to protect linkages and evolutionary processes.
3. Human communities must be considered part of the ecosystem, although human-induced impacts require monitoring to safeguard the integrity of ecological processes.
4. Development activities generally should not exceed defined levels of sustainability directly related to ecosystem health.
5. Sophisticated scientific knowledge and monitoring should be used in developing management objectives, measuring progress, and making corrective adjustments.
6. Management proposals should be evaluated using an ecologically derived time scale.

Failure to recognize and manage in accord with these propositions accounts for many of the intractable problems confronting federal land managers today and undercuts the development of sustainable resource management policies.

Ecosystem management can best be defined in terms of core governing principles. A primary goal of ecosystem management is to "protect or restore critical ecological components, functions, and structures in order to sustain resources in perpetuity" (Moote et al., 1994). This includes maintaining the ecological integrity of native ecosystems over broad spatial and temporal scales, including viable species populations, evolutionary processes, and a full range of ecosystem types (Grumbine, 1994). Because species and ecological processes transcend jurisdictional boundaries, ecosystem management requires coordination among governmental entities and other interested parties (Keiter, 1994). Because people are a part of nature, ecosystem management should promote sustainable natural resource policies to ensure viable communities and economic opportunities (Slocombe, 1993). Similarly, because human values influence any natural resource management policy, ecosystem management must accommodate human interests and afford widespread public involvement in planning and decision-making processes (Freemuth & Cawley, 1993; Shannon, 1993). Given the important role science plays in understanding natural systems, integrated and interdisciplinary scientific research is an important aspect of ecosystem management (Grumbine, 1994). Yet, because ecosystem management and the accompanying science is still experimental, ecosystem management institutions and processes should be adaptable to reflect changes in scientific knowledge as well as human values (Lee, 1993).

The NPS draft ecosystem management policy proposal encompasses most of these core principles. The agency's goal is "preserving, protecting, and/or restoring ecological integrity (composition, structure, and function) and also maintaining sustainable societies and economies" (National Park Service, 1994b). The policy statement then enumerates a set of related strategies to accomplish this goal:

1. Recognizing and employing multiple boundaries and scales.
2. Utilizing biodiversity and conservation biology science.
3. Merging cultural resources and traditions.
4. Incorporating social, cultural, economic, and political factors.
5. Utilizing scientific information for decision-making.
6. Establishing partnerships.
7. Developing interdisciplinary management styles.
8. Managing for long-term goals.
9. Employing adaptive and flexible management techniques.

Although other federal land management agencies use somewhat different terminology to define their own ecosystem management policies, they each reflect a similar commitment to long-term and sustainable resource management, preservation of biodiversity, and collaborative efforts across boundary lines (U.S. Forest Service, 1992; Bureau of Land Management, 1994).

Because the NPS operates under a powerful preservationist legal mandate (16 U.S.C. §1), it can—and should—emphasize the protection and restoration of native species and natural processes as a primary management goal. However, because national parks are part of larger ecological complexes, these goals cannot be fully accomplished without the active cooperation of neighbors. In many instances, this means that park managers must work with officials from the Forest Service or Bureau of Land Management (BLM), who are governed by quite different multiple-use mandates and obligated to produce natural resource commodities, like timber and minerals (Sax & Keiter, 1987). In other instances, this means that park managers must seek to collaborate with state and private landowners, whose ownership goals generally do not include biodiversity conservation (Sax, 1976). From the NPS perspective, therefore, ecosystem management involves establishing and maintaining working relationships with these diverse neighbors based on a shared commitment to protect national park resources and biological processes.

Notwithstanding their divergent statutory mandates, the federal land management agencies have each promulgated ecosystem management policies that are remarkably similar, reflecting a core concern for maintaining the integrity of ecological processes. While this common federal commitment to an ecosystem management regime should enable park managers to address resource problems more effectively, adjacent land managers are still confronted with powerful contrary political pressures, which could upset tenuous cooperative relationships. Ecosystem management, in other words, is not yet fully institutionalized on the public domain. Moreover, some private landowners, who often hold ecologically critical lands, have mounted a powerful counter-offensive based on property rights that is designed to undermine the concept entirely (Falen & Budd-Falen, 1994; Hage, 1989).

Ecosystem Management, Law, and Politics

Given its origins as an administratively conceived policy, ecosystem management cannot be divorced from the law or political reality. Although ecosystem management and the law are not ordinarily linked together, the practical reality is that federal land management agencies can only pursue ecosystem management policies within the limits of existing law. In other words, the law establishes the outermost parameters of agency-initiated ecosystem management policies. Failure to recognize this fact risks litigation to enjoin innovative initiatives or political retribution for bureaucratic adventurism. Moreover, during this transitional period, administratively conceived ecosystem management policies also must surmount additional legal, political, bureaucratic, and scientific obstacles.

The Legal Foundation

Existing law does not expressly endorse ecosystem management on the public domain. Instead, the authority to construct and implement ecosystem-based management policies must be extracted from laws that were framed for other purposes but which vest land managers with sufficient discretion to manage at the ecosystem level (Keiter, 1994). For the federal land management agencies, this means that the prevailing organic legislation, as further refined by overarching statutes like the Endangered Species Act (16 U.S.C. §§ 1531–43) and the National Environmental Policy Act (NEPA)(42 U.S.C. §§ 4321–61), establishes the framework within which ecosystem management priorities and relationships can be constructed. This section examines how these laws facilitate adoption of ecosystem management policies.

In the case of the NPS, ecosystem management policy can be grounded in statutory provisions that give the agency an unambiguous preservation mission. The Organic Act (16 U.S.C. §§ 1–18f) mandates that national parks are to be managed to conserve scenery, natural and historic objects, and wildlife, and to provide for public enjoyment, while ensuring that the parks are left "unimpaired for the enjoyment of future generations" (16 U.S.C. § 1). This nonimpairment requirement constitutes a clear substantive standard that obligates the agency to protect the natural integrity of park landscapes when human pressures come into conflict with resource values. Congress, however, through the appropriations process and otherwise, has often expressed a preference for visitor access and services rather than resource protection, and the NPS has generally been responsive to such political pressures, sometimes to the detriment of park ecology (Chase, 1986).

Although the Organic Act originally did not give the NPS explicit responsibility or authority to address activities occurring on adjacent lands that threaten park resources, the 1978 Redwood Amendments appear to contain such responsibility and authority. Section 1a-1 of the amended Organic Act obligates NPS officials to protect, manage, and administer national parks in light of the high public value and integrity of the national park system (16 U.S.C. § 1a-1). An outgrowth of the Redwood controversy, this mandate imposes a clear obligation on NPS officials to respond to threatening activities, whether internal or external to the parks, and to view their resource management responsibilities on an ecosystem scale. In cases where external activities are adversely affecting park resources, this means that the agency could exercise regulatory authority beyond park boundaries. Under existing legal precedent, federal land management agencies have the apparent authority to regulate activities occurring on adjacent lands that jeopardize federal resources, though the political will to extend federal regulatory power beyond the boundary line is often lacking (Keiter, 1989).

Moreover, the NPS Organic Act's preservation responsibilities effectively impose substantive coordination obligations on adjacent land managers, which should assist park managers to protect transboundary ecological resources. Because Forest Service and BLM managers operate under statutory planning requirements obligating them to "coordinate" with adjacent land management agencies (16 U.S.C. § 1604 (a); 43 U.S.C § 1712 (c)(9)), the NPS preservation mandate causes them to exercise restraint when contemplating development activities that could impact park ecological systems (Keiter, 1989). Although there is little judicial precedent directly supporting this interpretation, it is difficult to construe the statutory "coordination" requirement differently if ecosystem management is to have any meaningful content. In addition, NPS obligations under the Endangered Species Act and NEPA are plainly consistent with an ecosystem management approach to transboundary resource issues; both laws transcend jurisdictional boundaries and impose either substantive or procedural requirements on all federal land managers that require consideration of ecosystem-wide effects. Thus, even in the absence of an ecosystem management statutory mandate, park officials have the legal authority to involve themselves directly in transboundary problems and to protect park resources from harm.

In the case of designated wilderness lands, a similar conclusion is suggested by the governing legislation and related statutes. Although the Wilderness Act (16 U.S.C. §§ 1131–36) is not framed in terms of ecosystems, it contains an explicit preservation mandate requiring that wilderness areas be managed "unimpaired for future use and enjoyment as wilderness" (16 U.S.C. § 1131 (A)). The Act also imposes a general legal duty on federal land managers to protect and manage wilderness "so as to preserve its natural conditions" (16 U.S.C. § 1131 (c)), a statutory obligation closely paralleling the Redwood Amendments' requirement that NPS officials protect park resources from environmental harm (16 U.S.C. § 1a-1). One court has ruled that federal reserved water rights attach to designated wilderness areas (*Sierra Club v Block,* 1985), which essentially recognizes that wilderness areas are part of larger ecological complexes and are entitled to legal protection to ensure the area's ecological integrity. The ruling, however, was subsequently reversed on other grounds (*Sierra Club v Yeutter,* 1990), leaving unclear the scope of federal reserved water rights in wilderness areas. Moreover, other courts have interpreted the Wilderness Act to permit logging in wilderness areas to protect adjacent lands from insect damage (*Sierra Club v Lyng,* 1987; *Sierra Club v Lyng,* 1988), decisions at odds with the notion that wilderness is a setting where ecological processes are allowed to reign. Thus, despite its strong protective language, the Wilderness Act does not fully ensure that designated wilderness lands will be managed to protect natural processes or as part of a larger ecosystem (Rohlf & Honnold, 1988).

Other federal land management agencies are also governed by organic legislation that can be interpreted to support ecosystem management on the public lands. In the case of the Forest Service, which operates under a multiple-use mandate, the National Forest Management Act (NFMA)(16 U.S.C. §§ 1600–14), imposes detailed, interdisciplinary resource planning obligations (16 U.S.C. § 1604 (b)). Besides containing the only explicit federal statutory reference to biological diversity (16 U.S.C. § 1604(g)(3)(B)), the NFMA contains important environmental limitations on logging practices (16 U.S.C. § 1604 (g)) and requires coordinated planning with neighboring agencies or landowners (16 U.S.C. § 1604 (a)). In the case of the BLM, the Federal Land Policy Management Act (FLPMA)(43 U.S.C. §§ 1701–84) also establishes a multiple-use standard, which contemplates protecting ecological and environmental values as well as prohibiting permanent damage to the resource base (43 U.S.C. § 1702 (c)). FLPMA also provides for comprehensive, interdisciplinary planning (43 U.S.C. § 1712 (c)(2)), as well as coordination with adjacent land managers and communities (43 U.S.C. § 1712(c)(9)). In the case of the U.S. Fish and Wildlife Service, the national wildlife refuges are administered under governing legislation prohibiting secondary activities that are not "compatible with the major purposes for which such areas were established" (16 U.S.C. § 668dd(d)(1)(A)), which suggests that wildlife needs should receive priority in managing the refuges (Fink, 1994). In sum, although these organic statutes do not speak in explicit ecosystem management terms, they do establish a context or setting compatible with an ecosystem management approach.

Federal land management agencies also must adhere to environmental laws that are designed to protect critical ecological components, thus further supporting ecosystem management on the public domain. The Endangered Species Act contains an express federal commitment to protect species from extinction, connecting species conservation with habitat preservation (16 U.S.C. § 1531(b)). Applicable across the public domain, the Act enjoins federal agencies to conserve listed species and prevents them from jeopardizing protected species (16 U.S.C. § 1536(a)(1),(2)), and it prohibits anyone from taking a listed species (16 U.S.C. § 1538(a)(1)(B)). The courts have interpreted the statute strictly and applied its procedural mandates rigorously (e.g., *Thomas v Peterson,* 1985). NEPA (42 U.S.C. §§ 4321–61), which requires preparation of an environmental impact statement (EIS) whenever a major federal action significantly affecting the human environment is contemplated (42 U.S.C. § 4332(2)(C)), also provides indirect support for ecosystem management. Although NEPA is purely a procedural statute, the courts have ruled that its EIS requirements demand analysis and disclosure of the full ecological impacts of proposed actions, thus effectively obligating federal land managers to adopt an ecosystem perspective in their decision-making (Keiter, 1990). Along with other statutes designed to regulate environmental quality, these laws provide federal land managers with a statutory basis for viewing their management responsibilities in broader ecological terms that transcend existing boundary lines.

The law, therefore, supports ecosystem management on the public domain, and the Clinton administration has fostered administrative initiatives designed to bring it to fruition (Interagency Ecosystem Management Task Force 1995). However, the transition is not yet complete. Congress still has not endorsed the concept, and it is considering several measures that could severely undermine it. The federal agencies have focused most of their energies on coordination efforts without a corresponding commitment to establishing substantive goals or priorities. Many private landowners, who are under no legal obligation to engage in ecosystem management, are intent on invoking the constitutional takings doctrine to challenge any ecologically based regulatory efforts. In other words, despite a plethora of ecosystem management initiatives, serious obstacles still must be overcome before the concept is securely rooted in agency culture.

Obstacles to Change: Making the Transition

Although resting on a firm legal foundation, ecosystem management policy nonetheless faces significant legal, political, bureaucratic, and scientific obstacles. None of these obstacles is insurmountable, but they cumulatively make the transition to ecosystem management more difficult. They also will play a role in determining the final content and shape of ecosystem management policies within the individual federal agencies (Cortner et al., 1994).

Legally, several laws are not entirely compatible with ecosystem management concepts. Interagency coordination—a key aspect of ecosystem management—presents land managers with the difficult task of reconciling the preservation-oriented mandate of the NPS with the multiple-use mandates governing the Forest Service and BLM, which emphasize commodity production and not preservation (Cortner, 1994). Related resource development statutes, such as the General Mining Law of 1872 and the Taylor Grazing Act of 1934, grant access rights that often contradict ecosystem management goals (Wilkinson, 1992). Even the environmental protection laws, which are generally compatible with ecosystem management principles, contain provisions that can undermine ecological objectives, as in the case of the Endangered Species Act's focus on preserving single species (Grumbine, 1992). The Supreme Court's revitalization of constitutional takings doctrine imposes direct limitations on the regulatory authority that federal land managers might assert over adjacent private landowners, making it difficult to bring landowners to the bargaining table (Sax, 1993). Even if landowners wish to cooperate in ecosystem-based agreements that might restrict timber harvesting or other productive activities, this might present antitrust problems if the agreement constitutes a restraint on trade. In addition, conservative interest groups, such as the Wise Use Movement, are promoting a county land use planning process designed to compel federal land managers to adjust public land management priorities consistent with development-oriented county plans, but recent litigation suggests that this approach violates constitutional supremacy principles (Reed, 1993–1994).

One of the more difficult legal hurdles confronting interagency ecosystem coordination efforts is the Federal Advisory Committee Act (FACA) (5 U.S.C. App. at 1175), which has been invoked in an attempt to block such efforts (*Northwest Forest Resource Council v Espy*, 1994). FACA applies to any committee or similar group composed of federal and nonfederal members enlisted to provide advice or recommendations to federal agencies or officers (Bybee, 1994). The Act imposes detailed procedural requirements on such committees, including preparation of a written charter, prior Federal Register notice of meetings, balanced membership, and open meetings (5 U.S.C. App. 2). Failure to comply with the requirements has caused at least one court to enjoin federal officials from using a scientific committee's recommendations (*Alabama-Tombighbee Rivers v Department of the Interior*, 1994). Recent congressional legislation exempts, however, interagency committees composed of federal officials and state, tribal, or local governmental members from FACA requirements. The amendment does not apply to advisory committees with members representing private or nonprofit organizations, or just private citizens. Joint public–private coordination bodies will therefore continue to face extensive procedural hurdles, which may discourage all but the most dedicated ecosystem management proponents.

Politically, federal ecosystem management policies have yet to secure a firm foothold. Because Congress has not expressly endorsed ecosystem management, federal land management agencies must derive statutory authority for ecosystem-based initiatives from existing organic legislation and related environmental statutes. Responding to conservative property rights advocates, the 104th Congress has signaled that it has reservations about ecosystem management, particularly if it is viewed primarily as a biodiversity conservation strategy. The evidence includes Congress's refusal to endorse the National Biological Service, House passage of takings legislation, efforts to alter the Endangered Species Act, adoption of a salvage logging rider that opens national forests to increased timber harvesting, and rejection of mining law or range reform legislation. Moreover, congressionally mandated budget reductions are already impacting the land management agencies, which virtually ensures that fewer funds will be available to support the scientific and related studies that are required to understand complex ecological processes and to make informed ecosystem management decisions. Nonetheless, ecosystem management legislation has been introduced into Congress, in the form of a bill that would establish a congressional oversight committee to study the concept (S. 93, 141 Cong. Rec. S173-01), which suggests that even Congress may recognize the logic behind the concept. However, until Congress gives explicit endorsement to the ecosystem management concept, federal land management agencies must proceed with caution, recognizing that they are politically accountable to a potentially hostile Congress.

Bureaucratically, an administratively conceived ecosystem management policy presents several challenges. Federal land management agencies, including the NPS, traditionally have been insular institutions, protective of their own turf and suspicious of sister agencies. Federal land managers are quite sensitive to maintaining their own managerial discretion, which makes them reluctant to cede any power either to a coordinating body or another agency (Sax & Keiter, 1987). These same federal managers are also reticent about intruding into the managerial prerogatives of sister agencies or involving themselves in the affairs of local communities, where they often are perceived as outside interlopers (Sax & Keiter, 1987). In addition, general bureaucratic ossification and resistance to change make it difficult for many agency employees to accept the new imperatives and protocols of ecosystem management (Shannon, 1994), in part because they are not yet convinced that the shift to ecosystem management represents a permanent change. Yet, by administratively restructuring along ecoregion lines (National Park Service, 1994a), NPS leadership has sent a clear message that it is serious about this transition to ecosystem management. Similar messages are coming from the Forest Service, U.S. Fish and Wildlife Service, and BLM, which should help to surmount bureaucratic resistance to ecosystem management.

Nonetheless, real ecosystem management reform will require the land management agencies to establish substantive priorities between protecting ecological integrity and extracting resource commodities, and to engage in meaningful interagency coordination processes. Because ecosystem management requires flexibility and adaptiveness to respond to evolving scientific information, agency officials must also be prepared to show increased levels of responsiveness when data indicates that assumptions must be changed. In other words, besides establishing clear ecological goals to govern interagency coordination efforts, federal land managers also must adopt flexible strategies to pursue these goals—both of which are contrary to the way bureaucracies ordinarily function. In addition, although land managers may prefer to build interagency relationships around informal associations and arrangements, they should be encouraged to employ more formal memoranda of understanding and other written agreements to ensure that coordination efforts will survive beyond the tenure of current managers.

Because public land management policy draws heavily on public involvement, ecosystem management will require the agencies to reassess their relationships with the public as well as traditional constituencies. A major challenge will involve harmonizing scientific information with public values (Lee, 1993). Another challenge will involve determining precisely what the public values, which will require land managers to devise appropriate processes to elicit public input into potentially controversial decisions. At the same time, given the widespread public interest in all federal lands, agency officials will have to address the issue of how much weight to accord national versus local interests in establishing ecosystem management policies (Romm, 1994). Despite hopes that consensus can be achieved over ecosystem management priorities, the reality is that some mechanism for public appeal of ecosystem management decisions will be necessary to resolve irreconcilable conflicts. In sum, federal land management agencies must be prepared to redefine how the public and special interest groups participate in the ecosystem management dialogue (Cortner, 1994). For the NPS, which has never taken its NEPA and public involvement obligations too seriously, this will require carefully defining the public's role and its participation opportunities.

Scientifically, ecosystem management presents the federal agencies with difficult technical and organizational challenges. Basic data about public land resources are lacking, and scientists are still discovering new ecological relationships with significant implications for ecosystem management policy (U.S. General Accounting Office, 1994b). In many cases, the human role in ecological change is still being documented and clarified. Debate also continues over such matters as ecosystem boundaries, species population dynamics, and the interpretation of ecological data. Nonetheless, so long as managers are willing to adapt policies to changing information and circumstances, there is sufficient consensus among scientists to make ecosystem management a viable natural resource management policy.

As an organizational matter, interagency management initiatives predicated upon the ecological sciences are proving difficult to coordinate (Clark, 1993). Agency scientists traditionally have collected disparate and often incompatible data, making it difficult to compile meaningful ecosystem-wide ecological profiles. Recent geographic information system technology should help with this problem, once compatible data is assembled and entered into the system. Relatedly, the federal agencies have adopted different definitions of the relevant ecosystem for management purposes: The Forest Service focuses its management on terrestrial ecoregions, while the U.S. Fish & Wildlife Service is committed to managing watersheds (U.S. General Accounting Office 1994b). Such differences will make coordinated management difficult in locations where shared ecosystems overlap. Beyond science, land managers need reliable social-economic data to evaluate the impact of proposed ecosystem management policies on local communities, yet there is little useful data available (U.S. General Accounting Office, 1994b). Clearly, if the NPS and other federal land management agencies are to coordinate management policies at an ecosystem level, they must standardize, expand, and improve current systems of data collection and analysis.

Although daunting, none of these problems is insurmountable—a fact confirmed by the pace at which ecosystem management is being institutionalized within the federal land management agencies. What follows, therefore, constitutes an examination of complementary management strategies that might be employed to bring meaningful ecosystem management policies to full fruition on the public domain.

Ecosystem Management Strategies

Although the general principles governing ecosystem management are now widely accepted, implementation strategies are still being developed and refined. Because most ecosystem management definitions focus on maintaining the integrity of ecological processes, federal land managers must be prepared to address and participate in developments occurring beyond existing boundary lines. Thus, a major challenge confronting federal land management agencies is to establish both substantive and procedural standards to govern these interjurisdictional relationships. Two interrelated coordination strategies are necessary: a *process-based strategy* that employs voluntary cooperative arrangements to address ecosystem concerns, and a *substantive strategy* that uses legal authority to establish minimum standards for protecting shared ecosystems. An effective ecosystem management policy for national parks will require agency officials to pursue both strategies, employing them in a complementary fashion. Anything less risks diluting the ecosystem management concept into a purely procedural undertaking, with no assurance that ecosystem integrity will be protected.

Interagency Coordination: Structure and Process

Because ecosystem boundaries rarely correspond to national park boundaries, an effective ecosystem management policy requires that park officials interact with adjacent land managers and owners. Indeed, a critical dimension of ecological management is the establishment and maintenance of interagency partnerships and relationships. Federal land management agencies are now experimenting with a variety of coordinated ecosystem management initiatives (U.S. General Accounting Office, 1994b). These cooperative, consensus-based relationships are designed purposefully to reduce conflict and to identify common resource management objectives while preserving managerial discretion. To succeed, however, partnership participants must pay careful attention to preliminary organizational and procedural arrangements.

The NPS appears to be committed to consensus-based arrangements to promote its ecosystem management goals. These arrangements are intended to achieve cooperative solutions to transboundary ecological and other resource problems. As cooperative ventures, the arrangements mirror the general predisposition of most land managers, who place a high premium on maintaining neighborly relations to avoid conflict (Sax & Keiter, 1987). In fact, conflict resolution is not why most NPS managers have pursued a career in land and natural resource management. They would prefer to enter into amicable negotiations with an adjacent manager than to face the prospect of an adversarial confrontation over the legality (or illegality) of specific resource management decisions.

Numerous examples of cooperative, ecosystem-based management initiatives already exist. Although the genesis and official underpinnings of these initiatives differ widely, they each reflect an attempt to create an interagency structure or process to address transboundary resource management problems. In the Yellowstone region, the Greater Yellowstone Coordinating Committee (GYCC), which is composed exclusively of NPS and Forest Service managers, is responsible for coordinating federal land management policies throughout the region; it is based on a 1960 Interagency Memorandum of Understanding, which expressly retains each agency's management authority (Keiter, 1989). In northwestern Montana, the Crown of the Continent Ecosystem Center is an informal, voluntary affiliation involving Glacier National Park, Waterton Lakes National Park, the U.S. Forest Service, and others to promote research and education on ecosystem-wide problems in the Northern Rockies; it is not designed to play any particular managerial role. The related Flathead Basin Commission, which was established by Montana legislation (Mont. Code Ann. §§ 75-7-301 to -308) in the wake of a controversial British Columbia coal mining proposal, is composed of federal, tribal, state, and local officials with the goal of improving communication among the governmental entities responsible for managing land and resources in the Flathead River drainage. On the Colorado Plateau, the Canyon Country Partnership was created through an interagency partnership agreement to address common resource management problems confronting the NPS, Forest Service, BLM, and local officials, principally the booming recreational activities occurring around Moab, Utah (Bill Hedden, pers. comm.). Other examples of NPS involvement in innovative interjurisdictional initiatives include the High Sierras initiative in California, the California State-wide Biodiversity Conservation Agreement (Wheeler, 1993), the North

Cascades International Park Initiative, and the Southern Appalachians Model Biosphere Reserve Program (Gilbert, 1988). None of these coordination initiatives was mandated by federal law; each represents an ad hoc effort to promote consensus over ecosystem-wide resource problems.

The real attraction of consensus-based coordination strategies is that they are relatively easy to initiate and provide an opportunity to avoid confrontation. Indeed, few people object to coordination in theory. But when interagency coordination processes produce proposals for real reform in management priorities and practices, then objections begin to surface. Therefore, unless interagency coordination efforts are carefully conceived and executed, participants face lurking pitfalls that can scuttle even the most well-intentioned effort. And the larger the scope and potential impact of the coordination effort, the greater are the potential pitfalls ahead.

The problems are perhaps best illustrated by the Greater Yellowstone Coordinating Committee's (GYCC) ill-fated vision document process. The vision process was prompted by congressional concern over a lack of regional management coordination that placed the wide-ranging grizzly bear at risk (Congressional Research Service, 1986). The GYCC's vision document was designed to produce a set of guiding principles for interagency management establishing Greater Yellowstone as a "world class model" for integrated, coordinated natural resource management. The draft document did just that: It called for ecosystem management and envisioned "a landscape where natural processes are operating with little hindrance on a grand scale . . . a combination of ecological processes operating with little restraint and humans moderating their activities so that they become a reasonable part of, rather than incumbrance upon, those processes" (Greater Yellowstone Coordinating Committee, 1990). However, fearing a significant shift in policy, some residents and commodity groups denounced the document and enlisted local political officials to subvert the process (Goldstein, 1992). As a result, when the GYCC released the final document, it represented a pale replica of the original visionary statement, omitting any reference to ecological management and emphasizing instead the separate missions of the two federal agencies (Greater Yellowstone Coordinating Committee, 1991). Not surprisingly, the document appears to have become a relic and has not figured prominently in the resolution of recent issues of ecosystem-wide significance.

What immediate lessons can be gleaned from this experience? First, because the Greater Yellowstone vision process was initiated by congressional prodding and because the Yellowstone region boasts charismatic resources of national significance, the GYCC made a tactical error by not pursuing the vision document as a national rather than local initiative (Keiter, 1996). Second, the GYCC's limited composition, which consisted only of federal representatives, left it vulnerable to the charge that it did not represent local interests, and it was therefore unable to forge a mutually shared vision for the region or to garner necessary local public support (Freemuth & Cawley, 1993). Third, by consciously choosing to ignore the NEPA process in order to bring the vision process to fruition promptly, the GYCC left itself open to criticism that it did not adequately involve the public in the process (Goldstein, 1992). Fourth, the vision process suffered from lack of a clear problem definition at the outset, which meant that the GYCC lacked any consensus, either within itself or with outside constituents, about the exact nature of regional resource management problems (Clark & Minta, 1994).

More broadly, the Greater Yellowstone experience illustrates the importance of establishing clear ecosystem coordination guidelines from the outset. Key issues that should be addressed include: (1) who to involve in the coordination partnership; (2) what roles each participant will play; (3) how to define the common resource management problems; (4) what rules—both substantive and procedural—will govern the effort; and (5) how progress will be measured. To address these issues systematically, partnership members should consider preparing a formal agreement, perhaps a memorandum of understanding, charter, or similar document, which can also help to ensure regularity and continuity. While informal relationships among managers and agency employees are critical to establish trust initially, meaningful ecosystem-level reform will require more durable and formal arrangements. The agreement should define the partnership structure, including its membership composition and opportunities for public participation. Participants ordinarily should include representatives from the governing or management bodies with jurisdictional authority over the affected lands, which means most ecosystem management partnerships should include federal, state, and local government officials. In addition, participation opportunities should be afforded affected interest groups as well as the general public, with resource managers recognizing that public values must be reconciled with biological requirements through this process (Maguire 1994).

The agreement should also establish operating rules for the partnership, including how decisions will be reached and whether they will have a binding effect on members. In addition, it should clarify the partnership's goals by first identifying the relevant resource management problems and then establishing clear standards or criteria for resolving these problems (Clark 1993). Rather than simply coordinating for the sake of coordination, the agreement should identify common management goals, such as the protection of ecological integrity and processes, that will govern resolution of conflicts throughout the ecosystem (Keiter 1994). Moreover, partnership members should attempt to define how progress will be assessed, either through objective scientific criteria, related socioeconomic factors, or other indicators (Agee & Johnson 1988).

Without such a formal agreement, there is no assurance that the effort will survive beyond the tenure of the present participants, or that all participants will be committed to a common goal of maintaining ecosystem integrity. And without clearly specified goals and standards, ecosystem management will exist in name only.

Beyond Coordination: Law and Ecosystem Management Standards

Even when ecosystem management goals and processes are established through a formal agreement, difficult issues that go to the core of each agency's mission will undoubtedly surface and test the limits of coordination. Federal land management agencies have yet to confront the vexing question of how divergent statutory mandates are to be reconciled at the boundary line where ecosystems overlap (Cortner, 1994). Does or should the NPS's rigid preservationist mandate "trump" the Forest Service or BLM's flexible multiple-use mandate? While the answer may be clear when the agencies agree on the ecological or environmental ramifications of a particular decision, what is the result when the agencies disagree over the impacts or degree of risk involved? An ecosystem management strategy based upon interagency coordination principles suggests that these difficult questions should not be resolved merely by relying upon administrative discretion and authority. Rather, as we have seen, the law establishes important ecosystem-wide standards that address these problems. The NPS, therefore, can and should invoke the law to establish and define substantive coordination standards to protect ecological resources and processes.

The NPS, however, has traditionally been reluctant to rely upon law to define its role or responsibilities beyond park boundaries. Whether as a matter of agency culture, ingrained political sensitivities, or just deference to the prerogatives of adjacent land managers, NPS officials ordinarily have not aggressively asserted their legal authority, either when responding to threatening external activities or promoting ecosystem management initiatives (Sax & Keiter, 1987). In fact, the NPS has not been a particularly legalistic agency. It has only rarely initiated litigation, even to curtail external activities threatening vital park resources (Lockhart, 1988). It has promulgated very few binding regulations to define its natural resource management responsibilities, choosing instead to rely upon policy statements, which are more flexible and have no legally binding force. Perhaps because NPS's preservationist mission and implementation policies have been relatively noncontroversial, the agency has had few occasions to think in terms of its legal authority. But as the Yellowstone bison controversy illustrates so cogently, the agency's insularity is no longer assured, even in Yellowstone's expansive wilderness setting.

The era of ecosystem management presents the NPS with an opportunity to reexamine its traditional reluctance to use the law to accomplish natural resource management objectives. For several reasons, the NPS is in a uniquely strong position to rely upon the law as a tool for ecosystem protection. First, adjacent Forest Service or BLM managers, unlike their NPS counterparts, operate under multiple-use mandates, which provide them with the flexibility to forego or adjust activities that threaten common ecological resources. Second, under the Organic Act's clear preservationist mandate as well as its regulatory provisions, the NPS has the legal responsibility and authority to address matters beyond park boundaries to accomplish ecosystem management objectives. Third, the NPS also can invoke a plethora of related environmental laws, such as the ESA, the Clean Water Act, and National Forest Management Act, to constrain adjacent activities that might harm ecosystem resources or processes. In short, the law provides the NPS with a firm legal basis for redefining its relationship with neighboring land management agencies by promoting resource management objectives designed to ensure ecosystem integrity.

Despite the NPS's traditional aversion to law, it is the law—through litigation and otherwise—that has been regularly invoked to protect park resources and ecosystem integrity. Litigation directly against the agency, in the case of Redwood National Park, provided critical leverage to secure congressional passage of legislation expanding the park and thus stopping destructive upstream logging (Hudson, 1979). In the case of Everglades National Park, the agency's own involvement in litigation played a key role in forcing negotiation of a landmark federal-state-industry agreement that is designed to restore the Everglades ecosystem. Litigation by environmental groups, in the case of Glacier National Park, has forestalled threatened oil and gas development on Forest Service lands adjoining the park and thus secured critical wildlife migration corridors (Sax & Keiter, 1987). Even in the absence of litigation, participants in recent ecosystem management initiatives acknowledge that the coercive power of laws like the ESA provide both the catalyst and adhesion for successful interagency coordination efforts (Wheeler 1993). The law, therefore, should not be discounted as a key factor in promoting and shaping ecosystem-based management initiatives designed to protect national parks and other biological resources.

Nonetheless, it is unlikely that the NPS will soon become an aggressive litigative or regulatory agency. Federal agencies ordinarily do not sue each other. Disputes involving Department of the Interior agencies will generally be resolved through top-level political negotiations within the Department itself, just as disputes with non-Interior agencies will usually be addressed through similar political channels. However, this era of ecosystem management and carefully structured interagency coordination agreements based on clearly defined ecological standards should enable the federal land management agencies to resolve most problems before they escalate to a higher level. Because the NPS is mandated to protect park resources in an unimpaired condition, it can—and should—insist upon an ecosystem management standard that ensures ecosystem integrity. Failure to insist upon such a protective standard could place the NPS in violation of its principal legal obligation.

Similar considerations govern NPS relationships with private neighbors during this transitional ecosystem management period. Not only is the agency reluctant to litigate over threatening activities from adjacent private lands, but political considerations—often brought to bear by local congressional delegations—are likely to deter confrontations with prominent constituents. Nonetheless, NPS officials should be prepared to invoke the law as a framework for negotiations to protect park resources. Moreover, the law can and should be used to establish appropriate standards to govern public-private partnerships created for coordinated regional management purposes. Based upon its legal obligation to maintain park resources unimpaired, the NPS should insist that these cooperative relationships embrace the goal of maintaining ecosystem integrity. With the credible threat of exercising its regulatory authority beyond park boundaries or of initiating litigation to prevent damage to park resources, the NPS should have sufficient legal clout to bring even recalcitrant neighbors into an acceptable ecosystem management agreement.

In sum, the law provides the NPS with a legal basis for ensuring that the goal of maintaining the integrity of ecological resources and processes is incorporated into its ecosystem management relationships. By structuring cooperative agreements around its substantive nonimpairment standard, the NPS can help to ensure the ecological integrity of its own resources as well as those associated with overlapping ecosystems. Anything less would violate the agency's governing legal mandates, either opening it to litigation or forcing it to pursue litigation to protect park resources. Moreover, unless clear substantive standards are incorporated into ecosystem management relationships, partnerships with other agencies are in danger of collapsing into merely coordination exercises—all process and no substance. And that is clearly not what ecosystem management is about.

Conclusion

Returning to Yellowstone's bison controversy, it is evident that the matter must be addressed in its full ecological context. The controversy cannot be confined by artificial boundaries; it is affecting important park resources as well as the interests of adjacent landowners, just as resource development proposals outside the park inevitably affect the park's own resources. Something more than mere coordination will be required to reconcile the powerful concerns that are driving the participants toward intractable positions.

What the preceding discussion has been designed to suggest is that the law, which obligates the NPS to ensure the biological integrity of its bison population, provides standards that can and should be used to govern resolution of this matter as well as other regional resource management controversies. By structuring interagency coordination processes around these standards, the regional land and resource managers can employ ecosystem management principles to protect ecological processes and to reduce the level of conflict among governmental agencies and with private organizations and individuals.

During this transitional period, ecosystem management concepts are still being defined and developed. Key goals of ecosystem management are the protection of ecological processes and conservation of biological diversity, achieved in large part through interagency coordination. Yet because the federal land management agencies are governed by diverse laws that do not uniformly endorse these goals, a commitment to interagency coordination alone may not be sufficient to protect ecosystem integrity. Therefore, the NPS must be prepared to invoke its own and related legal mandates to ensure that ecosystem management relationships are based upon substantive standards that will, in fact, protect ecological resources across jurisdictional boundaries. Otherwise, this new era of ecosystem management is at risk of becoming yet another failed experiment in public land policy reform. Today's ecological pressures are too pressing to allow that to happen.

Literature Cited

Agee JK, Johnson DR. *Ecosystem Management for Parks and Wilderness.* Seattle: University of Washington Press, 1988.

Alabama-Tombighbee Rivers v Department of the Interior, 26 F3d 1103 (11th Cir 1994).

Boyce MS. Natural Regulation or the Control of Nature? In Keiter R, Boyce M (eds), *The Greater Yellowstone Ecosystem: Redefining America's Wilderness Heritage.* New Haven, CT: Yale University Press, 1991:183–208.

Budd KJ. Ecosystem Management: Will National Forests Be "Managed" into National Parks? In Keiter R, Boyce M (eds), *The Greater Yellowstone Ecosystem: Redefining America's Wilderness Heritage.* New Haven, CT: Yale University Press, 1991:65–76.
Bureau of Land Management. *Ecosystem Management in the BLM: From Concept to Commitment.* Washington, DC: U.S. Government Printing Office, 1994.
Bybee JS. Advising the President: Separation of powers and the Federal Advisory Committee Act. *Yale Law J* 1994;104:51–128.
Chase A. *Playing God in Yellowstone.* Boston: Atlantic Monthly Press, 1986.
Clark TW. Creating and using knowledge for species and ecosystem conservation: Science, organizations, and policy. *Persp. Biol Med* 1993;36:3–17.
Clark TW, Minta SC. *Greater Yellowstone's Future: Prospects for Ecosystem Science, Management, and Policy.* Moose, WY: Homestead Publishing, 1994.
Clark TW, Zaunbrecher D. The Greater Yellowstone Ecosystem: The Ecosystem Concept in Natural Resource Policy and Management. *Renew Resourc J*,1987;5(3):8–16.
Congressional Research Service. *Ecosystem Management: Federal Agency Activities.* Washington, DC. 1994. 94–339 ENR.
Congressional Research Service. *Greater Yellowstone: An Analysis of Data Submitted by Federal and State Agencies.* U.S. House Committee on Interior and Insular Affairs, Subcommittee on Public Lands and Subcommittee on National Parks and Recreation. Comm. Print No. 6. Washington, DC: U.S. Government Printing Office, 1986.
Cortner HJ. Intergovernmental Coordination in Ecosystem Management. In U.S. Senate Committee on Environment and Public Works. *Ecosystem Management: Status and Potential.* S. Rpt. 103–98. Washington, DC: U.S. Government Printing Office, 1994;229–242.
Cortner HJ, et al. *Institutional Barriers and Incentives for Ecosystem Management: A Problem Analysis.* Tucson, AZ: Water Resources Center, University of Arizona, 1994.
Falen FJ, Budd-Falen K. The right to graze livestock on the federal lands: The historical development of Western grazing rights. *Idaho Law Rev* 1993–1994;30:506–524.
Fink RJ. The National wildlife refuges: Theory, practice, and prospect. *Harv Environ Law Rev* 1994;18:1–135.
Forest Ecosystem Management Assessment Team. Forest Ecosystem Management: An Ecological, Economic, and Social Assessment. Washington, DC: U.S. Government Printing Office, 1993. Publ. 1993-793-071.
Freemuth J, Cawley RM. Ecosystem management: The relationship among science, land managers, and the public. *The George Wright Forum* 1993;10(2):26–32.
Gilbert VC. Cooperation in Ecosystem Management: In Agee JK, Johnson DR (eds), *Ecosystem Management for Parks and Wilderness.* Seattle: University of Washington Press, 1988:180–192.
Goldstein B. Can ecosystem management turn an administrative patchwork into a Greater Yellowstone Ecosystem? *Northwest Environ J* 1992;8:285–324.
Greater Yellowstone Coordinating Committee. *Vision for the Future: A Framework for Coordination in the Greater Yellowstone Area* (Draft). Billings, MT: U.S. Department of the Interior, National Park Service; U.S. Department of Agriculture; Forest Service, 1990.
Greater Yellowstone Coordinating Committee. *A Framework for Coordination in the Greater Yellowstone Area* (Final). Billings, MT: U.S. Department of the Interior, National Park Service; U.S. Department of Agriculture, Forest Service, 1991.
Grumbine RE. What is ecosystem management? *Cons Biol* 1994;8:27–38.

Grumbine RE. *Ghost Bears: Exploring the Biodiversity Crisis.* Washington, DC: Island Press, 1992.

Hage W. *Storm over Rangelands: Private Rights in Federal Land.* Bellevue, WA: Free Enterprise Press, 1990.

Hudson DA. Sierra Club v Department of the Interior: The Fight to preserve the Redwood National Park. *Ecol Law Quart* 1979;7:781–859.

Keiter RB. *Greater Yellowstone: Managing a Charismatic Ecosystem.* Logan, UT: Utah State University, Department of Natural Resources and Environmental Issues, 1996.

Keiter RB. Beyond the boundary line: Constructing a law of ecosystem management. *Univ Colorado Law Rev* 1994;65:293–333.

Keiter RB. NEPA and the emerging concept of ecosystem management on the public lands. *Land and Water Law Rev* 1990;25:43–60.

Keiter RB. Taking account of the ecosystem on the public domain: Law and ecology in the Greater Yellowstone Region. *Univ Colorado Law Rev* 1989;60:924–1007.

Keiter RB. On protecting the national parks from the external threats dilemma. *Land and Water Law Rev* 1985;20:355–420.

Keiter RB, Froelicher P. Bison, brucellosis, and law in the Greater Yellowstone Ecosystem. *Land and Water Law Rev* 1993;28:1–75.

Lee KN. *Compass and Gyroscope: Integrating Science and Politics for the Environment.* Washington, DC: Island Press, 1993.

Leopold A. *A Sand County Almanac.* New York: Oxford University Press, 1949.

Leopold AS, et al. Wildlife management in the national parks. *Trans N Am Wildl Nat Res Conf* 1963;28:29–44.

Lockhart WJ. External Park Threats and Interior's Limits: The Need for an Independent Park Service. In Simon D (ed), *Our Common Lands: Defending the National Parks.* Washington, DC: Island Press 1988:3–74.

Maguire LA. Science, Values, and Uncertainty: A Critique of the Wildlands Project. In Grumbine RE (ed), *Environmental Policy and Biodiversity.* Washington, DC: Island Press, 1994:267–272.

Moote MA, Burke S, Cortner HJ, Wallace MG. *Principles of Ecosystem Management.* Tucson, AZ: Water Resources Research Center, University of Arizona, 1994.

Nash R (ed). *Wilderness and the American Mind.* New Haven, CT: Yale University Press, 1982.

National Park Service. *Restructuring Plan for the National Park Service.* Washington, DC: National Park Service, 1994a.

National Park Service. *Ecosystem Management in the National Park Service* (Discussion Draft). Washington, DC: National Park Service, 1994b.

National Park Service. *State of the Parks 1980: A Report to Congress.* Washington, DC: National Park Service, 1980.

Northwest Forest Resource Council v Espy, 846 F. Supp. 1009 (D.D.C. 1994).

Noss RF, Cooperrider AY. *Saving Nature's Legacy: Protecting and Restoring Biodiversity.* Washington, DC: Island Press, 1994.

Reed SW. The county supremacy movement: Mendacious myth marketing. *Idaho Law Rev* 1993–1994;30:525–553.

Reese R. Greater Yellowstone: The national park and adjacent wildlands. *Montana Geographic Series* (2d ed) 1991;6:1–104.

Rohlf D, Honnold DL. Managing the balance of nature: The legal framework of wilderness management. *Ecol Law Quart* 1988;15:249–279.

Romm J. Jurisdictional Relations in Ecosystem Management: Some Observations from California. In U.S. Senate Committee on Environment and Public Works. *Ecosystem Management: Status and Potential.* S. Rpt. 103-98. Washington, DC: U.S. Government Printing Office, 1994:97–106.

Runte A. *National Parks: The American Experience* (2d ed). Lincoln, NE: University of Nebraska Press, 1987.

Sax J. Property rights and the economy of nature: Understanding Lucas v South Carolina Coastal Council. *Stanford Law Rev* 1993;45:1412–1455.

Sax J. Helpless giants: National parks and the regulation of private lands. *Michigan Law Rev* 1976;75:239–274.

Sax J, Keiter R. Glacier National Park and its neighbors: A study of federal interagency relations. *Ecol Law Quart* 1987;14:207–263.

Shannon MA. Coordination Among Federal Agencies: Cultures, Budgets, and Policies. In U.S. Senate Committee on Environment and Public Works. *Ecosystem Management: Status and Potential.* S. Rpt. 103-98. Washington, DC: U.S. Government Printing Office, 1994:209–228.

Shannon MA. Community Governance: An Enduring Institution of Democracy. In U.S. House Committee on Interior and Insular Affairs. *Multiple Use and Sustained Yield: Changing Philosophies for Federal Land Management.* Comm. Print No. 11. Washington, DC: U.S. Government Printing Office, 1993:219–246.

Sierra Club v Block, 622 F. Supp. 842 (D. Colo. 1985), vacated on other grounds sub nom. *Sierra Club v Yeutter,* 911 F2d 1405 (10th Cir. 1990).

Sierra Club v Lyng, 663 F. Supp. 556 (D.D.C. 1987).

Sierra Club v Lyng, 694 F. Supp. 1260 (E. D. Tex. 1988), aff'd on other grounds sub nom. *Sierra Club v Yeutter,* 926 F2d 429 (5th Cir. 1991).

Sierra Club v Yeutter, 911 F2d 1405 (10th Cir. 1990).

Slocombe DS. Environmental planning, ecosystem science, and ecosystem approaches for integrating environment and development. *Environ Manage* 1993;17:289–303.

Thomas v Peterson, 753 F2d 754 (9th Cir. 1985).

Thorne ET, Meagher M, Hillman R. Brucellosis in Free Ranging Bison: Three Perspectives. In Keiter R, Boyce M (eds), *The Greater Yellowstone Ecosystem: Redefining America's Wilderness Heritage.* New Haven, CT: Yale University Press, 1991:275–288.

U.S. Fish and Wildlife Service. *Grizzly Bear Recovery Plan.* Missoula, MT: 1993.

U.S. Forest Service. Memorandum from Dale Robertson, U.S. Forest Service Chief, to Regional Foresters. Washington, DC: 1992.

U.S. General Accounting Office. *National Park Service: Activities Outside Park Borders Have Caused Damage to Resources and Will Likely Cause More.* Washington, DC: 1994a. GAO/RCED-94-59.

U.S. General Accounting Office. Ecosystem Management: Additional Actions Needed to Adequately Test a Promising Approach. Washington DC: 1994b. GAO/RCED-94-111.

Wheeler DP. Foreword: A Strategy for the Future. *Stanford Environ Law J.* 1993;12:ix–xv.

Wilkinson CF. *Crossing the Next Meridian: Land, Water, and the Future of the West.* Washington, DC: Island Press, 1992.

Wright RG. Wildlife research and management in the national parks. Urbana, IL: University of Illinois Press, 1992.

II

Planning and Evaluating the Ecological Integrity of Parks and Protected Areas

Chapters 6–11 discuss how parks and protected areas can help protect a nation's biodiversity, the mechanisms for accomplishing and evaluating this mission, and the problems encountered when sufficient lands are not protected.

Chapter 6 questions whether sustainable use and sustainable development in the context of ecosystem management—concepts supported by several authors in the first section—are actually viable methods to preserve biodiversity. The author believes that large protected areas are essential to protect biodiversity and that they may function best when enveloped in a well-managed matrix of multiple-use zones. The fundamental question is "how large should these protected areas be?" Clearly, there is no one answer, as the author concludes, but in almost every region of the world much more land and water must be protected to maintain biodiversity.

The spatial databases depicting the distribution of plant communities and animal species over entire states, which are being developed by the Gap Analysis Program (GAP), offer a tremendous opportunity to identify assemblages of organisms that exist primarily on lands outside the networks of protected areas and to develop efficient ways to protect them before they become endangered. The GAP databases, which contain information at the large scales envisioned for protected areas in Chapter 6, provide a new and relatively untried venue for planning parks and evaluating the ecological suitability of lands to be included in parks. An application of how these data sets can be used to evaluate the conservation potential of new parks is described in Chapter 7.

However, examples of the actual establishment of new parks in the United States according to a systematic plan are rare. One example comes from Alaska.

There, the opportunity to establish new parks arose in the late 1970s as a result of expansive federal legislation to settle native land claim issues and because of the existence of large federal wilderness holdings in the state. A discussion of the methods used to expand the park system in that state and a retrospective view of the success of that effort are covered in Chapter 9.

It is clear, however, that planners need more than just biological information to adequately design and manage protected areas. This is particularly true when designing areas to protect species, like the grizzly bear, which often conflict with human values. In fact, as Chapter 8 concludes, biological factors are relatively unimportant determinants in the design of a grizzly bear preserve. Rather, it is knowledge of human values, behavior, densities, and distribution, both within and adjacent to protected areas, that should be the determining factors in the design of a successful grizzly bear preserve.

Of course, large, undisturbed natural areas are not the norm. In tropical and temperate forests, formerly continuous tracts have rapidly come to resemble islands of natural areas located in seas of modified habitat. The broad consequences of this phenomenon has led to extensive study of the function and design of habitat corridors reconnecting once continuous, but now isolated, blocks of undisturbed habitat. This issue is explored in detail in Chapter 10.

Chapter 11 presents the results of an eloquent study illustrating that human-caused habitat fragmentation and isolation are real and often decidedly negative phenomena faced by many animal species, and, in this study, by the California red-backed vole in particular. However, equally important, this study shows that fragmentation is not a universally negative phenomenon and that different species respond differently to fragmentation, isolation, and edge effects. The vulnerability of a species to the effects of fragmentation may be strongly related to its ability to tolerate the disturbed habitat matrix of the "sea" surrounding the undisturbed stands. This lead us to realize that land managers must focus not only on the undisturbed parks and corridors, as has been largely emphasized in the previous chapters, but they must also work to conserve and rehabilitate elements of the surrounding landscape.

6

Protected Areas: How Much Is Enough?

Reed F. Noss

For over a century, protected areas have been the cornerstones of biological conservation. They have been perceived as the last strongholds of wild nature, where the flora and fauna of a region could persist indefinitely without human meddling. But, in recent years, the role of protected areas in conservation has been questioned on several fronts. Conservation biologists have pointed out that reserves alone are unlikely to maintain viable populations of many species because they usually are too small and isolated from one another (Diamond, 1975; Wilcox, 1980; Harris, 1984; Noss & Harris, 1986). Other critics claim that parks and wilderness areas no longer play a useful role in reconciling conservation and development because they are elitist and anti-people. For example, the national park idea, when transferred to Africa and other developing countries, has conflicted with the needs and aspirations of local human communities (Harmon, 1987; Barnes, 1994). Because of these and other reasons, protected areas are becoming evermore difficult to establish in many parts of the world. Similarly, in the United States, the philosopher Callicott (1994/95), among others, contends that the wilderness idea is anachronistic, ecologically uninformed, ethnocentric, historically naive, and politically counterproductive. Although Callicott attacks the wilderness idea rather than wilderness areas, his critique comes at a time when politically inspired antagonism toward protected areas—and public lands in general—is becoming increasingly virulent.

These trends suggest that establishing new national parks, wilderness areas, and other reserves will be extremely difficult almost everywhere. But if not reserves, what then can we rely on to maintain our biological heritage? Most of the rational critics of protected areas offer some model of sustainable development and resource use as an alternative strategy. The idea is that people can learn to exploit natural resources without degrading ecosystems and losing biodiversity and, as economic conditions improve, people will be more willing to treat nature with respect (see Chap. 2). However, these hopeful assumptions appear naive. Experiments in sustainable development, sustainable use, multiple use, and ecosystem management have largely failed (Ludwig et al., 1993; Irvine, 1994), leading many biologists and conservationists to reassert the fundamental values of protected areas, while recognizing their limitations (Robinson, 1993; Noss & Cooperrider, 1994; Foreman, 1994/95; Noss, 1994/95).

An emerging consensus, at least among biologists, seems to be that protected areas are necessary but not sufficient to meet conservation objectives (see Meffe & Carroll, 1994). Yes, protected areas will be difficult to establish, but they seem to be an essential part of any successful conservation strategy. A revised biosphere-reserve model of interconnected reserves enveloped in multiple-use buffer zones or a well-managed matrix will apply to most regions and may offer the best hope of maintaining biodiversity over the long term (Noss & Cooperrider, 1994). The role of reserves in this new model is different and more interactive with surrounding human communities than in the past, but it is no less crucial. In almost every region of the world, much more land and water must be protected if biodiversity is to be maintained.

Despite the renewed recognition that protected areas play an essential role in conservation, the question "how much is enough?" remains controversial and fundamentally unanswered. For many years, the dominant research questions in conservation biology had to do with how large a single reserve must be to maintain its diversity over time and whether a single large reserve is better than several small reserves of equivalent total area. These questions now are seen as largely irrelevant to real-world conservation planning because conservation decisions are never that simple—many other factors, such as the type of surrounding land uses, the life histories and distributions of the species concerned, and the specific management practices employed, come into play (Noss & Csuti, 1994; Noss & Cooperrider, 1994). But the question of how much of a given region should be protected or managed—such that native biodiversity and ecological integrity are maintained—is still pertinent. So, too, are questions of how strictly regulated a reserve must be to qualify as a protected area, what distribution and spatial configuration of reserves works best, and how surrounding buffer or matrix lands should be managed. This chapter discusses these issues and provides some guidelines for constructing a reserve network adequate to maintain biodiversity. The discussion demonstrates that, although far from straightforward, the question "how much is enough?" is not intractable and must be approached empirically and with an open mind.

Science and Values

Science lacks ready formulas for determining how big reserves should be or how much land in a given region should be protected. In the first book published on conservation biology, Soulé and Wilcox (1980) stated, "[R]eserves should be large, manifold and dispersed (except that a few highly vagile species benefit from proximity of reserves). To such questions as 'how big?' or 'how many?,' the answers are 'as big as possible' and 'as many as possible.' "

Six years later, two scientists (Soulé & Simberloff, 1986) who had been on opposing sides of the small- versus large-reserves controversy came to consensus with a similar statement:

> Nature reserves should be as large as possible, and there should be many of them. The question then becomes how large and how many. There is no general answer. For many species, it is likely that there must be vast areas, while for others, smaller sites may suffice so long as they are stringently protected and, in most instances, managed. If there is a target species, then the key criterion is habitat suitability. Suitability requires intensive study, especially in taxa that contain species with narrow habitat requirements.

As McCoy (1983) noted, neither the island biogeographic method of extrapolating species-area equations nor the "method of addition," which simply calculates the minimum area that contains the number of species to be preserved in the sites being considered, has produced credible estimates of the minimum area required to meet conservation objectives. What is most often missing in such analyses is detailed data on the sites and the autecological requirements of the species in question (McCoy, 1983).

Thus, although hard answers are often demanded, scientists have been able to offer little explicit guidance on how much protected area should be established. Some specific recommendations have been offered by international commissions. The 1982 World Parks Congress in Bali called on governments to protect 10 percent of their lands, without offering any reasons for why this figure was selected. By this time, sustainable development had become the rallying cry of international conservationists, and parks were seen as needing redefinition within the context of national economies. In 1987, the Brundtland Commission of the United Nations defined *sustainable development* as "development that seeks to meet the needs and aspirations of the present without compromising the ability to meet those of the future" and reaffirmed the role of protected areas in achieving this goal (World Comm on Env & Dev, 1987). The Brundtland Commission was modest in its request, calling on all nations of the world to set aside 12 percent of their land base in protected areas (World Comm on Env & Dev, 1987). Since then, the 12 percent figure has been adopted as a target by many nations, states, and organizations around the world (e.g., the Endangered Spaces Campaign of World Wildlife Fund Canada [Hummel, 1989]), and the figure underscores the idea of adequately representing all ecosystem types in the country (Iacobelli et al., 1995).

Unfortunately, there is little evidence of any systematic process of analysis by the Brundtland Commission to arrive at the widely embraced 12 percent figure, and, thus, many question whether it is scientifically defensible. There is only the vague suggestion that "the total expanse of protected areas needs to be at least tripled if it is to constitute a representative sample of Earth's ecosystems" (World Comm on Env & Dev, 1987), citing McNeely and Miller (1982) for justification. Given that about 4 percent of the earth's land base was estimated to be in protected areas in 1987, a tripling would lead to 12 percent protection.

It is useful to evaluate any report or other effort in terms of its underlying values framework. Values, after all, are what determine goals and directions, in science as well as in politics. The singular focus of the Brundtland Commission was on humans, with interest in other species only to the extent that they contribute to meeting human needs and desires. The report explicitly advocated an expanding world economy fueled by technological innovation (Irvine, 1994). It is difficult to see how this development could be achieved without significant losses of biodiversity. Yet, implicit in the recommendation for 12 percent protection is the untested assumption that this level will be sufficient to provide the environmental services on which human society depends.

There is no scientific basis to the 12 percent and no reason to believe that internationally recognized conservation goals could be met by protecting only 12 percent of any region, if surrounding lands are developed intensively. The 12 percent figure seems to have arisen as a guess of what a politically reasonable target might be. Indeed, 12 percent would be a dramatic improvement over the present situation, in which only about 1.5 percent of the earth's surface, or 5.1 percent of national land area, is in protected areas of at least 1000 hectares in size (World Resources Institute et al., 1992). These estimates are by no means rigorous. Protected area proportions calculated from other figures in the same report (World Resources Institute et al., 1992) show a worldwide total for protected areas of 3.04 percent (Table 6.1). Despite uncertainty over the real amount of protected land, it surely is far less than 12 percent globally. However, whether 12 percent is enough to maintain critical ecological and evolutionary processes, to keep the air breathable and water drinkable, and to maintain viable populations of all earth's species are questions that were never asked explicitly by the Brundtland Commission. Rather, the 12 percent figure derives

Table 6.1 Numbers and combined areas of reserves at least 1000 hectares in size in IUCN protected area categories I–III (i.e., strict nature reserves, national parks, and natural monuments) by region, and percentage of region protected

Region	Number of reserves	Area (ha)	Region protected (%)*
North and Central America	610	170,344,290	7.25
South America	289	58,190,622	3.27
Africa	260	88,722,877	2.93
Asia (including Russia)	585	59,305,756	1.32
Europe	289	8,056,879	0.77
Australia and South Pacific	443	67,872,385	8.79
Antarctica	12	220,649	0.02
Total	2,488	452,713,458	3.04

IUCN = International Union for Conservation of Nature and Natural Resources (now, World Conservation Union).
*Calculated using regional areas from *Encarta Multimedia Encyclopedia,* 1993.
Source: Adapted from World Resources Institute, World Conservation Union, and United Nations Environment Programme. *Global Biodiversity Strategy* Washington, DC: World Resources Institute, 1992.

from a set of value assumptions that correspond to an anthropocentric worldview where humans are seen as the only legitimate objects of moral concern and where other species are valued only to the extent that they serve humans.

The point is that any answer to the question "how much is enough?" is necessarily value-laden. It will reflect the values and interests of the individuals and groups who frame the question and attempt to provide an answer. A scientific or rational process for determining the necessary extent of protected area is possible, but it cannot operate independently of values. Stated values can be incorporated into a set of explicit goals. Once these value-laden goals and objectives are agreed on—at least by those engaged in the analysis—then science can be applied relatively objectively to arrive at defensible answers to the question.

Noss (1992) and Noss and Cooperrider (1994) stated four objectives for regional conservation that contribute to the overarching goal of maintaining biodiversity and ecological integrity worldwide in perpetuity. These objectives are central to the strategy of The Wildlands Project (Foreman et al., 1992; Noss, 1992). The primary ethical assumption underlying these objectives is that species and other natural entities and processes are valuable for their own sake:

1. Represent, in a system of protected areas, all native ecosystem types and seral stages across their natural range of variation.
2. Maintain viable populations of all native species in natural patterns of abundance and distribution.

3. Maintain ecological and evolutionary processes, such as disturbance regimes, hydrological processes, nutrient cycles, and biotic interactions.
4. Design and manage the system to be resilient to short-term and long-term environmental change and to maintain the evolutionary potential of lineages.

An Empirical Evaluation Framework

How can we determine the area of protected land needed to achieve the goals and objectives just stated? First, we must recognize that the old debates about how big single reserves must be to maintain their diversity and whether single large reserves are better than several small reserves are, with few exceptions, defunct. Island biogeographic theory and species-area relationships offer an interesting intellectual background to real-world conservation planning but not much else. More important questions center around the area and configuration of habitat needed to maintain viable populations of particular target species and, ultimately, to support the entire native biota of a region (Karr, 1990; Woodley et al., 1993; Angermeier & Karr, 1994; Westra, 1994; Noss, 1995).

As noted by Murphy and Noon (1992), Noss (1992), Noss and Cooperrider (1994), Noss (1995), and Reid and Murphy (1995), iterative map-based conservation planning, using a variety of evaluation criteria, may be the most objective and comprehensive methodology available for determining an optimal reserve network. Preliminary maps serve as hypotheses to be tested; they are refined as analyses fail to confirm specific map properties or reveal new important areas (Murphy & Noon, 1992). Because of the uncertainty inherent in all databases and predictive models, a truly optimal reserve network is unknowable; however, an empirical approximation is possible when objectives and assumptions are explicit.

Geographic information systems (GIS) make multiple-criteria spatial analyses relatively straightforward. For determining an optimal network, each conservation criterion is represented as a layer in a GIS map overlay. Areas needed to fulfill each criterion in the region of interest are mapped, maps for all criteria are overlaid, and the final map represents the optimal reserve network (given the current state of knowledge). The proportion of the protected network within the region is the empirical answer to the question "how much is enough?" This answer is of political interest but is biologically superfluous; much more important is the spatial configuration of the reserve network and how the entire landscape is actually managed.

This empirical approach is highly case-specific and will result in different answers for different regions. Regions vary tremendously in terms of physical and biotic heterogeneity, area requirements of the extant fauna, land use, and other factors, such that 15 percent of one region may be adequate to maintain the extant fauna and flora, whereas 90 percent of another region is needed to accomplish the same objectives. Because putting together optimal reserve networks is a long-term endeavor, in which completion of networks may not be anticipated for several decades or more, the level of optimism about the future will necessarily influence answers. For example, 15 percent of a region in the midwestern United States may suffice to maintain the extant flora and fauna with current patterns of land use; however, abandonment of agricultural land, closure of roads, and reintroduction of extirpated species (e.g., bison, wolves) may be possible over the next century or two, such that 50 percent of the landscape in the year 2150 may be needed to fulfill the upgraded objectives. I therefore recommend a chronological sequence of reserve networks developed in phases, with each stage somewhat bigger or more ambitious than the preceding.

Ideally, regions should be defined on the basis of physiographic and/or biogeographic criteria rather than by political boundaries. Thus, the ecoregions of Omernick (1987) or Bailey et al. (1994) would be useful for analysis. The boundaries of regions or the regional classification used are really not all that important; in the final analysis, all that matters is that the classification make some ecological sense, that nothing is left out, and that opportunities for habitat connections between regions be examined.

The following section discusses the steps in the process of building information and data layers that ultimately will be overlaid to form the proposed, sequential reserve network (Fig. 6.1). At each step, factors and considerations that influence estimates of optimal protected area for a region are reviewed (see Noss & Cooperrider, 1994; Noss, 1995). It will be evident in this review that the process of reserve selection and design, although largely empirical and reasonably objective, is still as much an art as a science. It is a science in that scientific theories, models, methods, and data form the basis of all analyses. It is an art in that the process of combining, weighing, and evaluating various criteria and data layers requires human judgment and the "intuition" of experienced ecologists. The human brain, properly educated, is still superior to the most advanced supercomputers in these integrative, synthetic functions.

Represent Species or Habitat Types

Perhaps the best accepted biological conservation goal worldwide is representation of all elements of biodiversity in protected areas (Noss & Cooperrider, 1994). Assessments of species representation in reserve networks are common (Pressey et al., 1993). But, because data on individual species are often limited

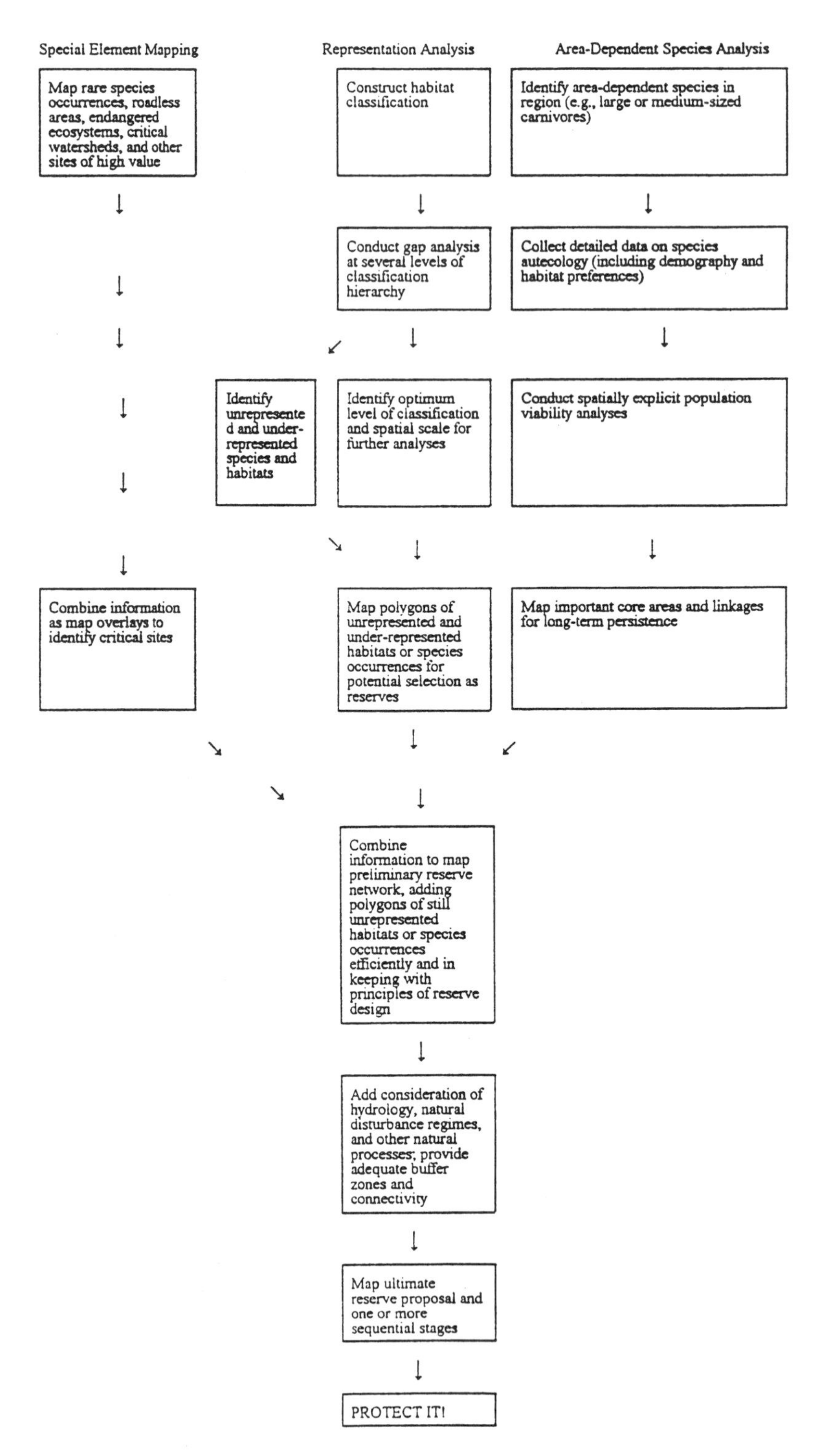

Figure 6.1 Building a reserve network of optimal scale. The flow-chart shows the steps in data analysis, reserve selection, and reserve design that lead to a defensible reserve network that will have a high probability of meeting ambitious but well-accepted conservation goals.

and patchily distributed—even for well-studied taxa, but especially so for more cryptic groups such as soil invertebrates, fungi, and bacteria—a "coarse filter" is often applied in which communities or habitats are the units of inventory and analysis (Noss, 1987a). The goal is to include examples of all communities, habitats, or other land classes in a reserve network and, by so doing, to assure that species associated with these habitats are also represented. As discussed later, the viability of populations in these representative habitat units and the integrity of the ecosystems are separate issues requiring additional and parallel analyses.

In the United States and Canada, habitat representation assessment has been called "gap analysis" (see Scott et al., 1993; Iacobelli et al., 1995). The "gaps" are habitat types (e.g., plant communities or physical landscape features) that are unrepresented or underrepresented in the current network of protected areas. The question of how much area must be protected in order to fill the gaps has no straightforward answer. It will depend on (1) the inherent abiotic and biotic heterogeneity of the region; (2) the fineness of classification or geographic subdivision ("splitting" versus "lumping") of habitats used in the analysis; (3) the *reservation threshold,* or proportion of a given habitat type that must be reserved in order to be considered represented; (4) the *replication threshold,* or number of sites in which a habitat type must be protected in order to be considered represented; and (5) the size of reserves in which a habitat type must be represented. As values for any of these five criteria increase, so generally will the estimate of the amount of land needing protection (Noss & Cooperrider, 1994; Pressey & Logan, 1994, 1995; Noss, 1995).

A landscape that is highly heterogeneous in topography, geology, soils, or other physical factors can be expected to have more patchy or variable species distributions than a landscape in the same region having a uniform physical environment. This is because each species is typically found in a specific and limited portion of an environmental gradient (Gleason, 1926). As a consequence of the turnover in species along environmental gradients, called *beta diversity* (Whittaker, 1972), representing only a portion of a gradient in a reserve will protect a limited subset of the species in the landscape. (See Chap. 2 for discussion of the types of diversity.) Many large protected areas in the western United States are primarily at high elevations and represent only the species and communities found within the upper portions of the elevation gradient; moreover, that portion of the gradient typically has the lowest habitat-specific diversity (*alpha diversity*) (Harris, 1984; Noss & Cooperrider, 1994; Caicco et al., 1995). Similarly, in Sweden, alpine landscapes are well represented in protected areas but species-rich river landscapes are poorly protected (Nilsson & Götmark, 1992).

We can predict that when beta diversity is high relative to alpha diversity—which is true in more heterogeneous landscapes—more total area will need to be protected in order to represent all species, habitats, or communities in reserves. The beta-alpha–diversity ratio may thus be useful for determining the need for protected areas in a region, in those rare cases where such data are available. More generally, because mapping of species distributions or even vegetation will be difficult in highly heterogeneous regions and all biological databases are likely to be incomplete, physical habitats may provide the best coarse filter. This approach is being tested in the Klamath-Siskiyou region of northwestern California and adjacent Oregon (Vance-Borland et al., 1995). In this extremely complex region—called the "Klamath knot" by geologists and long-recognized as a center of biodiversity (Whittaker, 1960, 1961; Wallace, 1983)—elevation, soil type, and, to a lesser extent, aspect most strongly influence the distribution of vegetation and (we assume) individual species. However, only about one-third of some 550 distinct physical habitats defined by these variables are represented in current congressionally withdrawn protected areas (Vance-Borland et al., 1995, 1996).

A few studies have attempted to determine how much area is needed to represent all ecosystem types or species of a region in protected areas. In an Australian river valley, 44.9 percent of the total area of wetlands is needed to represent each plant species at least once and 75.3 percent of the total area to represent all plant species and all wetlands types at least once (Margules et al., 1988). In deciduous forests of western Norway, 75 percent of the total area is needed to represent all plant species but only 20 percent to represent all bird species (Saetersdal et al., 1993). For islands in the Gulf of California, an incredible 99.7 percent of the total area is needed to represent all bird, mammal, reptile, and plant species at least once (Ryti, 1992). The plant species, many of which have very narrow distributions on these islands, contribute most of this estimate. Similarly, but not quite so extreme, 65 percent of the area in canyons in San Diego County, California, is needed to represent all bird, mammal, and plant species at least once (Ryti, 1992). In contrast, the state of Idaho has more uniform species distributions, as is typical of north temperate latitudes. Species composition changes dramatically along elevation gradients (high beta diversity) but not much from one mountain range to another (relatively low gamma diversity; see Whittaker, 1972). In Idaho, only 4.6 percent of the total area is needed to represent all vertebrate species at least once; 7.7 percent to represent all federally endangered, threatened, and candidate vertebrates and plants; and 8 percent to represent all 118 vegetation types (Noss & Cooperrider, 1994; Kiester et al., 1996).

Mathematical algorithms that emphasize efficiency, i.e., the most species, habitats, or other units represented in the minimum area, have been used for calculations of minimum area needed for complete representation. An optimal solution to the reserve-selection problem, in this sense, is one that is most efficient. These algorithms are iterative, and most incorporate the principle of complementarity (Kirkpatrick, 1983; Margules et al., 1988; Pressey et al., 1993). For example, the first site selected by the algorithm is the one with the most species or habitat types, whereas the second site does not necessarily have the second most species or habitats but, rather, has the largest number of species or habitats not found in the first site. There has been much debate recently about which algorithms are closest to optimal and whether optimality is the most important consideration in reserve selection (see Underhill, 1994; Pressey et al., in press; Csuti et al., in press). These discussions are highly technical and are not summarized here. The point is that objective and efficient methods exist for selecting a set of sites representing all species or habitats in a region. However, these algorithms have serious biological limitations (see Bedward et al., 1992b).

Simply representing a habitat type or species in an area of undetermined size is not biologically optimal, regardless of whether the algorithms used are mathematically optimal. Therefore, some studies have applied minimal criteria to determine the percentage of a species' range or area of a habitat type to be included in a protected areas network, the number of times each will be represented, and the size of reserve in which the species or habitat type will be represented. Using the criterion of Schonewald-Cox (1983) for the area required to maintain a population of a medium-sized mammalian predator, Kiester et al. (1996) considered a species to be protected in Idaho if it occurred in at least three protected areas of at least 10,000 hectares each. Species that met this criterion were screened out; analyses to determine the highest priority sites for protection were then limited to the set of "unprotected" species. In considering whether habitat types or other land classes were represented in Australian reserves, Pressey and Logan (1994, 1995) used only the presence of these types or classes, with levels of 1 percent, 5 percent, and 10 percent of the total area considered as alternative thresholds for protection. It is possible to choose any arbitrary threshold, but the amount of area required to achieve "adequate" representation will increase as the threshold increases.

The scale at which habitat types are delineated can have a profound effect on the effectiveness of the coarse filter strategy (Stoms, 1992). In interpreting results from studies using vegetation, physical habitat types, or other land classes for assessing representation, it is important to recognize that such types can be split or lumped almost endlessly. Generally, the amount of land required to represent every land class increases with the number of land classes in the classification. For instance, protecting examples of 50 habitat types in a region will require much less area than protecting examples of 500 habitat types in the same region. Existing reserves generally represent more coarse-scale than fine-scale classes, but the relationship depends on the reservation threshold (e.g., presence only, 1%, 5%, or 10%) (Pressey & Logan, 1994, 1995). Because coarse-scale land classes are internally heterogeneous, the chance of missing important physical and biological variations is high. However, at some threshold, progressively finer subdivision of classes will result in only slight increases in homogeneity or number of species represented (Pressey & Bedward, 1991; Bedward et al., 1992a; Pressey & Logan, 1995). Conservationists will need to determine, probably for each region individually, which level of classification hierarchy is optimal in terms of protecting the broadest range of physical and biological variation in a relatively efficient manner.

I recommend the following sequence for using representation analyses as part of a broader conservation evaluation (see Fig. 6.1).

1. Construct a hierarchical or subdivided classification of habitat types (biotic or abiotic) in the region of interest, or use a good existing classification.
2. Conduct a gap analysis to determine which habitats or species (if good data are available) are poorly represented in protected areas. Make the criteria for adequate representation reasonably conservative and stringent (e.g., at least 10% of the area of each habitat type protected), and conduct the analysis at several levels of classification hierarchy.
3. Determine the optimal scale of analysis.
4. Highlight on the map the polygons (i.e., defined map units of similar character) of unrepresented or underrepresented habitats or species occurrences—these are the options for completing representation, and at least some polygons of each type (e.g., equaling 10% of total area) must be represented in the final design.

Exactly which polygons or sites to select cannot be determined until after analysis and mapping of other criteria (which can be parallel and conducted concurrently with representation analyses), because polygons of many of the habitat types will be included in the map layers based on other criteria (e.g., roadless areas, rare species occurrences, travel corridors for large mammals). After mapping habitats that fulfill these other needs, it is possible to go back to the gap analysis and determine which polygons of each habitat type (based on adjacency, connectivity, and other design criteria, in addition to principles of efficiency and complementarity) should be added to the system to meet the representation requirements.

Protect Rare-Species Habitats

If biological databases are complete, all species (including the rarest) will be represented in a species-level representation analysis. However, in most cases biological data will be incomplete. One must use the coarse filter of habitat representation to capture (hopefully) most unmapped species, or use vegetation maps to predict species occurrences based on habitat relationship models (Scott et al., 1993). In either case, the coarse filter must be complemented by a fine filter that uses whatever rare-species site information is available. Biologists and naturalists have always been interested in the rare and the unusual, so a surprising amount of information may be obtainable on these species. In most states in the United States this information has been incorporated into the databases of the natural heritage programs (Jenkins, 1988). A reasonable goal is to protect all occurrences of each of the rarest species (e.g., those ranked as critically imperiled or imperiled at a global scale, G1 or G2, by criteria of The Nature Conservancy). However, many individual occurrences of rare plants, invertebrates, and some small vertebrates might be protected in rather small reserves or on private lands, if landowners are informed of their importance and are willing to cooperate in their conservation. For regional or broader conservation planning, the emphasis should be on identifying landscapes with concentrated occurrences of rare species (e.g., centers of endemism); these landscapes then become priority areas for new reserves (Terborgh & Winter, 1983; Noss & Cooperrider, 1994).

Level of endemism, or narrowness of species distributions in general, will affect the amount of area needed in reserves. As noted earlier, the proportion of the landscape needed to represent all species in regions with many rare and narrowly distributed species will be much more than in areas with more broadly distributed species. Hot spots of rarity or endemism may or may not coincide with hot spots of species richness for various taxa (Gentry, 1986, 1992). Recent studies have shown that hot spots of species richness for different taxa also may not coincide, at either the continental (Noss & Cooperrider, 1994) or regional scale (Prendergast et al., 1993; Saetersdal et al., 1993). For example, species-richness hot spots for butterflies, dragonflies, liverworts, aquatic plants, and breeding birds overlap poorly in Britain, and many rare species do not occur in the most species-rich areas (Prendergast et al., 1993). Thus, in order to ensure that the sites of greatest importance in terms of rare-species habitats are represented in a reserve network, there is no alternative to extensive field work and detailed mapping of the occurrences of these species.

Protect Other Valuable Habitats

In addition to sites with concentrated occurrences of rare species, other sites that should be included within reserve networks to complement the minimum set of sites necessary to ensure adequate representation include the following (Noss 1992, 1995; Noss & Cooperrider, 1994):

1. Roadless, undeveloped, or otherwise essentially wild areas of significant size (e.g., >1000 ha). Undeveloped areas, especially when inaccessible to humans, offer important refugia to species sensitive to human activities. This criterion is relative to the condition of the region. In heavily developed regions, undeveloped areas as small as a few hectares may be of high value because of their rarity.
2. Locations of rare or unusual plant or animal communities, endangered ecosystems defined by extent of decline since European settlement or by imminence of threat (Noss et al., 1995), depleted and important seral stages (such as old-growth forest), or animal concentration areas such as bird- or seal-breeding sites, ungulate winter range of calving grounds, snake- or bear-denning areas.
3. Resource hot spots such as sites of unusually high primary productivity, artesian springs, ice-free bays, mineral licks, etc.
4. Watersheds of high value for anadromous fishes or other elements of aquatic biodiversity.
5. Disturbance initiation or export areas (e.g., promontories where lightning often strikes and other areas necessary to accommodate natural disturbance regimes).
6. Sites of inherent sensitivity to development, such as watersheds with steep slopes or unstable soils, or aquifer recharge areas.
7. Sites recognized as sacred or otherwise important by indigenous peoples.
8. Sites that could be added to adjacent protected areas to form larger and more defensible reserves.

Sites can be scored on each of these criteria and criterion scores summed to yield total scores for each site. Sites can then be ranked in priority order by total score as has been done in many European conservation evaluations and in a recent Wildlands Project case study for the Oregon Coast Range (Noss, 1993a). When entered into a computerized spreadsheet, scores can be weighted in alternative ways and sites ranked according to which criteria are deemed most important (Duever & Noss, 1990). As noted earlier, the set of sites selected under these criteria will include many of the polygons of unrepresented or underrepresented habitats or species occurrences identified through a gap analysis, thus partially fulfilling representation objectives.

Maintain Sufficient Area and Connectivity for Wide-Ranging Animals

Representing a species or its habitat in a reserve network will not by itself ensure that the species will be able to persist there; thus, representation criteria must be complemented by considerations of population viability (Noss, 1992, 1995). Recent models of population viability have focused on loosely connected systems of local populations (metapopulations, considered generally), with time horizons of 100 to 500 years, and with predicted risk of extinction below 5 percent (see Shaffer, 1987; Beier, 1993). Modeling population viability for all species that may be potentially at risk of extinction in a region is virtually impossible and is not necessary to fulfill conservation objectives. We can assume that most species will be captured by the coarse filter of habitat protection, with the addition of rare-species hot spots and other sites of special importance, as discussed previously. For determining the optimal scale and configuration of a regional reserve network, spatially explicit population-viability analyses of the most area-demanding species complement data from other analyses. Large- and medium-sized carnivores, in regions where they still exist or can be reintroduced, are usually the optimal target species (Noss, 1992; Noss et al., 1996). Information on the requirements of these species is essential in order to design a reserve network that is truly a whole greater than the sum of its parts (Noss & Harris, 1986).

Large carnivores have large home ranges. For example, in the Rocky Mountains, individual, annual home ranges for black bears average about 150 square kilometers (Amstrup & Beecham, 1976; Beecham & Rohlman, 1994), for mountain lions, more than 400 square kilometers (Seidensticker et al., 1973), and for grizzly bears, nearly 900 square kilometers (Blanchard & Knight, 1991). Carnivores also typically have low values for reproduction, survival, and effective population size, among the major considerations in population viability analysis. These low values produce high estimates for the population sizes necessary to ensure persistence. The genetically effective population size (N_e) is usually far smaller than the actual population size (N) required for persistence in the wild. For instance, black bears may have an N_e of less than 70 percent of N (Chepko-Sade & Shields, 1987), whereas N_e calculations for grizzly bears can be as low as 24 percent of N (Harris & Allendorf, 1989).

As a consequence of these demographic and genetic characteristics, large- and medium-sized carnivores require a huge amount of area to maintain viable populations. For example, using average densities for grizzly bears in the

Rocky Mountains (4 bears per 259 km^2) and the 4 to 1 N: N_e ratio estimated by Harris and Allendorf (1989), Metzgar and Bader (1992) calculated that 129,500 square kilometers of wildland would be needed to maintain 2000 bears for an N_e of 500—the official recovery goal for the species. Establishing single, large reserves is usually impractical, but the prospects are more encouraging if one examines carnivore population viability in the framework of metapopulations: exchange between populations reduces the size of each subpopulation required for viability. Thus, the 129,500 square kilometers or more required to maintain a viable population of grizzly bears need not be protected in one place but, rather, across a broad region, so long as some interchange occurs between populations.

A metapopulation approach to conserving species requires that reserves and other habitats be physically connected so that individuals can disperse safely from one reserve to another. For biological, ethical, and aesthetic reasons, most conservationists prefer natural dispersal to artificial translocations of animals; translocations are often failures (Griffith et al., 1989). Fortunately, large carnivores have strong dispersal capabilities, with distances of 50 to 100 kilometers not unusual (Rogers, 1987; Blanchard & Knight, 1991; Pletscher et al., 1991; Lindsey et al., 1994) and hundreds of kilometers recorded for wolves (Ballard et al., 1987; Mech et al., 1995).

What constitutes adequate dispersal habitat for large carnivores has been poorly studied and is controversial (Noss et al., 1996). Connectivity might be provided either through relatively discrete habitat corridors or through buffer or matrix lands that permit safe movement among reserves or provide for a continuously distributed population. For large carnivores, connectivity is mainly an issue of circumventing barriers to animal movements (e.g., highways or cities) and minimizing human-caused mortality (e.g., from vehicles, hunting, or trapping) (Noss et al., 1996). The topic of how to design reserve networks with adequate connectivity for large carnivores and other species is complex (see Noss 1992, 1993b; Beier & Leo 1992; Noss et al., 1996) and is discussed in Chapter 10.

It is clear that considering the long-term needs of wide-ranging species will greatly increase estimates of how much protected area is enough. A model incorporating demographic and environmental stochasticity for cougars in southern California predicted that only 1000 to 2000 square kilometers of habitat is necessary to maintain a cougar population with a 98 percent probability of persistence for 100 years (Beier, 1993). However, consideration of long-term genetic integrity and evolutionary potential would lead to much higher estimates of required population size and area. The estimated 129,500 square kilometers required to maintain 2000 grizzly bears (N_e = 500) constitutes about 60 percent of the northern Rocky Mountains region of the United States. Similarly, the area needed to maintain an N_e of 500 Florida panthers would cover some 60 to 70 percent of the original range of the subspecies in the southeastern United States (Noss & Cooperrider, 1994); the area needed to maintain an N_e of 5000 would exceed the original range—which is not as ridiculous as it may sound, because the various subspecies of cougars do not differ much from one another and were probably linked by frequent gene flow. These figures are well above available estimates for the area required simply to represent all species or habitats in these regions (e.g., compare the 4.6 percent of Idaho needed to represent all vertebrate species, as noted earlier).

In regions that lack large carnivores and where carnivore reintroduction in the short term is not feasible, other species with fairly demanding area requirements and that are sensitive to fragmentation (e.g., neotropical migrant, forest-interior songbirds [Whitcomb et al., 1981; Robbins et al., 1989]) can be suitable indicator species for reserve design. Although the area and connectivity requirements of these species are modest compared to large carnivores, thousands of square kilometers of relatively intact forest may be required to ensure persistence. Again, consideration of the autecology of these species will provide information on the optimal configuration of reserve networks that is not provided by gap analysis or site-specific evaluations.

Maintain Ecosystem Health and Integrity

Ecosystems are more than species. Ultimately, the biodiversity of a region depends on the continued operation of natural climatic, geomorphological, hydrological, ecological, and evolutionary processes. Many of these processes require considerable space to function normally, i.e., to operate within the natural range of variability to which species have adapted over evolutionary time. For example, the hydrologically defined ecosystem of the Everglades originates in the chain of lakes just south of Orlando, Florida, and extends southward into Florida Bay. Disruption of this hydrological system, a direct consequence of human population growth, ill-considered engineering schemes, and the general failure of agencies and landowners to take a regional perspective in land-use planning, is probably the primary factor in the biological crisis of south Florida (Kushlan, 1979, 1983; Harris, 1990). In reply to the question, "How much is enough to maintain the hydrological integrity of the Everglades?," we would have to say "most of south Florida." In fact, Odum and Odum (1972) asked a similar question and concluded that about 50 percent natural and 50 percent developed land would optimize ecosystem services and economic and cultural well-being in south Florida. In Georgia, Odum (1970) estimated that a mix of land uses comprising about 40 percent natural, 10 percent urban-industrial, 30 percent food production, and 20 percent fiber production would optimize ecosystem services and produce a self-sufficient human-nature ecosystem. It is curious that factors of this kind were not considered by the Brundtland Commission, which was ostensibly concerned with similar matters of human-nature harmony, when it arrived at its 12 percent recommendation.

Several authors believe that disturbance regimes must be appraised in estimating the optimal amount of protected area in a region (Pickett & Thompson, 1978; Shugart & West, 1981; Baker, 1992; Noss, 1992). Disturbance is one of the central factors shaping the composition, structure, and function of ecosystems. Species in any area have evolved a set of responses to the natural disturbance regime, including adaptations for avoiding, tolerating, or exploiting disturbance events (White, 1979; Pickett & White, 1985). Reserves that are small relative to the spatial scale of disturbance (e.g., a 10,000 ha nature reserve representing a forest type that naturally experiences 10,000–100,000 ha fires every few decades) will offer insufficient refugia for species during disturbance events; therefore, opportunities for recolonization and recovery of disturbed areas will be limited. Many small reserves containing fire-dependent vegetation (e.g., tallgrass prairie, longleaf pine) will not receive lightning strikes often enough to burn naturally and will require perpetual programs of prescribed burning (Noss & Cooperrider, 1994).

As with the other criteria for determining the optimal scale of reserve networks, there are no ready formulas for calculating how large a reserve or a reserve network must be to sustain a natural disturbance regime. Shugart and West (1981) suggested that a landscape would need to be 50 to 100 times larger than the average disturbance patch in order to maintain a "quasi-equilibrium" of seral stages over time. Other authors have disputed whether a quasi-equilibrium or steady state is even possible or desirable (Baker, 1989), but most agree that larger reserves have a lower probability of radical shifts in landscape structure that could threaten species persistence (Turner et al., 1993). Hence, once again bigger is better. Connectivity among reserves could facilitate the spread of disturbances, for better or worse (Simberloff et al., 1992), but may also provide a "fire escape" function and allow organisms either to escape disturbance events or recolonize areas after disturbance. Buffer zones could be used to help control the spread of disturbances into human-populated areas (Baker, 1992). For example, firewood collecting or other active fuel reduction activities within buffer zones could reduce the threat of fire in adjacent suburban areas.

Consideration of the natural processes of landscapes suggests that reserves be designed around natural features. It has been recommended that reserves encompass entire watersheds and not leave out headwater areas or sever drainages. Reserve boundaries should also avoid severing areas of active terrain (e.g., colluvial fans, glaciers, seismic zones) or geological features, such as sinkholes, eskers, and end moraines (Theberge, 1989; Noss, 1995). When reserve boundaries are adjusted to encompass these natural features and associated processes, considerably more area may be added to the reserve network.

In conclusion, thoughtful consideration of hydrology, geomorphology, natural disturbances regimes, and other patterns and processes of natural landscapes will usually lead to additional area in reserve network designs, beyond that necessary for representation of habitats. In some cases, the area could even exceed that needed for long-term population viability of large mammals. Precisely how much area to add and where to add it to accommodate natural processes and maintain ecological integrity cannot be calculated with any confidence; rather, these questions force one into the "art" of reserve design, which depends on the knowledge and intuition of ecologists familiar with the geomorphology, hydrology, vegetation, and disturbance-recovery dynamics of the landscape.

Provide Adequate Buffers

If a reserve network is expansive enough, it may be essentially self-buffering. But, in the early stages of putting a network together, many protected areas—especially smaller ones—may need to be insulated from the intensively used landscape in order to offer sufficient protection to species sensitive to the direct and indirect impacts of human activities. To offer one example, a recent study in Ontario, Canada, found the diversity and abundance of neotropical migrant songbirds in forest patches surrounded by residential development much lower than in forest patches with few or no nearby houses (Friesen et al., 1995). Possible explanations for the reduced numbers of songbirds in residential areas include the abundance of house cats and gray squirrels, which prey on birds or their eggs. Buffer zones with low housing density around natural areas may help protect sensitive species (Noss & Cooperrider, 1994).

There are other good reasons for establishing buffer zones. As noted earlier, they may be useful for controlling the spread of disturbances either into or out of reserves and, perhaps also, for thwarting the invasion of exotic organisms. They are also places where low-density human habitation and gentle land use might serve as models for relearning how to live harmoniously with nature. As for many of these criteria, the optimum width or size of buffer zones cannot be determined by any rigorous, quantitative technique. Decisions will once again depend on the knowledge and experience of ecologists and others who know the land.

Add Polygons of Habitats Still Underrepresented

By this stage in the reserve network-building process, many or most of the unrepresented and underrepresented habitat types or species occurrences identified in previous analyses will have been captured by consideration of rare species habitat, area requirements of wide-ranging species, natural processes, and other criteria. Polygons of habitat types still not adequately represented in preliminary map overlays can be added to the network at this time. Selection algorithms can help improve the efficiency of selections. More important, however, are the principles of reserve design—including bigness, redundancy, and connectivity (see Noss & Cooperrider, 1994)—that should be considered in choosing among alternative polygons for inclusion in the network to meet minimal representation requirements (e.g., 10% of the area of each habitat type represented).

Consider Land Parcels

A final consideration has to do with the actual availability of lands for purchase and, of course, the funding to acquire these lands. Private lands are available for sale as parcels that will probably show little relationship to the polygons or sites identified through the empirical reserve selection process described in the preceding pages (Strittholt & Boerner, 1995). Similarly, public lands are often divided into management units or agency jurisdictions whose boundaries, in the short term at least, are relatively fixed. It may not be feasible to protect only a portion of a management unit (e.g., because of fencing or law enforcement constraints).

Therefore, for both private and public lands, parcels or management units will determine what is actually available to protect. Some private landowners will refuse to sell or will ask unreasonable prices. Some public agencies have multiple-use mandates and grandfather clauses on activities, such as mining or grazing, that do not allow them to offer the kind of restrictive management conservationists would like to see. Thus, the real world of land protection opportunities will seldom match the ideal world defined by science-based analyses. All the conservationist can do in such cases is try to make the match as close as possible. In the short term, less land will be available for protection (much less actually affordable) than desired. In some areas, however, the mismatch between habitat polygons and ownership parcels will mean that more land than is actually needed will have to be acquired. For example, a kettlehole bog loaded with rare species may occupy only 5 out of 500 acres in a land parcel, but the owner may be willing to sell all or nothing. These complications reduce the efficiency of reserve selection and affect real-world answers to the question "how much is enough?"

Combine Overlays and Set Priorities for Phased-in Design

When all the steps previously described have been completed and all the factors that might determine how much protected area is needed have been considered, the map layers can be overlaid to produce the "final" reserve network design for the region. Of course, the design is never really "final"—it must be continually updated and refined with new data in order to remain credible. But the area of the combined overlay map represents the proportion of the region that, given the current state of knowledge, should be protected in order to have a high probability of attaining the stated conservation goals. This design is the best available answer to the question "how much is enough?" It is what conservationists might hope to accomplish, given optimistic assumptions about human population, resource use, and civilization in general, in 50 to 200 years, depending on the region. Full recovery of the habitats and populations within the protected area could take longer, possibly 500 or 1000 years.

Because the ultimate, most ambitious proposal cannot be implemented at once, it is desirable to produce a chronological sequence of maps showing progressively more area being protected over time. Thus, the sites that are at greatest risk of significant biotic impoverishment—those that have the most to lose if not protected soon—must be identified as priorities in the first phase of protection. Subjective judgment will be required, but it should not be too difficult for a panel of experts, at least some of whom were involved in the previous analyses, familiar with the region's biota to rank sites and connections in terms of urgency of protection. Candidate sites for immediate protection might include one of the last populations of an endemic plant species that is threatened by development, the only known occurrence of a particular plant community in the region, the last large roadless area in a national forest that is proposed for logging, or a habitat connection critical to maintaining a population of cougars over the next 10 or 20 years. Other sites, such as corridors needed to restore genetic linkages between populations so that their long-term evolutionary potential is maintained, might not need full protection for 50 years. Perhaps the site can be logged now, and in 50 years the forest will have recovered enough to be suitable as movement habitat. Building homes in such an area is a more permanent alteration and should be discouraged. Similarly, a site needed to represent an unusual geological feature or other physical habitat, but which contains no rare or sensitive species, can wait a few decades for protection (unless of course, it is targeted for mining or other activity that would irrevocably alter its character).

Figure 6.2 shows proposed reserve network progressions for Florida. This proposal is preliminary and will be refined with further work. Perhaps the most

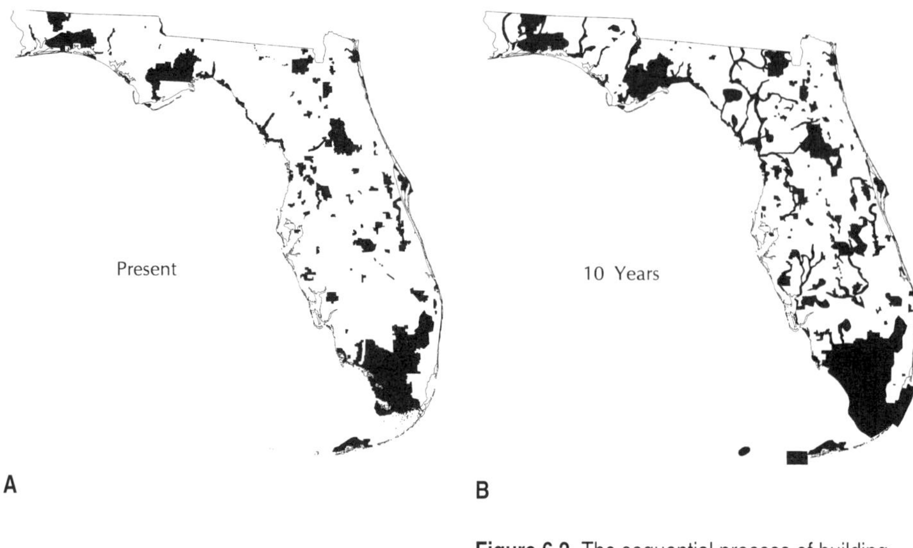

Figure 6.2 The sequential process of building a reserve network, illustrated for Florida. The sequence shows (A) managed areas existing today, which constitute 3,206,050 hectares or 22 percent of the state; (B) what might exist in about 10 years if current land protection initiatives are successful, totaling 5,692,923 hectares or 39 percent of the state; (C) what might exist in 100 years if more ambitious protection initiatives are successful, totaling 8,658,244 hectares or 59 percent of the state. Importantly, the managed areas existing today are not all functioning optimally as reserves. Some of them (e.g., the three national forests) are managed for timber production and are riddled with roads. Thus, changing management practices is as crucial as adding areas to the network. (Data from Noss [1987b], Cox et al. [1994], and The Nature Conservancy/Florida Natural Areas Inventory [unpublished].)

noteworthy feature of this sequential proposal is its audacious optimism. Instead of the process we are used to seeing in chronological series of maps or photographs—declining forest cover, fragmentation, cancer-like growth and metastasis of cities—here we see a process of recovery, of nature returning. Thus, these proposals provide hope for the future, an antidote to despair. Our experience in The Wildlands Project is that these ambitious proposals inspire activists and scientists alike to keep on working for ecological recovery, no matter how bleak the political situation may seem today. It is this unflinching optimism that confuses and exasperates the opponents of conservation.

The process of selecting priority areas for near-term protection and then actively acquiring or defending these sites should never become divorced from the broader, long-term goal of recovering the full sweep of native biodiversity and wildness in a region. Too often conservationists get caught up in "brush fires" and political emergencies and lose sight of the bigger vision. In the short term, compromise is essential if one is to accomplish anything. But, if compromise comes too readily or leads to irreplaceable losses of biodiversity, then it thwarts long-term goals.

Reserves or Something Else?

This question deserves a book of its own. If we conclude, from the type of analysis described in this chapter, that some 50 percent of a region will ultimately need protection in order to achieve our conservation goals, what does "protection" mean? Does it mean fee-simple acquisition or congressional designation as inviolate reserves? Or could that 50 percent of the region include managed forests, grazed rangelands, or other multiple-use buffer or transition areas where some resource exploitation would continue? What specific kinds and intensities of human activities should be allowed in various zones? Answers to these questions, as for most others posed in this chapter, will be highly case-specific. A human activity, such as livestock grazing or selection forestry, that is perfectly compatible with conservation objectives in one area may be destructive in another. As used in this book, the term *protected area* refers to areas that would qualify as type I, II, or III (i.e., strict nature reserves, national parks, national monuments, or equivalents) in the World Conservation Union (IUCN) classification. The types of human uses permitted within these categories vary tremendously from site to site. The key concept is compatibility with conservation objectives. This has to be assessed flexibly site-by-site but taking care, in the face of uncertainty, to risk erring on the side of protecting too much. In a preliminary proposal for the Oregon Coast Range (Noss, 1993a) I made specific recommendations for human activities that should be encouraged and those that should be discouraged in three protection zones and the matrix. The areas proposed for strictest protection were *not* human-exclusion zones; nonmotorized recreation, nature study, nonmanipulative scientific research, and environmental education were encouraged, as was active restoration (e.g., thinning of tree farms, removal of roads, stream rehabilitation).

Conclusion

The best available data and analyses suggest that considerable areas of land must be maintained in a reasonably natural and healthy condition in order to fulfill the well-accepted conservation goals of representing all kinds of ecosystems, maintaining viable populations of all native species, sustaining key ecological and evolutionary processes, and maintaining landscapes and biota that are resilient to environmental change. Specific answers to the question "how much is enough?" are not defensible unless the goals and objectives of conservation are explicit and the analyses are scientifically valid, and even then all figures are highly uncertain. Nevertheless, available data support my earlier estimates that somewhere between 25 percent and 75 percent of a given region (or an average of 50%) would need to be protected in order to meet well-established conservation goals (Noss, 1992; Noss & Cooperrider, 1994). "Protected" does not usually mean human-free or unmanaged, but it means the needs of the native biota (emphasizing those most sensitive to human activities) come first.

The best available answer to "How much is enough?," based on biocentric values and scientific analyses, may not be the best answer politically. This is where the strategy of phasing in protection over a period of decades comes into play. Today, whether or not it is prudent to display the final reserve network proposal—the one deemed necessary to fulfill all stated goals and objectives—to policymakers and the general public is an open question. This problem has been much debated among proponents of The Wildlands Project. My current opinion is that it is sensible in most regions to show the entire sequence, from what we have today in terms of protected areas, through one or more intermediate stages, to the ultimate proposal. This sequence is shown in Figure 6.2 for the development of a reserve network in Florida. My previous published proposals for Ohio, Florida, and the Oregon Coast Range (Noss, 1987b, 1993a) showed only the ultimate proposal. The maps predictably shocked and enraged some people, but they inspired many others. The proposals stimulated others to undertake more detailed analyses and produce more modest and politically credible proposals (e.g., Cox et al., 1994; Strittholt & Boerner, 1995) that will contribute to the same biological objectives and can be seen as steps toward the more ambitious proposals.

Other conservationists would argue that the ultimate proposals should never be shown to the public because of the potential for stimulating political backlash. In some regions, they may be right. But I believe that conservationists everywhere would do well to question "politically correct" estimates of how much protected area is enough, such as the Brundtland Commission's indefensible 12 percent, and then advocate scientifically informed, empirical approaches of the type outlined in this chapter. Goals and objectives of the conservation strategy should be stated openly and the data made available to anyone who wants to analyze them. The results of analyses that proceed from biocentric goals and objectives can be expected to point to enormous protection needs and to specific proposals that are unpopular in many circles. However, the alternative is to protect less than necessary to maintain the full richness of life—an option I consider ethically repugnant and unacceptable.

Literature Cited

Amstrup SC, Beecham JJ. Activity patterns of radio-collared black bears in Idaho. *J Wildl Manage* 1976;40:340–348.

Angermeier PL, Karr JR. Biological integrity versus biological diversity as policy directives. *BioScience* 1994;44:690–697.

Bailey RG, Avers PE, King T, McNab WH (eds). *Ecoregions and Subregions of the United States* (Map). Reston, VA: U.S. Department of Agriculture Forest Service and U.S. Department of Interior Geological Survey, 1994.

Baker WL. Landscape ecology and nature reserve design in the Boundary Waters Canoe Area, Minnesota. *Ecol* 1989;70:23–35.

Baker WL. Landscape ecology of large disturbances in the design and management of nature reserves. *Landscape Ecol* 1992;7:181–194.

Ballard WB, Whitman JS, Gardner CL. Ecology of an exploited wolf population in south-central Alaska. *Wildl Monogr* 1987;98:1–54.

Barnes RFW. Sustainable Development in African Game Parks. In GK Meffe, CR Carroll (eds), *Principles of Conservation Biology.* Sunderland, MA: Sinauer Associates, 1994:504–511.

Bedward M, Pressey RL, Keith DA. Homogeneity analysis: Assessing the utility of classifications and maps of natural resources. *Australian J Ecol* 1992a;17:133–139.

Bedward M, Pressey RL, Keith DA. A new approach for selecting fully representative reserve networks: Addressing efficiency, reserve design and land suitability with an iterative analysis. *Biol Cons* 1992b;62:115–125.

Beecham JJ, Rohlman J. *A Shadow in the Forest, Idaho's Black Bear.* Moscow, ID: University of Idaho Press, 1994.

Beier P. Determining minimum habitat areas and habitat corridors for cougars. *Cons Biol* 1993;7:94–108.

Beier P, Loe S. A checklist for evaluating impacts to wildlife movement corridors. *Wildl Soc Bull* 1992;20:434–440.

Blanchard BM, Knight RR. Movements of Yellowstone grizzly bears. *Biol Cons* 1991;58:41–67.

Caicco SL, Scott JM, Butterfield B, Csuti B. A gap analysis of the management status of the vegetation of Idaho (U.S.A.). *Cons Biol*1995;9:498–511.

Callicott JB. A critique of and an alternative to the wilderness idea. *Wild Earth* 1994/95;4(4):54–59.

Chepko-Sade BD, Shields WM. The Effects of Dispersal and Social Structure on Effective Population Size. In BD Chepko-Sade, ZT Halpin, (eds), *Mammalian Dispersal Patterns.* Chicago: University of Chicago Press, 1987:287–321.

Cox J, Kautz R, MacLauglin M, Gilbert T. *Closing the Gaps in Florida's Wildlife Habitat Conservation System.* Tallahassee, FL: Florida Game and Fresh Water Fish Commission, 1994.

Csuti B, et al. A comparison of reserve selection algorithms using data on terrestrial vertebrates in Oregon. *Biol Cons* (in press).

Diamond JM. The island dilemma: Lessons of modern biogeographic studies for the design of natural preserves. *Biol Cons* 1975;7:129–146.

Duever LC, Noss RF. A Computerized Method of Priority Ranking for Natural Areas. In RS Mitchell, CJ Sheviak, DJ Leopold (eds), *Ecosystem Management: Rare Species and Significant Habitats.* Proceedings of the 15th Annual Natural Areas Conference. Bulletin No. 471. Albany, NY: New York State Museum, 1990:22–33.

Encarta Multimedia Encyclopedia. Bothel, WA: Microsoft Corporation, 1993.

Foreman D. Wilderness areas are vital. *Wild Earth* 1994/95;4(4):64–68.

Foreman D, Davis J, Johns D, Noss RF, Soulé, M. The Wildlands Project mission statement. *Wild Earth* (Special Issue): 1992;3–4.

Friesen LE, Eagles PFJ, MacKay RJ. Effects of residential development on forest-dwelling neotropical migrant songbirds. *Cons Biol* 1995;9:1408–1414.

Gentry AH. Endemism in Tropical Versus Temperate Plant Communities. In ME Soulé (ed), *ConservationBiology: The Science of Scarcity and Diversity.* Sunderland, MA: Sinauer Associates 1986:153–181.

Gentry AH. Tropical forest biodiversity: Distributional patterns and their conservational significance. *Oikos* 1992;63:19–28.

Gleason HA. The individualistic concept of the plant association. *Bull Torrey Botanical Club* 1926;43:463–481.

Griffith B, Scott JM, Carpenter JW, Reed C. Translocation as a species conservation tool: Status and strategy. *Science* 1989;245:477–480.

Harmon D. Cultural diversity, human subsistence, and the national park ideal. *Environ Ethics* 1987;9:147–158.

Harris LD. *The Fragmented Forest: Island Biogeography Theory and the Preservation of Biotic Diversity.* Chicago: University of Chicago Press, 1984.

Harris LD. An Everglades Regional Biosphere Reserve. *Florida Fish Wildl News* 1990;4(2): 12–15.

Harris RB, Allendorf FW. Genetically effective population size of large mammals: An assessment of estimators. *Cons Biol* 1989;3:181–191.

Hummel M (ed). *Endangered Spaces: The Future for Canada's Wilderness.* Toronto: Key Porter, 1989.

Iacobelli T, Kavanagh K, Rowe S. *A Protected Areas Gap Analysis Methodology: Planning for the Conservation of Biodiversity.* Toronto: World Wildlife Fund Canada, 1995.

Irvine S. The cornucopia scam: Contradictions of sustainable development. Part I: Ignoring the limits to growth. *Wild Earth* 1994;4(3):73–81.

Jenkins RE. Information Management for the Conservation of Biodiversity. In EO Wilson (ed), *Biodiversity.* Washington, DC: National Academy Press, 1988:231–239.

Karr JR. Biological integrity and the goal of environmental legislation: Lessons for conservation biology. *Cons Biol* 1990;4:244–250.

Kiester AR, Scott JM, Csuti B, Noss RF, Butterfield B. Conservation prioritization using GAP data. *Cons Biol* 1996;10 (in press).

Kirkpatrick JB. An iterative method for establishing priorities for the selection of nature reserves: An example for Tasmania. *Biol Cons* 1983;25:127–134.

Kushlan JA. Design and management of continental wildlife reserves: Lessons from the Everglades. *Biol Cons* 1979;15:281–290.

Kushlan JA. Special species and ecosystem preserves: Colonial water birds in US national parks. *Environ Manage* 1983;7:201–207.

Lindsey FG, et al. Cougar population dynamics in southern Utah. *J Wildl Manage* 1994;58:619–624.

Ludwig D, Hilborn R, Walters C. Uncertainty, resource exploitation, and conservation: Lessons from history. *Science* 1993;260:17, 36.

Margules CR, Nicholls AO, Pressey RL. Selecting networks of reserves to maximize biological diversity. *Biol Cons* 1988;43:63–76.

McCoy ED. The application of island-biogeographic theory to patches of habitat: How much land is enough? *Biol Cons* 1983;25:53–61.

McNeely J, Miller K (eds). *National Parks Conservation and Development: The Role of Protected Areas in Sustaining Society.* Proceedings of the World Congress on National Parks. Washington, DC: Smithsonian Institution Press, 1984.

Mech LD, Fritts SH, Wagner D. Minnesota wolf dispersal to Wisconsin and Michigan. *Am Midl Nat* 1995;133:368–370.

Meffe GK, Carroll CR (eds). *Principles of Conservation Biology.* Sunderland, MA: Sinauer Associates, 1994.

Metzgar LH, Bader M. Large mammal predators in the northern Rockies: Grizzly bears and their habitat. *Northwest Environ J* 1992;8:231–233.

Murphy DD, Noon BR. Integrating scientific methods with habitat conservation planning: Reserve design for Northern Spotted Owls. *Ecol Appl* 1992;2:3–17.

Nilsson C, Götmark F. Protected areas in Sweden: is natural variety adequately represented? *Cons Biol* 1992;6:232–242.

Noss RF. From plant communities to landscapes in conservation inventories: A look at The Nature Conservancy (USA). *Biol Cons* 1987a;41:11–37.

Noss RF. Protecting natural areas in fragmented landscapes. *Nat Areas J* 1987b;7: 2–13.

Noss RF. The Wildlands Project: Land conservation strategy. *Wild Earth* (Special Issue) 1992;10–25.

Noss RF. A conservation plan for the Oregon Coast Range: Some preliminary suggestions. *Nat Areas J* 1993a;13:276–290.

Noss RF. Wildlife Corridors. In DS Smith, PC Hellmund (eds), *Ecology of Greenways.* Minneapolis: University of Minnesota Press, 1993b:43–68.

Noss RF. Wilderness—now more than ever: A response to Callicott. *Wild Earth* 1994/95;4(4):60–63.

Noss RF. *Maintaining Ecological Integrity in Representative Reserve Networks.* Toronto and Washington, DC: World Wildlife Fund Canada, World Wildlife Fund, 1995.

Noss RF, et al. Conservation biology and carnivore conservation. *Cons Biol* 1996;10 (in press).

Noss RF, Cooperrider A. *Saving Nature's Legacy: Protecting and Restoring Biodiversity.* Washington, DC: Defenders of Wildlife, Island Press, 1994.

Noss RF, Csuti B. Habitat Fragmentation. In GK Meffe, CR Carroll (eds), *An Introduction to Conservation Biology.* Sunderland, MA: Sinauer Associates, 1994.

Noss RF, Harris LD. Nodes, networks, and mums: Preserving diversity at all scales. *Environ Manage* 1986;10:299–309.

Noss RF, LaRoe ET III, Scott JM. *Endangered Ecosystems of the United States: A Preliminary Assessment of Loss and Degradation.* Biological Report 28. Washington, DC: National Biological Service, 1995.

Odum EP. Optimum population and environment: A Georgia microcosm. *Curr Hist* 1970;58:355–359.

Odum EP, Odum HT. Natural areas as necessary components of Man's total environment. *Proc N Am Wildl Nat Res Conf* 1972;37:178–189.

Omernick JM. Ecoregions of the conterminous United States. *Ann Assoc Am Geograph* 1987;77:118–125.

Pickett STA, Thompson, JN. Patch dynamics and the design of nature reserves. *Biol Cons* 1978;13:27–37.

Pickett STA, White PS. *The Ecology of Natural Disturbance and Patch Dynamics.* Orlando, FL: Academic Press, 1985.

Pletscher DH, et al. Managing wolf and ungulate populations in an international ecosystem. *Trans N Am Wildl Nat Res Conf* 1991;56:539–549.

Prendergast JR, et al. Rare species, the coincidence of diversity hotspots and conservation strategies. *Nature* 1993;365:335–337.

Pressey RL, et al. Beyond opportunism: Key principles for systematic reserve selection. *Trends Ecol Evol* 1993;8:124–128.

Pressey RL, Bedward M. Inventory and Classification of Wetlands: What for and How Effective? In R Donohue, B Philips (eds), *Educating and Managing for Wetlands Conservation.* Canberra: Australian National Parks and Wildlife Service, 1991: 190–198.

Pressey RL, Logan VS. Level of geographic subdivision and its effects on assessments of reserve coverage: A review of regional studies. *Cons Biol* 1994;8:1037–1046.

Pressey RL, Logan VS. Reserve coverage and requirements in relation to partitioning and generalization of land classes for Western New South Wales. *Cons Biol* 1995;9: 1506–11517.

Pressey RL, Possingham HP, Margules CR. Optimality in reserve selection algorithms: When does it matter and how much? *Biol Cons* (in press).

Reid TS, Murphy DD. Providing a regional context for local conservation action. *Bio-Science* (Suppl) 1995:84–90.

Robbins CS, Dawson DK, Dowell BA. Habitat area requirements of breeding forest birds of the Middle Atlantic states. *Wildl Monogr* 1989;103:1–34.

Robinson JG. The limits to caring: Sustainable living and the loss of biodiversity. *Cons Biol* 1993;7:20–28.

Rogers LL. Effects of food supply and kinship on social behavior, movements, and population growth of black bears in northern Minnesota. *Wildl Monogr* 1987;97:1–72.

Ryti RT. Effect of the focal taxon on the selection of nature reserves. *Ecol Appl* 1992;2: 404–410.

Saetersdal M, Line JM, Birks HJB. How to maximize biological diversity in nature reserve selection: Vascular plants and breeding birds in deciduous woodlands, western Norway. *Biol Cons* 1993;66:131–138.

Schonewald-Cox CM. Conclusions. Guidelines to Management: A Beginning Attempt. In CM Schonewald-Cox, SM Chambers, B MacBryde, WL Thomas (eds), *Genetics and Conservation: A Reference for Managing Wild Animal and Plant Populations.* Menlo Park, CA: Benjamin/Cummings, 1983:141–145.

Scott JM, et al. Gap analysis: A geographical approach to protection of biological diversity. *Wildl Monogr* 1993;123:1–41.

Seidensticker JC, Hornocker MG, Wiles WV, Messick JP. Mountain lion social organization in the Idaho Primitive Area. *Wildl Monogr* 1973;35:1–60.

Shaffer M. Minimum Viable Populations: Coping with Uncertainty. In M Soulé (ed), *Viable Populations for Conservation.* Cambridge: Cambridge University Press, 1987:69–86.

Shugart HH, West DC. Long-term dynamics of forest ecosystems. *Am Sci* 1981;69: 647–652.

Simberloff, D, Farr JA, Cox J, Mehlman DW. Movement corridors: Conservation bargains or poor investments? *Cons Biol* 1992;6:493–504.

Soulé ME, Simberloff D. What do genetics and ecology tell us about the design of nature reserves? *Biol Cons* 1986;35:19–40.

Soulé ME, Wilcox BA. Conservation Biology: Its Scope and Challenge. In ME Soulé, BA Wilcox (eds), *Conservation Biology: An Ecological-Evolutionary Perspective.* Sunderland, MA: Sinauer Associates, 1980:1–8.

Stoms DM. Effects of habitat map generalization in conservation planning. *Photo Eng Rem Sens* 1992;58:1587–1591.

Strittholt JR, Boerner REJ. Applying biodiversity gap analysis in a regional nature reserve design for the Edge of Appalachia, Ohio (USA). *Cons Biol* 1995;9:1492–1505.

Terborgh J, Winter B.A method for siting parks and reserves with special reference to Colombia and Ecuador. *Biol Cons* 1983;27:45–58.

Theberge JB. Guidelines for drawing ecologically sound boundaries for national parks and nature reserves. *Environ Mangage* 1989;13:695–702.

Turner MG, et al. A revised concept of landscape equilibrium: Disturbance and stability on scaled landscapes. *Landscape Ecol* 1993;8:213–227.

Underhill LG. Optimal and suboptimal reserve selection algorithms. *Biol Cons* 1994;70:85–87.

Vance-Borland K, Noss RF, Strittholt J, Carroll C, Nawa R. A biodiversity conservation plan for the Klamath-Siskiyou region. *Wild Earth* 1995/96;5(4):52–59.

Wallace DR. *The Klamath Knot.* San Francisco, CA: Sierra Club Books, 1983.

Westra L. *An Environmental Proposal for Ethics: The Principle of Integrity.* Lanham, MD: Rowman & Littlefield, 1994.

Whitcomb RF, et al. Effects of Forest Fragmentation on Avifauna of the Eastern Deciduous Forest. In RL Burgess, DM Sharpe (eds), *Forest Island Dynamics in Man-dominated Landscapes.* New York: Springer-Verlag, 1981:125–205.

White PS. Pattern, process, and natural disturbance in vegetation. *Bot Rev* 1979;45: 229–299.

Whittaker RH. Vegetation of the Siskiyou Mountains, Oregon and California. *Ecol Monog* 1960;30:279–338.

Whittaker RH. Vegetation history of the Pacific Coast states and the "central" significance of the Klamath Region. *Madrono* 1961;16:5–23.

Whittaker RH. Evolution and measurement of species diversity. *Taxon* 1972;21:213–251.

Wilcox BA. Insular Ecology and Conservation. In ME Soulé, BA Wilcox (eds), *Conservation Biology: An Ecological-Evolutionary Perspective.* Sunderland: Sinauer Associates, 1980:95–117.

Woodley S, Kay J, Francis G. *Ecological Integrity and the Management of Ecosystems.* Ottawa: St. Lucie Press, 1993.

World Commission on Environment and Development. *Our Common Future.* New York: Oxford University Press, 1987.

World Resources Institute (WRI), World Conservation Union (IUCN), and United Nations Environment Programme (UNEP). *Global Biodiversity Strategy.* Washington, DC: WRI, 1992.

7

Evaluating the Ecological Suitability of Lands for Parks and Protected Areas Using Gap Analysis Databases

R. Gerald Wright
J. Michael Scott

The great variability of the lands included in the U.S. National Park System probably is obvious to most readers of this book. In addition to the clear physical distinctions among the landscapes in various categories of areas in the national park system—natural and cultural, urban and wilderness, battlefield and birthplace, recreational and spiritual—there are qualitative differences as well. As described in Chapter 1, in the years since the creation of Yellowstone National Park and the National Park Service (NPS) in 1916, the national park system has experienced tremendous growth. The expansion of the system has been most dramatic in the last two decades during which the number of units in the system was increased by almost 30 percent, bringing the total number of parks to 366. This growth has prompted criticism of the park system; some believe that it now includes areas of marginal significance, which diminish the value of all parks in the system. It also has been argued that additions of marginal quality to the park system may compromise opportunities to add other new areas that have significant resources now lacking in the system or to expand existing units to better protect resources in and adjacent to them (Ridenour, 1994).

From its inception, the goal of the NPS has been to acquire only the most outstanding lands and resources, with "national significance" as the primary criterion. The broad interpretation historically given to this criterion has contributed to the great diversity of landscapes now included in the system. As indicated in Chapter 1, it was not until 1972 that the NPS undertook a formal effort to classify the natural regions of the United States and to identify the physiographic zones not represented by a unit in the system (NPS, 1972) (see Table 1.1). However, professional guidelines for evaluating national significance have not always been foremost in the minds of those responsible for establishing new parklands (NPS 1990; Mackintosh, 1991). Factors such as opportunity, aesthetics, local economic impacts, surrounding land uses, and political support are often more important in park establishment than resource considerations and system-wide plans (Wright, 1984; Pressey, 1994; Wright et al., 1994). Thus, a park establishment bill backed by an influential constituency and lacking significant outside opposition is apt to be passed by Congress without great regard for the opinions of scientists, historians, or other professional specialists. Likewise, from the beginning, the leaders of the NPS sought to enlarge the public and political constituency of the agency by acquiring more parks in more places. These aims have made most NPS managers reluctant to vigorously resist popular park proposals if questioned only by their professional advisors; they have also tended to dissuade managers from pursuing proposals that would establish new parks or enlarge or reshape existing parks when these actions are potentially controversial (Conservation Foundation, 1985).

Political considerations aside, one reason that the NPS has had difficulty evaluating park proposals is that its planners and scientists have generally lacked the necessary resource data, particularly in a spatial context, to evaluate adequately the ecological significance of new or expanded areas (Freemuth, 1991). With the advent of the Gap Analysis Program (Scott et al., 1993), this situation is changing.

The Gap Analysis Concept

Gap analysis provides an overview of the distribution and conservation status of components of biodiversity, such as vegetation types and species. It seeks to identify "gaps" (i.e., vegetation types or species that are not found on existing protected lands) that may be filled through the establishment of new reserves or changes in land management practices. The process uses digital maps of vegetation types, species distribution, land ownership, and land management status that are overlaid in a geographic information system (GIS) to identify individual or groups of vertebrate species and vegetation types that are unrepresented or underrepresented in existing protected areas.

The Gap Analysis Program (GAP) thus provides—often for the first time in many states—accurate, state-wide digital resource maps. The potential application of these databases is virtually unlimited. In this chapter, we focus on their use in evaluating the ecological importance (i.e., an ability to contribute to the conservation of unprotected biological resources) of five areas in Idaho and Oregon. The areas chosen have, in recent years, been proposed by various advocacy groups for national park designation or as an expansion area to an existing park (MacCracken & O'Laughlin, 1992). In order to provide an ecologically defensible spatial context for this analysis, the ecoregion concept was employed. *Ecoregions* are geographic areas that exhibit similarities in the mosaic of environmental resources, ecosystems, and effects of humans (Omernik, 1987).

Preservation of at least 10 percent of the land area occupied by each plant-cover type in a particular ecoregion was established as the minimum criterion needed to protect unprotected biological resources (Miller, 1984; Wright et al., 1994). Each of the five proposals was evaluated in terms of its contribution to this conservation goal. We recognize that 10 percent is a highly arbitrary figure and that, as explicitly pointed out by Noss in Chapter 6, the amount of area needed to protect a given resource is dependent on the type of resource and the ecosystem in question. Because, in this chapter, we are evaluating a concept, the concept can be applied using any chosen value.

Plant-cover type or the alliance level of classification is an intermediate level of vegetation classification (Jennings, 1993). *Plant-cover type* is defined as a group of plant communities having the same primary species and similar physiognomy. Plant communities are, in turn, an assemblage of plant species that interact at the same time and place, have a defined species composition and physiognomy, and usually are named by combining the name of the species that dominates the canopy layer with the name of the species that dominates the lower vegetation layers (Jennings, 1993).

The conservation goal of 10 percent preservation contains two major assumptions, both of which have not yet been tested. The first assumption is that all defined plant-cover types have equal importance. Obviously, this may not be true and to thoroughly evaluate this assumption would involve an analysis of the risk associated with the loss of each cover type throughout its entire range. The second assumption is that plant-cover types, because they are easily derived and more widely available, can be used as surrogates for information on other biological resources, including the distribution of animal species (Pressey, 1990). We recognize that most species have specific micro-habitat requirements that cannot be mapped feasibly for large areas. However, we have been encouraged by the results of independent comparisons between gap analysis predictions and observed species occurrences. Scott et al. (1993) and Edwards et al.

(1996) reported accuracy rates in excess of 70 percent. These analyses show that the gap data can, at the very least, provide a useful first step in identifying species distributions. However, these analyses should be considered a "coarse filter" that must be complemented by a species-specific strategy to conserve uncommon or sensitive species (Noss, 1987).

Areas Studied

Craters of the Moon National Monument

This unit would be created by expanding by approximately sixfold the existing 21,686-hectare Craters of the Moon National Monument. The area includes significant examples of plains volcanism and supports extensive undisturbed (i.e., ungrazed) stands of sagebrush-wheatgrass (*Artemisia* spp., *Agropyron* spp.) vegetation (NPS, 1989).

Hells Canyon National Park, Chief Joseph National Preserve, Snake River Breaks National Recreation Area

This 640,420-hectare proposal is centered around the current Hells Canyon National Recreation Area (NRA) and includes lands in Idaho and Oregon (Hells Canyon Preservation Council, 1992). The canyon is the deepest river-carved gorge in North America. Because of extensive elevation gradients and diverse topographic features, the area supports many different ecosystems as well as significant archaeological resources.

Sawtooth Mountains National Park and Recreation Area

This 410,764-hectare proposal is located in central Idaho in an area of high elevation, rugged mountains, spectacular scenery, high mountain lakes, and abundant wildlife. It is an area that has been favorably compared to Grand Teton National Park, and attempts to designate the area as a national park date back to 1911.

Owyhee Canyonlands National Park

The lands in this proposal, which have been under study for wilderness designation by the Bureau of Land Management (BLM), encompass a relatively isolated, high-elevation desert plateau, which is dissected by the scenic canyons of the Owyhee River and its tributaries.

Table 7.1 Park proposals for Idaho and Oregon by ecoregion

Proposal	Ecoregion	State
Craters of the Moon National Park and Preserve	Snake River Basin	Idaho
Hells Canyon National Park, Chief Joseph National Preserve, and Snake River Breaks National Recreation Area	Blue Mountains	Idaho and Oregon
Sawtooth Mountains National Park and National Recreation Area	Northern Rockies	Idaho
Owyhee Canyonlands National Park	Snake River Basin	Idaho
Steens Mountain National Park	Snake River Basin	Oregon

Steens Mountain National Park

Steens Mountain is a large fault-block mountain that rises from the plains of eastern Oregon. Its sharp elevational gradients create a series of diverse plant communities unique for eastern Oregon. The 155,780-hectare area currently is managed as the Steens Mountain Recreation Area by the BLM.

Methods

Four databases, defining ecoregions, plant-cover types, land ownerships, and land status, were used in the analysis.

1. *Natural ecoregions,* as described by Omernik (1987), defined the spatial bounds of analysis. Idaho contains portions of six ecoregions and Oregon nine, with four of these ecoregions found in both states. The proposals fell within three of the eleven ecoregions common to both states; the Blue Mountains, Northern Rockies, and Snake River Basin/High Desert (Table 7.1).
2. The *plant-cover types,* which occupy each proposal area and its respective ecoregion, were identified from vegetation maps of Idaho and Oregon compiled by GAP (Scott et al., 1993) and were delineated at scales of 1 : 500,000 and 1 : 100,000, respectively. Caicco et al. (1995) defined 26 state-wide "natural" plant-cover types for Idaho. "Natural" implies that the classification excluded lands in agriculture, land in urban or industrial use, bare ground, and open water but not alien species. The 26 cover types were based on a consolidation of 68 different, natural plant communities. We defined 64 natural plant-cover types for Oregon based on 292 plant communities. The Oregon cover types were consolidated and matched to

those in Idaho, resulting in a combined classification of 30 natural cover types, of which 4 were unique to southern and eastern Oregon and were not included in the analyses as they occurred outside the ecoregions studied.

3. *Land ownership* databases compiled for both states at a scale of 1 : 100,000 by GAP were used in the analyses. Ownerships by all federal and state agencies were combined into one class, resulting in two broad categories of land ownership, private and public.
4. *Land status* maps at a scale of 1 : 100,000, also compiled by GAP, were used for both states. These maps identified three land classes: protected public lands, unprotected public lands, and private lands.
   a. *Protected public lands* include those in which most resource extraction, development, and habitat manipulation is prohibited by law or agency policy.
   b. *Unprotected public lands* are those in which consumptive resource use and development are permitted according to various laws and regulations.
   c. *Private lands* are lands owned privately. Because the opportunities to effect land management changes on public lands generally are far greater than on private lands, and because the great majority of lands involved in the proposals were in public ownership, our analyses were limited to public lands.

Results

Most lands in each ecoregion of the two states are publicly owned, ownership being divided primarily between the BLM and U.S. Forest Service (USFS) (Table 7.2). Only a relatively small proportion of these lands are fully protected, with most occurring in USFS wilderness areas. The Northern Rockies ecoregion is unique, with over 22 percent of the land area in protected status, most in the Frank Church River-of-No-Return Wilderness.

Each ecoregion is relatively diverse in terms of the plant-cover types found within it. Each of the ecoregions considered contains 24 of the 26 plant-cover types analyzed. The Blue Mountains lack representation by the mountain hemlock (*Tsuga mertensiana*) and western red cedar (*Thuja plicata*) types, the Snake River Basin lack mountain hemlock and the montane forest type, and the Northern Rockies lack the lodgepole/aspen (*Pinus contorta/Populus tremuloides*) and wet meadow types. The distribution of the areas occupied by each plant-cover type within a given ecoregion, however, is highly skewed, with relatively few types dominating the land area of the particular ecoregion. For example, in the Blue Mountains, two types (ponderosa pine [*Pinus*

Table 7.2 Status of publicly owned lands in Idaho and Oregon by ecoregion

Ecoregion	Public lands (ha)	Percent	Protected lands (ha)	Percent
Idaho				
Snake River Basin	4,937,700	83	392,636	6.6
Blue Mountains	465,980	60	52,810	6.8
Northern Rockies	7,213,185	88	1,836,085	22.4
Oregon				
Snake River Basin	6,454,478	65	261,072	4.0
Blue Mountains	5,793,935	70	422,696	7.2
Northern Rockies	—	—	—	—

ponderosa] and juniper woodlands [*Juniperus* sp.]) occupy 59 percent of the ecoregion, in the Snake River Basin, four types (big and mixed sagebrush [*Artemisia* sp.], shrub steppe, and salt desert shrub) occupy 76 percent of the land area, and in the Northern Rockies, four types (western red cedar, Douglas fir [*Pseudotsuga menziesii*], ponderosa pine, and subalpine parklands) occupy 52 percent of the area.

Most cover types in each ecoregion either are unprotected or inadequately protected according to our ten percent conservation goal (Table 7.3). In general, plant-cover types that are protected tend to occupy more land area and are more widespread in their respective ecoregions. Conversely, those types that are unprotected have a more restricted distribution and, in total, occupy less than 1 percent of the land area in any of the ecoregions.

Because the lands included in each proposal, if accepted, will be classified as protected, each proposal adds to the total land area in protected status within its respective ecoregion. The proposals contain from 4 to 20 of the plant-cover types found in their respective ecoregions (Table 7.4). However, because most of these cover types occur over relatively small areas, each proposal contributes little to the protection of inadequately protected types. Overall, in these

Table 7.3 Number of plant-cover types receiving different levels of protection in the ecoregions studied*

Ecoregion	Unprotected	Protected (<10%)	Protected (≥10%)
Blue Mountains	6	8	10
Snake River Basin	8	12	4
Northern Rockies	4	8	12

**Note:* The conservation goal in this study is 10 percent protection.

Table 7.4 Added protection given to inadequately protected plant-cover types (n) by each proposal

	Number of plant-cover types		
Park proposal	No added protection	Added protection (<10% of land area protected)	Added protection (≥10% of land area protected)
Craters of the Moon (n = 20)	14	6	0
Hells Canyon (n = 14)	5	7	2
Sawtooth Mountains (n = 12)	8	3	1
Owyhee Canyonlands (n = 20)	16	3	1
Steens Mountain (n = 20)	12	6	2

proposals, sufficient land is protected to reach the conservation goal for only five previously unprotected plant-cover types.

Discussion

It is logical that each proposal contains, owing to its specific geographic location and size, only a portion of the total number of plant-cover types classified for its particular ecoregion. Thus, it is not surprising that each proposal makes only a limited contribution to achieving the total conservation goal for its respective ecoregion. In addition, the dominant plant-cover types within each of the proposals are, for the most part, those already adequately protected elsewhere in the ecoregion. In general, the inadequately protected plant-cover types are relatively uncommon and widely dispersed in their respective ecoregions. The unprotected types tend to be found in a small number of distinct areas and, thus, the probability of their occurring within any one geographic area (i.e., those covered by the proposals) is low. The one exception to this is the Sawtooths proposal, the cover types of which are quite representative of those in the Northern Rockies ecoregion. This fact suggests that achieving a conservation goal, such as protecting 10 percent of all plant-cover types in a geographic region, may require that planners and scientists focus specifically on the locations of uncommon types.

Wright et al. (1994) examined the effect of modifications to the boundaries of the Craters of the Moon and Owyhee Canyonlands proposals on the conservation of underprotected plant communities. They found that because the proposals are largely surrounded by lands of similar character, modifying the proposals and/or expanding the boundaries would have limited value in terms of achieving the conservation goal.

Combining data across ecoregions would change the picture slightly. For example, while no alpine communities are protected in the Snake River Basin, 25 percent of the lands in this vegetation type are protected in the Northern Rockies ecoregion. However, efforts to protect representative samples of regionally distributed plant-cover types are best evaluated when tied to a relatively homogeneous geographical unit. The ecoregion has been considered an appropriate unit of spatial analysis and has been used in other studies and management programs to define gaps in the protection of biological and physiographic resources (NPS, 1972; Austin & Margules, 1986).

This approach provides one method for setting priorities among various proposals, based on their contribution to the protection of biological resources. The analysis also illustrates that the planning process for new parks (and other reserves), needs to be an interdisciplinary effort. The proposals were based primarily on visual and recreational resources, as well as dissatisfaction with current management programs affecting the resources. The proposals would probably do an adequate job of protecting the scenic and recreational attributes of each area. However, the relatively small contribution they make toward conserving the biological elements of their respective ecoregions reflects the need for more attention to ecological criteria in proposal selection and design. Implicit in this suggestion is the idea that areas set aside for protection should be representative of the nation's natural and cultural heritage. In the United States, this role is typically played by the national parks. While most natural geographic features and phenomena can be represented in parks, some argue that judging all representative natural geographic features worthy of park status oversteps the traditional concept of parks as places for public enjoyment (Foresta, 1984).

The approach used in this chapter was the converse of the traditional gap analysis methodology in that it was constrained by a focus on preselected areas. As a result, it was less likely (i.e., only by coincidence) to encounter unique resources. In addition, the failure of the proposals to significantly add to the conservation of biological diversity in Idaho and Oregon was due as much to the resources found in each proposal as it was to the large proportion of currently protected public lands. With more than 5 million hectares of roadless, undeveloped lands in Idaho and Oregon undergoing examination for wilderness status (Merrill et al., 1995), the potential to provide greater protection to resources at risk is great.

Literature Cited

Austin MP, Margules CR. Assessing Representativeness. In MB Usher (ed), *Wildlife Conservation Evaluation.* London: Chapman and Hall, 1986:45–67.

Caicco SL, Scott JM, Butterfield B, Csuti B. A gap analysis of the management status of the vegetation of Idaho. *Cons Biol* 1995;9:498–511.

Conservation Foundation. *National Parks for a New Generation.* Washington, DC: The Conservation Foundation, 1985.

Edwards TC, Desler E, Foster D, Moisen GG. Adequacy of wildlife habitat relations models for estimating spatial distributions of terrestrial vertebrates. *Cons Biol* 1996;10:263–270.

Foresta RA. *America's National Parks and Their Keepers.* Washington, DC: Resources for the Future, 1984.

Freemuth JC. *Islands Under Siege: National Parks and the Politics of External Threats.* Lawrence, KS: University of Kansas Press, 1991.

Hells Canyon Preservation Council. Information package for the Proposed Hells Canyon/Chief Joseph National Park and Preserve. Joseph, OR: Hells Canyon Preservation Council, 1992.

Jennings MD. Natural terrestrial cover classification: Assumptions and definitions. *Gap Analysis Technical Bulletin 2.* Moscow, ID: University of Idaho Cooperative Fish & Wildlife Research Unit, 1993.

MacCracken JG, O'Laughlin J. A national park in Idaho? Proposal and possibilities. *Idaho Forest, Wildlife, and Range Policy Analysis Group Report 7.* Moscow ID: University of Idaho, 1992.

Mackintosh B. *The National Parks: Shaping the System.* Washington, DC: National Park Service, 1991.

Merrill T, Wright RG, Scott JM. Using ecological criteria to evaluate wilderness planning options in Idaho. *Environ Manage* 1995;19:815–825.

Miller KR. The Bali Action Plan: A Framework for the Future of Protection Areas. In J McNeely, KR Miller (eds), *National Parks, Conservation, and Development.* Washington, DC: Smithsonian Institution Press, 1984:756–764.

National Park Service. *Part Two of the National Park System Plan: Natural History.* Washington, DC: National Park Service, 1972.

National Park Service. *Reconnaissance Survey, Expansion of Craters of the Moon National Monument.* Washington, DC: National Park Service, 1989.

National Park Service. *Resource Topics for Parklands: Criteria for Parks.* Washington, DC: National Park Service, 1990.

Noss RF. From plant communities to landscapes in conservation inventories: A look at The Nature Conservancy. *Biol Cons* 1987;41:11–37.

Omernik JM. Ecoregions of the coterminous United States. *Ann Assoc Am Geograph* 1987;77:118–125.

Pressey RL. Preserve selection in New South Wales: Where to from here? *Australian Zool* 1990;26:70–75.

Pressey RL. Ad hoc reservations: Forward or backward steps in developing representative reserve systems. *Cons Biol* 1994;8:662–668.

Ridenour JM. *National Parks Compromised: Pork Barrel Politics and America's Treasures.* Merrillville, IN: ICS Books, 1994.

Scott JM, et al. Gap analysis: A geographic approach to protection of biological diversity. *Wildl Monogr* 1993;123:1–41.

Wright RG. Wildlife resources in creating the new Alaskan parks and preserves. *Environ Manage* 1984;14:349–358.

Wright RG, MacCracken JG, Hall J. An ecological evaluation of proposed new conservation areas in Idaho: Evaluating proposed Idaho national parks. *Cons Biol* 1994;8:207–216.

8

Designing and Managing Protected Areas for Grizzly Bears: How Much Is Enough?

David J. Mattson
Stephen Herrero
R. Gerald Wright
Craig M. Pease

Grizzly bears (*Ursus arctos*) have survived in the southern part of their current North American range only because of stringent protection and the availability of habitat inaccessible to humans. Hunting regulations and the U.S. Endangered Species Act have contributed substantially to grizzly bear survival, along with de facto protection by roadless areas and the organic acts of provincial and national parks (Keiter, 1991; U.S. Fish and Wildlife Service, 1993; Keiter & Locke, 1996). However, as much as nearly all involved concede these points, there is major disagreement over the protection grizzlies will need for long-term survival in parts of their range that are more heavily impacted by humans (Shaffer, 1992; Primm, 1993; Mattson & Craighead, 1994).

The debate has focused on the size and distribution of populations, the level and effectiveness of protection accorded bears within their occupied range, and the associated need for connectivity among populations (see Wilcove et al., 1986). In other words, how many bears, distributed over how large an area, with what level of interchange among "populations," and exhibiting what level of long-term fluctuations in growth, do humans want? At one level, this discussion is about the degree to which short-term human prerogatives are curtailed and the areas where these restrictions occur. Therefore, part of the debate could be resolved by identifying factors clearly relevant to answering questions about the size, juxtaposition, and protection of bear ranges; at the very least, it would provide a common framework for discussion.

Much of protected-area design is predicated on existing or potentially induced heterogeneity in the vital rates of target species (see Howe et al., 1991). In theory, species densities vary with the rates of births, deaths, immigration, and emigration, depending largely on habitat productivity and the densities of predators. Prior to the arrival of Europeans, the distribution of North American grizzlies reflected broad-scale patterns of climate and vegetation, possibly modified by competition with black bears (*Ursus americanus*) and low rates of mortality caused by indigenous humans (Storer & Tevis, 1955; Herrero, 1978; Brown, 1985). Grizzlies otherwise had no known major predators.

There is little doubt that the current persistence of grizzly bears at lower latitudes is largely determined by human predation, modified by the effects of food abundance on recruitment (Bunnell & Tait, 1981; Servheen, 1990; Stringham, 1990; McLellan, 1994). The decline of grizzly bear populations is clearly linked to human-caused mortality, which continues to account for virtually all deaths of grizzly bears older than 1 year in the southern Canadian Rockies and in the contiguous United States. Out of 174 grizzlies that were radio-marked and died in this area between 1974 and 1996 (Russell et al., 1979; Dood et al., 1986; Craighead et al., 1988; Aune & Kasworm, 1989; McLellan, 1989a; Nagy & Gunson, 1990; Raine, 1991; Wakkinen & Zager, 1991; Mace & Manley, 1993; Kasworm & Thier, 1994; Wielgus & Bunnell, 1994; Montana Department of Fish, Wildlife, and Parks, unpublished data), 85 to 94 percent were killed by humans (the range in percentages depends on whether unknown causes of death were included and whether the calculation was pooled or averaged over studies). The few demographic studies from this area have also concluded that survivorship, especially of females, outweighs the effects of fecundity on population growth and density (Knight & Eberhardt, 1985; McLellan, 1989b; Eberhardt et al., 1994).

Protection of grizzlies thus needs to reflect the importance of human–grizzly bear interactions. In particular, good management and reserve designs depend on a robust understanding of factors that control the frequency and lethality of contact between humans and grizzly bears (Mattson et al., 1996) and especially the spatial characteristics of this contact. Ideally, we would know what attributes of grizzly bear habitat increase the likelihood of human–grizzly bear contact and which classes of bear are most affected; equally important, we would know why. We would also know how human behavior affects human-bear interactions and their outcome, and how human behavior is modified by the presence of facilities such as roads, trails, and residences.

Having said this, human values are clearly of overriding importance in the design of protected areas for grizzly bears (Mattson et al., 1996). We decide how much we value grizzlies relative to resources that would otherwise be available from their habitat. In other words, humans inescapably define the benefits and opportunity costs of saving grizzly bears and, thus, the risk we are willing to take with this species (Shrader-Frechette, 1991; Kellert, 1994a). This human-centered accounting generates the policies that guide choices governing what, where, and when human activities occur, and are defined for our purposes in terms of spatially explicit modifications of human behavior (if any) intended to conserve grizzly bears (see Clark & Kellert, 1988).

For these reasons, this chapter is devoted to two major topics: (1) the spatial dimensions of human–grizzly bear relationships and factors that influence their dynamics, and (2) a conceptual model for protected-area design that does not just reflect grizzly bear behavior and related features of the biophysical environment but also considers human behavior and the constraints imposed by human values. We hope that this information will establish useful points of reference and thus promote more fruitful discussions about protected-area design, not only for grizzly bears but also for other medium- to large-size carnivores. To this end, we conclude the chapter by generalizing some points to the management of other carnivores as well as some guidelines for the design of carnivore protection in areas with substantial foreseeable or existing human impacts.

The Spatial Dimensions of Human–Grizzly Bear Relationships

Much of the research on grizzly bear habitat relationships has been motivated by concerns over human impacts and safety. How much is bear habitat use affected by the presence of humans or their facilities? Or, how do other human-induced habitat modifications affect grizzly bear cover, security, and food? Regardless of the specific question, these studies have typically relied on the relative spatial distribution of locations from radio-marked bears for their answers and have registered human presence in terms of physical surrogates such as roads, trails, campsites, and towns (Mattson, 1990; McLellan, 1990); it is too difficult or has been deemed less important to track individual humans. Less often, the research design has employed transects that provided absolute estimates of sign density. In either case, the behavior and distribution of live bears have been emphasized.

Perhaps more to the point, spatially explicit mortality risk has also been investigated. What is the probability of a given bear dying as a function of habitat attributes, most importantly, those related to human presence? For various reasons, relatively little research on this topic has been done for bears despite its obvious relevance; interestingly, elk researchers have vigorously pursued the issue because of its tie to hunter success and behavior (Christensen et al., 1991). However, whether we consider behavior or survival to be the primary mediator of human–grizzly bear interactions, both are important parts of the equation. In the following sections, we consider behavior and survival separately but then integrate these factors for a more holistic consideration of habitat–human–grizzly bear interactions, using the Yellowstone ecosystem as an example.

Behavior as a Spatial Phenomenon

There is a relative wealth of information concerning the spatial responses of grizzlies to humans or human facilities. At the population level, a number of studies have shown that grizzlies typically underused areas within 100 to 500 meters of roads, but in one study avoided areas as far away as 914 meters (Archibald et al., 1987; Mattson et al., 1987; McLellan & Shackleton, 1988; Aune & Kasworm, 1989; Kasworm & Manley, 1990). Construed a different way, Mace and Manley (1993) found that grizzlies in their Montana study area underused habitat where road densities exceeded 1.6 kilometers per square kilometer. Interestingly, this underuse by bears did not vary appreciably with use by humans or road design and was exhibited at very low levels of traffic (0.5–1.9 vehicles per hour) (Archibald et al., 1987; McLellan & Shackleton, 1988).

Underuse of areas near campgrounds and town-sites was even more extreme. Use of habitat within 400 to 2000 meters of campsites or cabins by grizzlies and brown bears was 40 to 67 percent less than expected by the available area (Elgmork, 1983; Gunther, 1990). In Yellowstone National Park, underuse of habitat near major recreational developments was typically evident out to 4–5 km (but in spring only out to 1 km) and was 46 to 94 percent less than expected by area and food abundance, depending on the season, food (e.g.,ungulates or cutthroat trout [*Onchorhychus clarki*]), and site (Mattson et al., 1987; Reinhart & Mattson, 1990; Mattson & Knight, 1992).

All of these studies were compromised potentially by uncontrolled biases. Most important was the possible "avoidance" of human facilities as an artifact of bears underusing inherently less attractive habitats that were fortuitously correlated with roads and town-sites. Human-influenced habitat may also receive full, but undocumented, use at night (Schleyer, 1983; Harting, 1985; McLellan & Shackleton, 1988; Nadeau, 1989). These biases most likely affect studies based strictly on daytime sampling of radio-marked bears, coupled with analyses that do not control for the spatial abundance of foods. Transect-based studies of grizzlies using ungulate carcasses (Mattson & Knight, 1992) and spawning cutthroat trout (Reinhart & Mattson, 1990) in Yellowstone National Park avoided both pitfalls and suggested relatively strong avoidance of roads and other human developments. Consistency of results among studies from diverse study areas, coupled with the nonexistence of studies showing bear use greater than or equal to that expected at random near humans, increase our confidence that, wherever killed by humans, grizzlies will not fully use habitat near human facilities.

A HYPOTHESIS. These behavioral results could imply that grizzly bears exhibit an irreducible avoidance of humans. Yet, there are numerous observations of grizzlies foraging within a few meters of humans during daylight hours (i.e., habituating to the human presence) (Herrero, 1985). Bears are clearly able to tolerate humans, presumably as a means of accessing food or finding security from other potentially threatening bears (McCullough, 1982; Herrero, 1985; Mattson et al., 1987, 1992; McLellan & Shackleton, 1988). Habituation therefore seems to affect disproportionately subadult males and females with young, who are plausibly at risk from older males (Tracy, 1977; Warner, 1987; Dau, 1989; Mattson et al., 1992; Fagen & Fagen, 1994).

We know or have good reason to believe that habituated bears are more likely than wary bears to be killed by humans (Meagher & Fowler, 1989; Mattson et al., 1992). Habituated bears are more vulnerable to poaching, more often in conflict with humans, or simply viewed as more of a threat to human safety (Herrero, 1985). Yet, because habituated bears tend to concentrate nearer to humans (within 2–4 km of human facilities in Yellowstone National Park), they are also candidates for making fullest use of human-influenced habitats (Mattson et al., 1992).

Taken together, these results suggest that the behavioral responses of grizzly bears to humans at the population level are largely artifacts of the rate at which humans kill bears and the degree to which this mortality is selective against habituated animals. In other words, we expect greater "avoidance" of human facilities if bears able to use this human-influenced niche are killed faster than they are recruited or simply at a higher rate than wary bears. This type of selective mortality would provide surviving bears more foraging opportunities in remote areas because of reduced overall bear densities. In summary, this hypothesis predicts that habitat "impairment" and the related need for grizzly bear habitat secure from humans devolve to the rate at which we kill bears and to the degree of our tolerance for bears that tolerate humans. In any case, the focus is on human-caused mortality rather than some presumed fundamental inability of grizzlies to use habitat near humans.

Mortality as a Spatial Phenomenon

In the relatively few instances when researchers have examined the joint distributions of human facilities (for lack of records from humans themselves) and known grizzly bear mortalities, there has been a positive association. Nagy et al. (1989) found that 75 percent of bear mortalities occurred within 1 kilometer of all-weather roads in their Alberta study area; Aune and Kasworm (1989) and Dood et al. (1986) found similar concentrations within 1.0 kilometer (63%) and 1.6 kilometers (48%), respectively, of roads in central and northwestern Montana. Unfortunately, none of these studies treated the analyses in terms of how the observed pattern departed from patterns expected at random, and other researchers have merely commented that most grizzly bears killed by humans were shot from or near roads (McLellan & Shackleton, 1988). One of the strongest associations between human access and, in this case, brown bear mortality comes from Chichagof Island, Alaska, where deaths of legally hunted bears were annually highly correlated ($r^2 = 0.86$) with miles of road during a period of road expansion (Titus & Beier, 1992).

Despite these convergent results, it could be argued that there is a bias toward detecting grizzly bear deaths nearer roads, regardless of their frequency, simply because, at a road, it is more likely humans will discover a bear if it dies. Presumably, this bias would be stronger for illegal mortality, which predictably goes unreported by the perpetrator, and for observations of unmarked dead bears. By these standards, the results from Alberta, Montana, and Alaska, based largely on reports of legal hunting kills, would be relatively robust. However, we still do not know how much illegal, unreported poaching goes on and to what extent this mortality is dissociated from access. Clearly, this topic warrants further inquiry.

Although most of these studies are potentially biased or statistically inconclusive, and none has benefited from comprehensive analyses of the several variables likely to influence grizzly bear mortality, they are consistent with history, theory, and more numerous and conclusive observations of the human–grizzly bear conflict. That is, observations of bear mortality concentrated near roads do not contradict some well-supported expectations. A number of studies at several locations in Alaska, Montana, and Wyoming have shown that human–grizzly bear conflict is positively correlated with annual or seasonal changes in human activity or frequency of human and bear contact, especially in areas where human activity is largely unregulated (i.e., excluding areas such as McNeil River Falls, Alaska) or where the bear population is protected from hunting (e.g., most national parks) (Martinka, 1982; Kendall, 1983; Keating, 1986; Dalle-Molle & Van Horn, 1989; Smith et al., 1989; Albert & Bowyer, 1991; Mattson et al., 1992; Fagen & Fagen, 1994). History also has

Table 8.1 The proportion of total area and recorded grizzly bear mortalities, and numbers of these mortalities prorated to the affected area, for strata defined by the level and nature of human impacts and the control of firearms (i.e., park versus nonpark lands), for the Yellowstone National Park ecosystem, 1975–1994

Strata	Proportion of Yellowstone ecosystem[a]	Proportion of recorded mortalities[b] ($n = 179$)	Mortalities per 1000 km^2
Town-sites and park developments[c]	0.09	< 0.30[d]	28.4
Primary (paved) roads[e]	0.12	= 0.13	9.0
Secondary roads[f]	0.12	< 0.25	17.2
U.S. Forest Service roadless areas	0.38	> 0.09	4.9
Yellowstone backcountry	0.29	> 0.23	2.6

[a]Defined by the most peripheral recorded grizzly bear mortalities (see Fig. 8.1).
[b]Including known, probable, and possible (see Craighead et al., 1988).
[c]Major human facilities where food services and overnight accommodations are provided (see Fig. 8.1), including areas within 4 km (the zone of influence on bear behavior [Mattson et al., 1992]).
[d]Indicates the relationship of proportional mortality to proportional area; protected ($df = 4$, $G = 121.8$, $P < .001$) multiple comparisons (Bonferroni confidence intervals) at $\alpha = 0.05$.
[e]Paved all-weather roads, open during the bears' active season, including areas within 2 km (Mattson et al., 1992).
[f]Gravel roads passable to most vehicles during the bears' active season, including areas with 1 km.

clearly shown that most grizzly bears that died between 1850 and the mid-1980s were killed by humans (Storer & Tevis, 1955; Brown, 1985) who we know or suspect did not range far from roads and trails (see Thomas et al., 1976; Lucas, 1980, 1985).

Results from the Yellowstone ecosystem suggest that mortality has varied substantially depending on agency jurisdiction and nearness to human facilities. Roughly 33 percent of the total habitat available to grizzlies is impacted substantially by humans; this is evidenced most graphically by the disproportionate 68 percent of recorded grizzly mortality from 1975 through 1994, for which humans are responsible (taken from Craighead et al. [1988] and Montana Department of Fish, Wildlife, and Parks, unpublished data). When prorated to zones of influence defined by behavioral responses of grizzlies to town-sites and roads, unit-area mortality was greatest near town-sites, and 5.8 to 11.0 times greater than the lowest rates in U.S. Forest Service roadless areas and U.S. Park Service backcountry, respectively (Table 8.1). Mortality in front-country areas was more often due to agency control of a "hazardous" bear, while mortality in the National Forests, in roadless areas and near secondary roads, was largely due to conflict with hunters and livestock, respectively (Table 8.2). Mortality in Yellowstone's backcountry was further distinguished from all other strata by a high proportion of "natural" cub mortality (Table 8.3).

Table 8.2 Proportional distribution of recorded grizzly bear mortality among causes, by strata (see Table 8.1 for definitions), for the Yellowstone National Park ecosystem, 1975–1994

Cause	Developments ($n = 53$)	Primary roads ($n = 24$)	Secondary roads ($n = 45$)	Yellowstone backcountry ($n = 16$)	U.S. Forest Service roadless areas ($n = 41$)
Poaching[a]	0.208a[b]	0.125ab	0.133ab	0.000b	0.073ab
Hunter/outfitter-related[c]	0.076bc	0.167bc	0.244b	0.000c	0.683a
Management[d]	0.528a	0.125b	0.089b	0.000b	0.000b
Livestock-related[e]	0.000b	0.000b	0.400a	0.000b	0.049b
Natural	0.076b	0.333b	0.089b	0.812a	0.171b
Other human-caused[f]	0.113a	0.250a	0.044a	0.188a	0.024a

[a] Illegal mortality not directly associated with hunting or livestock.
[b] Proportions followed by the same letter in rows are not different at $\alpha = 0.05$, based on Bonferroni confidence intervals calculated on both proportional distributions, pairwise, by strata. Multiple comparisons were protected (df $= 20$, G $= 172.8$, $P < .001$), and all strata differed from each other in aggregate.
[c] Mortality resulting from chance encounters with hunters, mistaken identity, conflict over hunter kills, or conflict at camps of outfitters that catered to hunters.
[d] Bears removed by managers because of concern for human safety or for "humane" reasons.
[e] Mortality resulting from conflict over livestock, primarily sheep.
[f] Includes accidental deaths due to research captures, collision with motor vehicles, etc.

These results illustrate why it makes sense to describe grizzly bear habitat fragmentation in terms of the increased mortality risk associated with human facilities and differences in jurisdictions (Mattson & Reid, 1991). By this reckoning, fragmentation is an increasing threat to grizzly bear populations in their southernmost ranges. The Yellowstone ecosystem is already substantially fragmented (Fig. 8.1), while most grizzly bear ranges in southern British Columbia and Alberta are at risk (see Horejsi, 1989; McCrory et al., 1990)—situations exacerbated by the common juxtaposition of human facilities with primary grizzly bear habitat concentrated in narrow transverse valleys (Purves et al., 1992).

SOURCES AND SINKS. There is a strong spatial component to grizzly bear mortality that is closely associated with human access and the unregulated presence of firearms. This spatial heterogeneity is perhaps most usefully described as a source-sink structure, with areas near human facilities constituting the sinks (Knight et al., 1988; Doak, 1995). As an obvious consequence, grizzly population growth rate will decline as the ratio of sink (human-impacted) habitat to source (remote) habitat increases. More importantly, depending on the rate of movement by bears between source and sink habitats, the decline in averaged population growth rate may accelerate relative to the increase in areas impacted by humans (Wilcox & Murphy, 1985; Doak, 1995).

Parks are commonly perceived as grizzly bear population sources. This structure has been postulated for southern Alberta, associated with the Rocky Mountain parks complex (Nagy & Gunson, 1990), and is also thought to exist along some boundaries of Glacier (eastern) and Yellowstone (western) National Park in the United States. As has been shown, backcountry areas of Yellowstone National Park do seem to have the lowest unit-area mortality rates in the ecosystem, approximately half being recorded in roadless areas under the jurisdiction of the U.S. Forest Service. By contrast, town-sites or other major human developments on private and public lands are the most lethal to bears, followed by areas affected by secondary roads. Relative to proportional area, there was also a pronounced difference (2.3 times) between recorded grizzly bear mortality inside and outside Yellowstone National Park (i.e., 23% of total mortality and 41% of total area in the park [log-likelihood test, $G = 26.6$, $P < .001$]). This was despite the application of stringent protections to all federal lands under the U.S. Endangered Species Act, in contrast to an even greater dichotomy of protection between park and nonpark lands in Canada (Keiter & Locke, 1996).

Several researchers have observed that large ranges render grizzlies extremely vulnerable to *population sinks,* whether these are defined as settled areas outside park boundaries or simply as areas impacted by major human

Table 8.3 Proportional distribution of recorded grizzly bear mortality among bear sex-age cohorts, by strata (see Table 8.1 for definitions), for the Yellowstone National Park ecosystem, 1975–1994

Grizzly bear sex-age cohort	Developments ($n = 53$)	Primary roads ($n = 24$)	Secondary roads ($n = 45$)	Yellowstone backcountry ($n = 16$)	U.S. Forest Service roadless areas ($n = 41$)
	a[a]	ab	ac	b	c
Subadult male[b]	0.315a[c]	0.250ab	0.136ab	0.062b	0.146ab
Subadult female[b]	0.111a	0.167a	0.114a	0.062a	0.000a
Adult male	0.167ab	0.125b	0.250ab	0.062ab	0.293a
Adult female	0.241a	0.208a	0.227a	0.188a	0.293a
Cubs[d]	0.148b	0.208ab	0.114b	0.625a	0.122b
Other[e]	0.018ab	0.042ab	0.159a	0.000b	0.146a

[a]Strata identified by the same letter did not differ in aggregate by protected (df = 20, G = 46.7, $P = .001$) multiple comparisons at $\alpha = 0.05$.

[b]Animals < 5 years old, unless accompanied by dependent young.

[c]Proportions followed by the same letter in rows are not different at $\alpha = 0.05$, based on Bonferroni confidence intervals calculated on both proportional distributions, pairwise, by strata.

[d]Dependent young, typically $\leq 1\frac{1}{2}$ years old.

[e]Animals not identified to cohort.

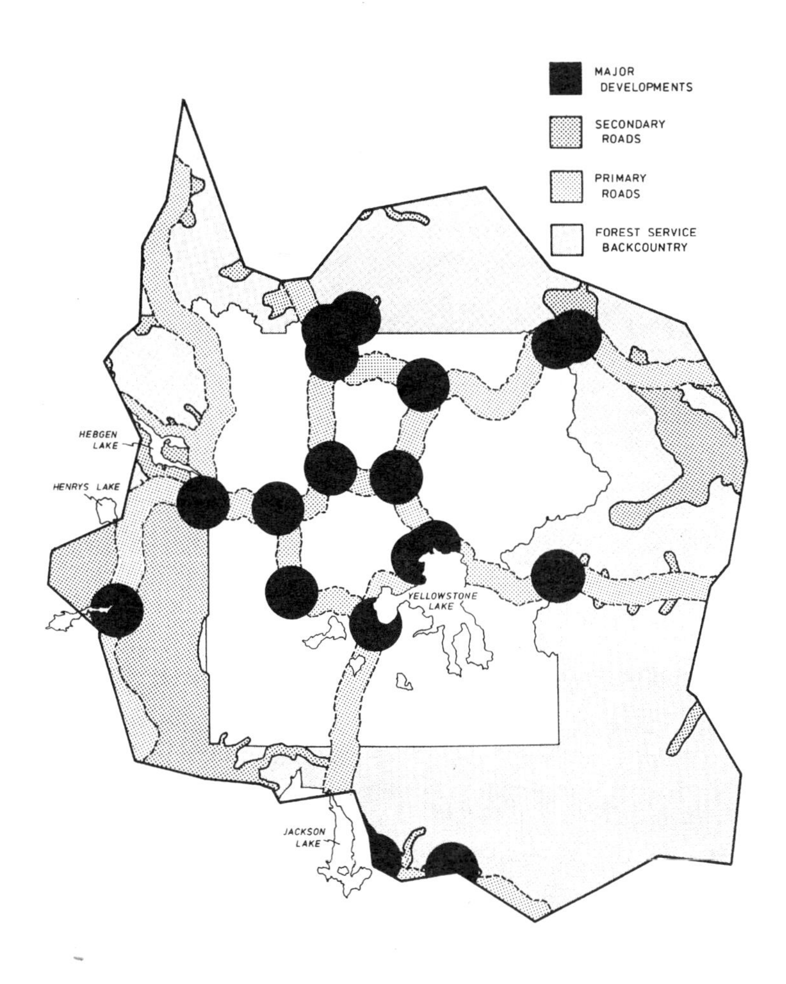

Figure 8.1 Stratification of the Yellowstone National Park ecosystem by the nature of road access and the legal presence of assembled firearms, configured by zones of influence on grizzly bear behavior (see Table 8.1 for definition of strata). Unshaded areas in the core coincide with Yellowstone National Park backcountry.

development (Bunnell & Tait, 1980; Knight et al. 1988; McCrory et al., 1990). For example, bears that spent relatively little time outside of Yoho and Kootenay National Parks in British Columbia were nonetheless quite vulnerable to legal bear hunters (Raine, 1991). In effect, the rate of bear movement between source and sink habitat is often quite high. While mitigating against unpredictable nonlinear responses in population growth rate, this exchange between source and sink habitats could make populations sensitive to each incremental increase in the area of human impacts (Doak, 1995). By implication, heterogeneity in vital rates sufficient to maintain a stable source-sink structure may require very large areas free of substantial human-caused mortality, depending on overall population growth rate (Doak, 1995; see Schonewald-Cox & Bayless, 1986).

An Integrated View: Yellowstone National Park as an Example

Grizzly bear research from the Yellowstone ecosystem gives us the opportunity to construct an integrated view of relationships among grizzly bears, their foods, and humans that may have general application to southern portions of the grizzly bear range. The Yellowstone data set is of sufficient duration and breadth to provide some room for inference, albeit with a dose of speculation. Radio-marked grizzlies relevant to this discussion have been tracked since 1975, and their relocations analyzed relative to a map of habitat types and human features. The abundance of key foods was also monitored for much of this period, and attention has been given to the consequences of bear behavior.

During this study, the survivorship of bears habituated to humans or conditioned to human foods was much lower than bears judged to be wary. Proportionally, 3.1 times as many radio-marked habituated bears were killed from 1975 to 1990 compared to their wary counterparts (Mattson et al., 1992). Because these habituated bears accounted for most habitat use within 2 kilometers of roads and 4 kilometers of major recreational developments, it is likely that the higher mortality of these bears also accounted for the population-level underuse of habitat in these corresponding zones (Mattson et al., 1987; see above).

Not all bear sex and age classes were equally prone to habituation. Adult males, on average, were the wariest of all bears, while subadult males and adult females with dependent young tended to range closest to human facilities and were as likely as subadult females to be habituated (Mattson et al., 1992). Humans were exceptionally intolerant of habituated subadult males and killed most of these animals as they were recognized. Thus, apparently few habituated subadult males survived to be habituated adults. At the same time, subadult males and females with young seemed to avoid the surviving wary males distributed in typically remote areas, presumably because these males posed a threat to them or to their dependent young (Mattson et al., 1987, 1992).

Thus, the human tendency not just to kill habituated bears but virtually eliminate habituated subadult males had several plausible consequences. Remote foraging sites were apparently preempted by the surviving wary adult males, who displaced other bears into chronically underused human-influenced habitats, where these bears habituated to humans as a means of accessing otherwise unavailable resources. The common availability of foods at human facilities no doubt increased the frequency and intensity of habituation, as these bears also conditioned to human foods (Herrero, 1985).

This situation arguably placed adult females in increased jeopardy by effectively reducing the availability of habitat remote from humans. From the perspective of fitness strategies, they were in a catch-22 between what may have been for them the palpable, evolutionarily potent risk of losing offspring to other adults (Bunnell & Tait, 1981), and the latent, evolutionarily recent risk of losing their own lives to humans. Although not yet conclusively demonstrated, the demographic consequences were probably significant, given the sensitivity of population growth rate to survivorship of adult females (Knight & Eberhardt, 1985).

Apparently, these conditions were aggravated by the juxtaposition of human facilities and key bear habitats. Not surprisingly, roads, recreational developments, and town-sites were concentrated in lower-elevation valley bottoms as was prime grizzly bear spring range (Mattson et al., 1987). Adult females also happened to make disproportionately intense use of spring habitats (Mattson et al., 1987), likely increasing the probability that they would lose their fear of humans as they sought out and used ungulate carcasses on elk and bison winter ranges. Although human-caused mortality occurred primarily during late summer and early fall, it is reasonable to expect that potentially fatal patterns of behavior could be initiated or reinforced at any time during the nondenning season.

In contrast, human facilities typically were far removed from the richest fall habitats (Mattson et al., 1987), where, during some years, bears foraged on seeds in high-elevation whitebark pine stands (*Pinus albicaulis*) (Mattson & Reinhart, 1994). While this disposition was ostensibly favorable to grizzlies, bears still died in their greatest numbers during the fall foraging season. However, this mortality occurred largely during years when whitebark pine seeds were unavailable and bears sought out alternate foods that typically occurred at lower elevations, nearer to, if not closely associated with, human facilities (Mattson et al., 1992). Thus, it was not sufficient for humans to be remote from the richest habitats when food crops produced in these areas failed. Perhaps as important, grizzly bear mortality was dependent on the presence of humans in habitats containing various alternate foods. Because predicting these alternate foods is fraught with uncertainty (Mattson et al., 1991a) and because they are often astride park boundaries, resolution of this problem is inherently difficult.

The choice of bears to kill and the placement of human facilities thus appear to be central parts of grizzly bear management, although likely secondary to decisions regarding the overall level of bear mortality and human activity. Killing "problem" habituated bears probably enhances human safety (Herrero,

1985; Herrero & Fleck, 1990), but it very likely reduces a population's ability to fully use available habitat and could indirectly increase the vulnerability of adult females to human-caused mortality. It probably is also insufficient to merely judge the impacts of human activities in terms of habitats that are, on average, most heavily used by bears during hyperphagia. Spring habitat and infrequently productive fall habitats are of equal importance to an assessment.

Designing Protected Areas

So far, we have emphasized the effects of human facilities on grizzly bear mortality and distribution, and the extent to which these effects might vary depending on the site and bear behavior. We have not addressed questions of larger import related to the size, security, and connectivity of protected areas. Although we cannot provide answers, in this section we identify some relevant considerations and relate them to existing protected area strategies.

Given an interest in grizzly bear conservation, protected areas are relevant usually where the bears' range has been or is becoming fragmented by human settlement. In areas typical of Alaska and northern Canada, there are few resident humans, and currently conservation is primarily a function of managing legal hunts and apprehending poachers. The design and establishment of protected areas is thus most immediately relevant to conservation of grizzly bears in Alberta and British Columbia and in the few areas of the contiguous United States where grizzlies survive (McCrory et al., 1990; U.S. Fish and Wildlife Service, 1993); it will become more important in the North as resource development continues.

A Conceptualization

Protected-area size is theoretically a function of population goals, range sizes and range overlap, impairment by human activity and behavior (Schonewald-Cox & Buechner, 1991), lethality of the surrounding matrix (Schonewald-Cox & Bayless, 1986; Franklin, 1993), habitat variability (Goodman, 1987; Thomas, 1994), and the shape of the protected area (Schonewald-Cox & Bayless, 1986). In other words, protected areas need to be larger if the desired population is large, composed of wide-ranging animals that occupy highly variable habitat impacted by substantial human use, and if the protected area is surrounded by a deadly matrix from which the bears need to be buffered. Conversely, protected areas theoretically can be smaller where range sizes and population goals are smaller, the habitat better protected and less variable, and the surrounding matrix more benign. As illustrated by Kootenay and Yoho National Parks in southeastern British Columbia (McCrory et al., 1990; Raine, 1991), all else being equal, linear protected areas leave bears more vulnerable than areas with a lower edge-to-interior ratio.

SOME PRIMARY DETERMINANTS. It could be argued that population goals, more than anything else, determine the size and design of protected areas for grizzly bears (Fig. 8.2). Protected areas would vary substantially depending, for example, on whether 50 or 2000 bears were deemed sufficient. In the same way, societal decisions regarding acceptable risks to populations and relevant time frames have major theoretical implications to protected area design. Managing at a very small risk of extinction within 1000 years obviously entails substantially greater levels of protection than managing for a moderate risk of extinction within 100 years (compare Shaffer, 1992, and U.S. Fish and Wildlife Service, 1993). Despite assertions to the contrary, these decisions are fundamentally expressions of human values, possibly influenced by a cost-benefit analysis that considers biological information (Kellert, 1994a).

Carrying capacity is another important variable in protected area design (see Fig. 8.2). Lower unit-area carrying capacity obviously engenders the need for greater space, all else being equal. However, the usefulness of this concept is compromised by its ambiguity and the fact that it has a frequently ignored temporal dimension (McNab, 1985). Carrying capacity could be described in primal terms; i.e., given the existing vegetation, how many grizzlies would have lived in an area prior to the arrival of Europeans? Inevitably, however, this type of construct needs to be translated as "habitat effectiveness" or "habitat capability," by accounting for diminishments attributable to humans. Sidestepping these semantics, the area required to support a given number of bears will depend on both habitat productivity and the human presence (Weaver et al., 1986). Less productive habitat and a chronically lethal human presence will dictate a need for greater space as a means of providing more inherently lower-density feeding opportunities in areas secure from human impacts.

Humans, therefore, are perhaps best characterized as a habitat feature. Like vegetation types, human presence will vary temporally and spatially. However, unlike vegetation, which is understood primarily as a surrogate for population productivity, the human presence is most fruitfully understood as a surrogate for mortality, as described earlier. In simple terms, habitat variation is gainfully understood as fecundity varying with food abundance and mortality varying with the numbers and behavior of humans. Although not explicit in current grizzly bear cumulative-effects models, this conceptualization is implicit to the way habitat effectiveness is calculated and to the acknowledgment that separate

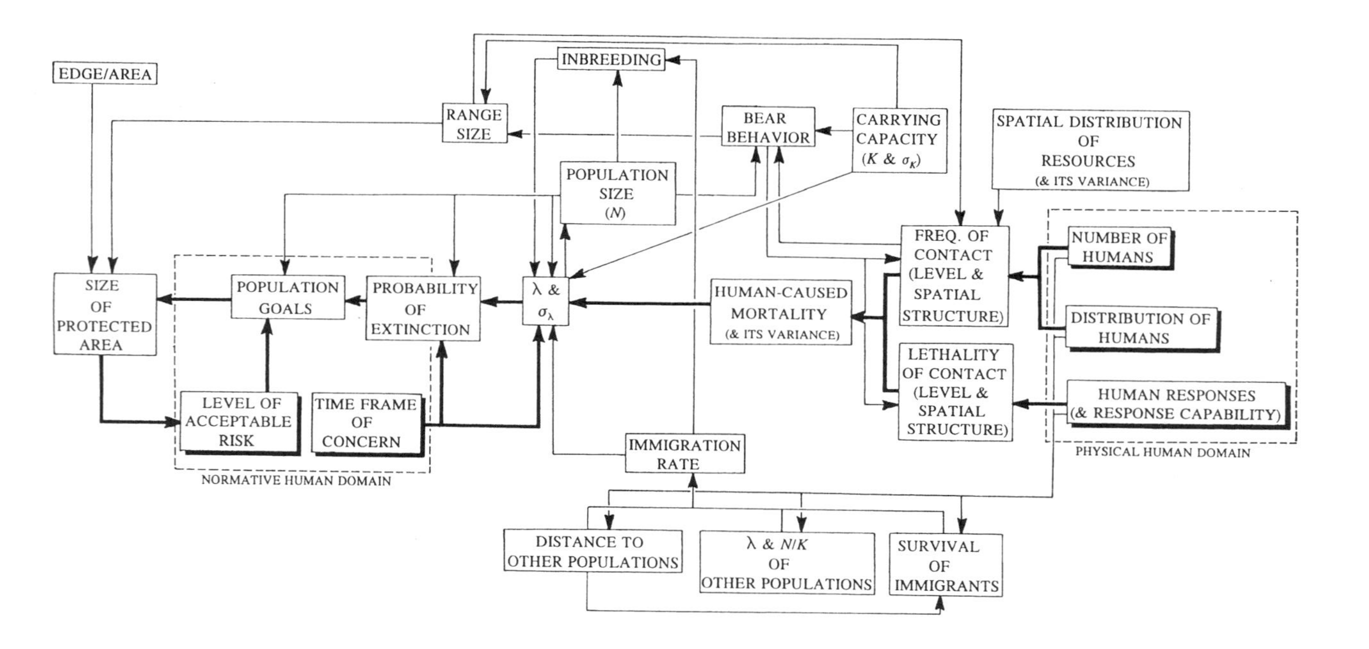

Figure 8.2 A simplified conceptual model of the relationship between attributes of grizzly bear protected areas and influential factors related primarily to human and grizzly bear behavior. Key manipulable factors are denoted by broken boxes and major relationships by heavier lines. Human-related factors are segregated as physical and normative (related to values or cultural norms) elements. Population growth rate is denoted by λ and variance, in places, by σ.

calculations of mortality risk as a function of the human presence are needed (Weaver et al., 1986).

While information on the spatial distributions of humans and foods is important, it is not sufficient for deciding how much protection a bear population needs to achieve specified goals. The abundance and quality of food may vary dramatically, not only from one year to the next but among decades or centuries (Hamer & Herrero, 1987a; Mattson et al., 1991a; Mattson & Reid, 1991). Similarly, both the numbers and behavior of humans may vary so that the frequency and lethality of contact between bears and humans changes (see Fig. 8.2). In theory, bear populations in areas subject to greater variation in human and biophysical domains need greater protection as well as prerogative on a larger area to ensure longer-term survival (Goodman, 1987; Thomas, 1994).

These considerations are graphically illustrated by historical changes in the Yellowstone National Park ecosystem and the Canadian ecosystem containing the four Rocky Mountain National Parks (i.e., Banff, Jasper, Yoho, and Kootenay). Habitat conditions are dynamic and relatively unpredictable in both areas. For example, in the Yellowstone ecosystem, fires recently burned a substantial part of available grizzly bear habitat (Schullery, 1989) during the same period that bears were beginning to use an entirely new (at least in recent history) grizzly bear food (army cutworm moths [*Euxoa auxiliaris*]) (Mattson et al., 1991b). Fire control along the east front of the southern Canadian Rockies has similarly led to the widespread diminishment of a key bear food (buffalo-berry [*Sheperdia canadensis*]) that is dependent on periodic stand-replacement fires (Russell et al., 1979; Hamer & Herrero, 1987b). Considerable annual variation in abundance of key foods is further nested within these longer-term changes. More to the point, this environmental variation has affected bears and, in the Yellowstone ecosystem, has led not only to possible changes in grizzly bear fecundity (Mattson & Reinhart, 1994) but also dramatic variation in conflict with humans (Mattson et al., 1992).

Recent theoretical work suggests that catastrophes and endemic environmental variation are important in defining extinction risks for currently stable populations (e.g., Lande, 1993). Parenthetically, it is worth noting that both types of variation are probably secondary in importance to the deterministic effects of increasing numbers of humans in grizzly bear habitat (Mattson et al., 1996). Nonetheless, annual human-caused mortality in or near protected areas is quite varied, and, prorated to the number of humans, has no doubt changed with time. While recognizing the importance of deterministic grizzly bear–population declines, it makes sense to recognize variance in the human-related causes and design, accordingly. If other buffers are absent, it seems logical to buffer grizzlies from variations in habitat conditions that lead to variation in risk of human-caused mortality. It also makes sense to at least anticipate the possible consequences of catastrophes, most logically defined in terms of dramatic adverse changes in either the numbers or behavior of humans. Although grizzlies may be physically able to accommodate most annual variation in the distribution and abundance of foods (Herrero, 1978; Stirling & Derocher, 1990), they are exceptionally ill-equipped to buffer themselves from temporal and spatial variations in risk posed by humans.

THE HUMAN DIMENSION. It is evident from these considerations that the key variables in designing protected areas for fragmented grizzly bear populations are the numbers, distribution, and behaviors of humans, both in and around reserves, and management time frames and levels of acceptable risk (see Fig. 8.2). While important, it would be difficult to argue that variation in habitat productivity, aside from effects on the frequency of encounters between humans and bears, equals the effects of those factors more directly tied to the rate at which humans kill grizzlies. In part, this follows from the apparent greater sensitivity of grizzly bear–population growth to survivorship rather than fecundity (Knight & Eberhardt, 1985; McLellan, 1989a; Eberhardt et al., 1994), and the demonstrated ability of humans to eliminate grizzlies, regardless of their reproductive potential (Storer & Tevis, 1955; Brown, 1985).

The size and characteristics of protected areas, therefore, will largely be a function of human values and behavior, although the role of "carrying capacity" as well as the need to understand how human numbers and behavior affect grizzly bear vital rates should not be trivialized. In this light, protection is ultimately defined in terms of frequency and lethality of contact with humans (see Fig. 8.2) (Mattson et al., 1996). Thus, the need to control humans varies with the number of humans in bear habitat, their distribution, and their behavior. More specifically, it is relevant whether humans are dispersed or aggregated, near or far from important bear habitat, armed or not, and prone to intolerance and fear of grizzlies.

Theoretically, grizzlies could coexist with relatively large numbers of humans if the humans were unarmed, tolerant of injury and competition for common resources, and aggregated in the poorest grizzly bear habitat. If this scenario were pervasive, the issue of providing grizzly bears with areas secure from humans might be moot. We do not know the theoretical or practical limits of human tolerance, but we do know that grizzlies can become highly habituated to humans and that unarmed humans do not pose much immediate risk to a grizzly bear. Clearly, many issues—and limitations—of grizzly bear conversation are rooted in human values and behavior.

There is contradictory evidence and some dispute over the extent to which human values can be changed in a short-term tactical sense (Clark & Reading, 1994; Kellert, 1994a). Additional information often is used to justify or rationalize set values and opinions (Reading & Kellert, 1993). On the other hand, education can modify the expressed values of students (Caro et al., 1994). Regardless, we know of no situation when, as a consequence of education programs by advocacy groups or government agencies, base values have changed toward the management of grizzly bears and their habitat. At best, such efforts might serve to mobilize interested people and reduce unintended conflicts between bears and humans by providing people with the information needed to produce an already desired outcome.

Although it may not be possible to change human values, human behavior can be influenced by the extent to which people are involved in decision-making processes. Greater acknowledgement of individual concerns often serves to alleviate conflict and can lead to more successful management (Gregory & Keeney, 1994; Wondolleck et al., 1994). In the case of grizzly bears, antagonism toward protective measures and, possibly, the proclivity to kill bears illegally might be reduced by recruiting acceptance, if not goodwill, through collaborative development of management strategies (Kellert 1994b). Although this type of endeavor would have to be bounded at some level by broad policy objectives, there is good reason to believe that a consensus-building approach may have the greatest short-term impacts on human behavior (see Cohn, 1988; Nowak, 1995). Unfortunately, there has been little impetus from policymakers for this kind of approach to grizzly bear management, and interaction between managers and stakeholders has largely consisted of a formal exchange of "information" and "concerns."

These reasons, as well as limited authority to change public values, have led to a management style that emphasizes modification of human behavior through coercion (i.e., laws, regulations, and their enforcement), access restriction, and control of firearms (as in most parks) (Dood et al., 1986; Nagy & Gunson, 1990; U.S. Fish and Wildlife Service, 1993). In a proximal sense, this management style has focused on sanitizing human facilities to reduce their attractiveness to bears and regulating direct human-caused mortality. Very little attention, if any, has been given to limiting the overall number of people in grizzly bear habitat or to the major capital investments, such as roads and accommodations, that often facilitate human presence. In the final analysis, because education seems to have limited tactical use—especially when confronted by millions of transient tourists—because regulation of human numbers seems to be challenging politically, and because disarming humans is feasible politically only in a few areas, "protection" of grizzly bears has come to be defined in terms of coercion and access restriction. Protected areas for grizzly bears are correspondingly defined primarily in terms of prohibitive regulations and few roads.

Corridors and Connectivity

Connectivity among populations is generally thought to enhance long-term survival. Relatively high levels of exchange, in theory, allow for "risk averaging," characterized by periodic rescue of declining populations (see Brown & Kodric-Brown, 1977) by emigration from growing populations and prevention of inbreeding by maintenance of genetic heterozygosity. Conversely, Simberloff and Cox (1987) point out that interchange among populations might transmit catastrophic diseases and dilute local genetic adaptations in species, which tend to maintain fragmented ranges under natural conditions. Therefore, as a bottom line, these authors contend that connectivity is not always a good thing.

But how does connectivity affect bears? There is no evidence that grizzly bear populations were characteristically fragmented or exhibited anything resembling a classic metapopulation structure prior to European settlement. There is no evidence that grizzlies exhibit major local genetic adaptations, as would be expected by their more-or-less continuous distribution, large ranges, phenotypic plasticity, and flexible behavior (F. Allendorf, personal communication). Unlike some other carnivores (see Young, 1994), brown bears are not threatened by any known virulent diseases anywhere in the world. In short, there are no obvious biological reasons why connectivity would be bad for grizzlies and numerous reasons to think it would be good (Shaffer, 1992; U.S. Fish and Wildlife Service, 1993). Theoretically, it would be especially beneficial to increase exchange among small, otherwise vulnerable, populations, especially if connections were made to larger populations (Harrison, 1991).

Corridors are commonly conceived as strips of connecting habitat that not only are attractive to a species but also are sufficiently secure to allow for safe transit. In this sense, corridors provide a means for individuals to travel among populations. However, this conception is useful only when populations are within a range of lifetime movements (Fahrig & Merriam, 1994). There is limited information about how far grizzlies will disperse. In the Yellowstone area, juvenile males have been known to relocate 45 to 105 kilometers from maternal ranges through relatively friendly habitat (Blanchard & Knight, 1991). Populations isolated by greater distances of hostile habitat, like Yellowstone's and possibly the bears in Washington State's North Cascades, would not benefit from traditionally conceived corridors. Successful movement to these more isolated populations instead would depend, in particular, on the establishment and survival of adult females in intervening habitat, which would function as a sequence of demographic stepping stones. In these cases, connectivity becomes a matter of creating conditions where females can survive and reproduce; the exact demographic requirements logically depend on the distance between "populations." In essence, separate populations would be united (Shaffer, 1992).

Conclusions and Implications

Hopefully this discussion clarifies that there is no single size, configuration, or suite of attributes for areas designed to protect grizzly bears. The primary determinants of size seem to be (1) specified or operational time frames and levels of acceptable risk; (2) rates of human-caused mortality and the extent to which they are selective against habituated bears; (3) unit-area carrying capacity, in the loose sense that it has been defined here; and (4) net exchange with other bear populations. Of these key factors, only one falls largely within the biological domain, while the others are determined primarily by human values, behavior, densities, and distribution within protected areas and the intervening "hostile" matrix. It therefore is logical that the design of protected areas for grizzlies should consider pertinent information about humans and the attributes of human–grizzly bear interactions that lead to bear deaths (Mattson et al., 1996). By this reckoning, bear biology is secondary to human biology.

Compartmentalization: A Key Strategy

A key strategy has been implicit to this entire discussion, assuming that there is some level of irreducible conflict between grizzlies and humans and related demonstrable risks to human safety. This scheme, which we call compartmentalization, follows from our analyses and is a more formalized approach, convergent with the fragmented evolution of grizzly bear management during the last two decades. Simply put, compartmentalization seeks to segregate a given number of humans and grizzlies, as a means of minimizing contact between the two species. Implementation of this strategy depends on (1) minimizing the attractiveness to bears of areas occupied by humans, whether this attractiveness derives from human activity or native foods (Herrero, 1985; Herrero & Fleck, 1990); (2) aggregating as much human use as possible within areas that are inherently unattractive to bears (Herrero et al., 1986; McCrory et al., 1986); (3) coercing bears that enter these human aggregations to leave as soon as they are detected (Greene, 1982; McCullough, 1982); and (4) excluding humans from high-value grizzly bear habitat during the seasons when bears are most active (Martinka, 1982). When implemented in concert with firearms control, this strategy promises to maximize short-term compatibility between grizzlies and humans.

National parks in Canada and the United States have come closest to implementing this strategy. Thus, we see grizzlies surviving during their active season in areas visited by several million people. Virtually all visitors remain aggregated in relatively few areas, sanitation is usually stringent, and areas of frequent bear use or known high-quality habitat are closed as needed or by a regular schedule, which allows for areas permanently allocated to high levels of human use, regardless of attractiveness to bears. Extremes of this strategy are evident at places like McNeil River Falls, Alaska, where the number and temporal and spatial distributions of humans are tightly controlled, albeit in extremely attractive bear habitat, but with minimal apparent impacts on bear behavior or survival (Aumiller & Schoen, 1991).

However, conflicts continue to plague the national parks, partly because many human developments were placed in attractive bear habitat at a time when minimizing conflict with grizzlies was not a priority. This is especially true of places such as Lake and Grant Villages in Yellowstone National Park, located on top of cutthroat trout spawning streams (Reinhart & Mattson, 1990); Old Faithful and Mammoth Villages, located in prime ungulate winter ranges (Mattson & Knight, 1992); and the Town of Banff and the Trans-Canada Hiway in Banff National Park, located in the middle of a travel corridor and otherwise productive habitat (Purves et al., 1992). In addition, the mere presence of such human multitudes seems to precipitate unavoidable conflict.

There are thus obvious limits to compartmentalization, even in parks. However, predictably, these limits are reached sooner in nonpark areas, where humans are freely armed and many are engaged in activities inherently prone to disperse them on the landscape, i.e., hunting, backcountry recreation, ranching, and most industrial activities. In these cases, we are confronted with somehow limiting the numbers of people or modifying their behavior to such an extent that their presence poses little threat to grizzlies. For reasons discussed previously and in other sources (Mattson & Craighead, 1994; Mattson et al., 1996), these tactics are exceptionally difficult to implement. Parks will therefore likely remain cornerstones of grizzly bear conservation (Martinka, 1982; McCrory et al., 1990), especially in areas with high densities of resident humans or where grizzlies still are hunted legally (Herrero, 1994).

Possible Improvements

HUMAN DIMENSIONS. It is vital for people involved in grizzly bear conservation, and especially the biologists and scientists, to understand that population goals and acceptable risks and time frames are societal choices that reflect the relative value placed on grizzlies by humans. None of these issues is inherently "biological," although biological information may be germane to human deliberations, especially as it elucidates the consequences of different choices (Kellert, 1994a). Biological information commonly is used to support a given position by highlighting specific implications, but this information still does not equate to the advocated values. Thus, the single most important variable in protected-area design is likely social and not biological. The process of designing protected areas for grizzlies therefore may rely as much on the efficiency and utility with which we garner consensus or specify guidance in law as on our ability to generate and use information about grizzly bears (Mattson et al., 1996).

Protected-area design and grizzly bear management, in particular, would therefore benefit from more systematic elucidation of relevant norms, either as a useful protocol for involving local stakeholders in the specification of goals or as a useful direct specification in high-level policy documents (Mattson & Craighead, 1994; Mattson et al., 1996). Some have contended that grizzly bear management already accounts for local social considerations. However, we believe that it is important not to confuse the minimization of risk to individual managers with the optimal expression of human values at the local and national levels. It could be argued that grizzly bear management has too often expressed more of the former and less of the latter, this rationalized as being attentive to "social realities" (Primm, 1993; Mattson & Craighead, 1994).

Moreover, managers would have more options and could be more effective if they had greater knowledge of socially acceptable techniques for modifying human behavior (Clark & Reading, 1994), with the intent of harmonizing humans and grizzly bears. Although other means may exist, one promising approach integrates stakeholders more closely in the development of management objectives and strategies (see earlier discussion; Gregory & Keeney, 1994) and even the design and implementation of research (Mattson et al., 1996). This not only increases the positive investment of people most likely to directly or indirectly affect bear survival, but also increases the likelihood that their concerns will be addressed in a constructive preventive manner (Wondolleck et al., 1994). Most important, this type of process holds the promise of increasing local acceptance of grizzly bears and the related necessary restrictions on human activity (Kellert, 1994b). Ironically, unless legislation in the United States is revised so that citizens can be more directly involved in management of federal lands, this potential reconciling tactic may be unavailable to managers.

BIOLOGICAL DIMENSIONS. Our understanding of spatial variation in grizzly bear mortality is not yet very robust. For example, no one has explicitly accounted for either movement between sources and sinks or the effects of ancillary variables. Existing information also does not clarify whether park boundaries or human-impacted areas, regardless of jurisdiction, better distinguish sinks and sources; the utility of either stratification will likely depend on jurisdictional dichotomies in legal protection. By implication, further analysis and modeling would simultaneously consider both stratifications of the landscape. Conservation efforts would benefit substantially from a more detailed understanding of how specific human activities affect grizzly bear survival and, taken together, ultimately how they affect population growth (Mattson et al., 1996). As a bottom line, ideally planners would be able to predict the demographic consequences of changing the types, numbers, and locations of human facilities, or changing management regimes associated with jurisdictional lines.

The existing empirical research for grizzly bears is a weak base for developing the spatially explicit models and predictions that most scientists and managers consider central to applying scientific results. Doak (1995) modeled the Yellowstone population as a fairly simplistic source-sink structure, yet elucidated some important management implications. However, to go beyond this type of first approximation, the available data must be subjected to more rigorous and comprehensive spatial analyses, and, perhaps, we will need to design additional field studies around explicit hypotheses concerning the spatial component of grizzly bear mortality. All of this research is inherently limited, however, by relatively small numbers of grizzly bears and the correspondingly small numbers of dead animals on which we can base any model or statistical test. So while this is a vital issue, we will have continuing difficulty generating relevant but reliable information (Mattson et al., 1996).

Furthermore, designers and managers of protected areas would benefit from being able to express biophysical landscapes in terms of grizzly bear fecundity. Together with an understanding of how grizzly bear mortality varies with the human presence, we would then be able to anticipate grizzly bear population dynamics as a function of dynamic landscapes. The difficulties of such an undertaking are obvious. There first needs to be some common currency for describing landscapes of the composition and productivity of varied plant species. We then need to relate this currency to observed fecundity, presumably through some analysis of ecological energetics. Finally, if there are density-dependent responses in reproduction at the landscape level, we need at least to comprehend the magnitude and nature of this effect. At best, researchers have made some small uncoordinated steps in this direction, but there is a long way yet to go.

Implications to Other Carnivores

We conclude this chapter with some observations on whether or how considerations in the design and management of protected areas for grizzly bears extrapolate to other medium- to large-sized carnivores such as black bears, American martens (*Martes americana*), fishers (*M. pennanti*), lynx (*Lynx canadensis*), cougars (*Felis concolor*), wolverines (*Gulo gulo*), and wolves (*Lupus canis*). Commonalities logically follow from (1) the extent to which humans cause mortality; (2) the extent to which densities are affected by variations in prey induced by humans either directly or through modification of habitat structure; (3) the extent to which there are economic incentives for humans to kill the carnivore; and (4) the extent to which the carnivore threatens human safety or assets such as livestock and crops.

By these standards, cougars probably pose nearly as much threat to human safety as do grizzlies. Wolves and cougars (Dixon, 1982; Paradiso & Nowak, 1982), along with grizzlies and black bears (Mattson, 1990), are sometimes a threat to livestock. Together with wolverines (Banci, 1994), these four large carnivores seem to die largely because humans kill them, at least under the same types of conditions where protected areas are important to conservation of grizzlies (Dixon, 1982; Paradiso & Nowak, 1982; Brown, 1983). Although humans may have additional substantive impacts on cougars and wolves through management of ungulates, it is likely that direct interactions with humans, and associated human behavior, are of great importance to the design of protected areas for all four larger carnivores. Many of the considerations regarding human–grizzly bear interactions are therefore relevant to black bears, wolves, cougars, and wolverines, although habituation to humans may not be of as much importance to mortality of the nonbear species.

Lynx, American martens, and fishers pose virtually no direct threat to human safety or assets. All are potentially subject to heavy trapping because of the sometimes high economic value of their pelts, and all can be dramatically affected by changes in the structure of their habitat, which are manifested primarily through changes in prey tied closely to older, typically closed-canopy, forests (see Powell, 1993; Buskirk et al., 1994; Ruggiero et al., 1994). Because of these dissimilarities to grizzly bears, the considerations in this chapter probably have minimal or, at most, only very general relevance to conservation of these species. A different model would better tend to the importance of habitat-prey dynamics and the economics of trapping and timber harvest.

Literature Cited

Albert DM, Bowyer RT. Factors related to grizzly bear-human interactions in Denali National Park. *Wildl Soc Bull* 1991;19:339–349.

Archibald WR, Ellis R, Hamilton AN. Responses of grizzly bears to logging truck traffic in the Kimsquit River Valley, British Columbia. *Int Conf Bear Res Manage* 1987; 7:251– 257.

Aumiller L, Schoen J. McNeil River: Managing for wildlife viewing. *Magazine of the Alaska Department of Fish and Game* 1991;23:42–43.

Aune KA, Kasworm W. Final report—East Front grizzly studies. Helena, MT: Montana Department of Fish, Wildlife and Parks, 1989.

Banci V. Wolverine. In LF Ruggiero, et al. (eds), *The Scientific Basis for Conserving Forest Carnivores: American Marten, Fisher, Lynx, and Wolverine in the Western United States.* Fort Collins, CO: U.S. Department of Agriculture, Forest Service General Technical Report (RM-254) 1994:99–127.

Blanchard BM, Knight RR. Movements of Yellowstone grizzly bears. *Biol Con* 1991;58: 41–67.

Brown, DE (ed). *The Wolf in the Southwest: The Making of an Endangered Species.* Tucson, AZ: The University of Arizona Press, 1983.

Brown DE. *The Grizzly Bear in the Southwest.* Norman, OK: University of Oklahoma Press, 1985.

Brown JH, Kodric-Brown A. Turnover rates in insular biogeography: Effect of immigration on extinction. *Ecol* 1977;58:445–449.

Bunnell FL, Tait DEN. Bears in models and in reality—implications to management. *Int Conf Bear Res Manage* 1980;3:15–23.

Bunnell FL, Tait DEN. Population Dynamics of Bears—Implications. In CW Fowler, TD Smith (eds), *Dynamics of Large Mammal Populations.* New York: John Wiley & Sons, 1981:75–98.

Buskirk SW, Harestad AS, Raphael MG, Powell RA (eds). *Martens, Sables, and Fishers: Biology and Conservation.* Ithaca, NY: Cornell University Press, 1994.

Caro TM, Pelkey N, Grigione M. Effects of conservation biology education on attitudes towards nature. *Cons Biol* 1994;8:846–852.

Christensen AG, Lyon LJ, Lonner TN (compilers). Proceedings of the elk vulnerability symposium. Bozeman, MT: Montana State University, 1991.

Clark TW, Kellert SR. Toward a policy paradigm of the wildlife sciences. *Renew Res J* 1988; (Winter):7–16.

Clark TW, Reading RP. A Professional Perspective: Improving Problem Solving, Communication, and Effectiveness. In TW Clark, RP Reading, AL Clarke (eds), *Endangered Species Recovery: Finding the Lessons, Improving the Process.* Washington, DC: Island Press, 1994:351–370.

Cohn JP. Culture and conservation. *BioScience* 1988;38:450–453.

Craighead JJ, Greer KR, Knight RR, Pac HI. Grizzly bear mortalities in the Yellowstone ecosystem 1959–1987. Bozeman, MT: Montana Department of Fish, Wildlife, and Parks, 1988.

Dalle-Molle JL, Van Horn JC. Bear–People Conflict Management in Denali National Park, Alaska. In M Bromley (ed), *Bear–People Conflicts: Proceedings of a Symposium on Man-*

agement Strategies. Yellowknife: Northwest Territories Department of Renewable Resources, 1989:121–128.

Dau CP. Management and Biology of Brown Bears at Cold Bay, Alaska. In M Bromley (ed), *Bear–People Conflicts: Proceedings of a Symposium on Management Strategies.* Yellowknife: Northwest Territories Department of Renewable Resources, 1989:19–27.

Doak DF. Source-sink models and the problem of habitat degradation: General models and applications to the Yellowstone grizzly. *Cons Biol* 1995;9:1370–1379.

Dood AR, Brannon R, Mace R. Final programmatic environmental impact statement: The grizzly bear in northwestern Montana. Helena, MT: Montana Department of Fish, Wildlife, and Parks, 1986.

Dood AR, Pac HI. Five-year update of the programmatic environmental impact statement: The grizzly bear in northwestern Montana. Helena, MT: Montana Department of Fish, Wildlife, and Parks, 1993.

Eberhardt LL, Blanchard BM, Knight RR. Population trend of the Yellowstone grizzly bear as estimated from reproductive and survival rates. *Can J Zool* 1994;72:360–363.

Elgmork K. Influence of holiday cabin concentrations on the occurrence of brown bears (*Ursus arctos* L.) in south-central Norway. *Acta Zool Fennica* 1983;174:161–162.

Fagen JM, Fagen R. Interactions between wildlife viewers and habituated brown bears, 1987–1992. *Nat Areas J* 1994;14:159–164.

Fahrig L, Merriam G. Conservation of fragmented populations. *Cons Biol* 1994;8:50–59.

Franklin JF. Preserving biodiversity: Species, ecosystems, or landscapes? *Ecol Appl* 1993;3:202–205.

Goodman D. How do any species persist? Lessons for conservation biology. *Cons Biol* 1987;1:59–62.

Greene RJ. 1982. An application of behavioral technology to the problem of nuisance bears. *The Psychological Record* 32:501–511.

Gregory R, Keeney RL. Creating policy alternatives using stakeholder values. *Manage Sci* 1994;40:1035–1048.

Gunther KA. Visitor impact on grizzly bear activity in Pelican Valley, Yellowstone National Park. *Int Conf Bear Res Manage* 1990;8:73–78.

Hamer D, Herrero S. Grizzly bear food and habitat in the Front Range of Banff National Park, Alberta. *Int Conf Bear Res Manage* 1987a;7:199–214.

Hamer D, Herrero S. Wildfire's influence on grizzly bear feeding ecology in Banff National Park, Alberta. *Int Conf Bear Res Manage* 1987b;7:179–186.

Harrison S. Local extinction in a metapopulation context: An empirical evaluation. *Biol J Linn Soc* 1991; 42:73–88.

Harting AL, Jr. Relationships between activity patterns and foraging strategies of Yellowstone grizzly bears. Bozeman, MT: Montana State University M.Sc. Thesis, 1985.

Herrero SM. A comparison of some features of the evolution, ecology, and behavior of black and grizzly/brown bears. *Carnivore* 1978;1:7–17.

Herrero S. *Bear Attacks: Their Causes and Avoidance.* New York: Nick Lyons Books, 1985.

Herrero S. Canadian national parks and grizzly bear ecosystems: The need for interagency management. *Int Conf Bear Res Manage* 1994;9:7–22.

Herrero S, Fleck S. Injury to people inflicted by black, grizzly or polar bears: Recent trends and new insights. *Int Conf Bear Res Manage* 1990;8:25–32.

Herrero S, McCrory W, Pelchat B. Using grizzly bear habitat evaluations to locate trails and campsites in Kananaskis Provincial Park. *Int Conf Bear Res Manage* 1986; 6:187–194.

Horejsi BL. Uncontrolled land-use threatens an international grizzly bear population. *Cons Biol* 1989;3:220–223.

Howe RW, Davis GJ, Mosca V. The demographic significance of 'sink' populations. *Biol Cons* 1991;57:239–255.

Kasworm WF, Manley TL. Road and trail influences on grizzly bears and black bears in northwest Montana. *Int Conf Bear Res Manage* 1990;8:79–84.

Kasworm WF, Thier TJ. Cabinet-Yaak ecosystem grizzly bear and black bear research: 1993 progress report. Missoula, MT: U.S. Fish and Wildlife Service Grizzly Bear Recovery Coordinator's Office, 1994.

Keating KA. Historical grizzly bear trends in Glacier National Park, Montana. *Widl Soc Bull* 1986;14:83–87.

Keiter RB. Observations on the future debate over "delisting" the grizzly bear in the greater Yellowstone ecosystem. *The Envir Prof* 1991;13:248–253.

Keiter RB, Locke H. Law and large carnivores in the Rocky Mountains: Evaluating the legal framework for species conservation. *Cons Biol* 1996;10 (in press).

Kellert SR. A Sociological Perspective: Valuational, Socioeconomic, and Organizational Factors. In TW Clark, RP Reading, AL Clarke (eds), *Endangered Species Recovery: Finding the Lessons, Improving the Process.* Washington, DC: Island Press, 1994a: 371–389.

Kellert SR. Public attitudes towards bears and their conservation. *Int Conf Bear Res Manage* 1994b;9:43–50.

Kendall KC. Trends in grizzly/human interactions in Glacier National Park, Montana (Presented paper). Grand Canyon, AZ: 6th International Conference on Bear Research and Management. 1983.

Knight RR, Blanchard BM, Eberhardt LL. Mortality patterns and population sinks for Yellowstone grizzly bears, 1973–1985. *Wildl Soc Bull* 1988;16:121–125.

Knight RR, Eberhardt L L. Population dynamics of Yellowstone grizzly bears. *Ecol* 1985;66:323–334.

Lande R. Risks of population extinction from demographic and environmental stochasticity and random catastrophes. *Am Nat* 1993;142:911–927.

Lucas RC. Use patterns and visitor characteristics, attitudes and preferences in nine wilderness and other roadless areas. Ogden, UT: U.S. Department of Agriculture, Forest Service Research Paper (INT-253), 1980.

Lucas RC. Visitor characteristics, attitudes, and use patterns in the Bob Marshall Wilderness complex, 1970–1982. Ogden, UT: U.S. Department of Agriculture Forest Service Research Paper (INT-345), 1985.

Mace R, Manley TL. South Fork Flathead River grizzly bear project: Progress report for 1992. Helena, MT: Montana Department of Fish, Wildlife, and Parks, 1993.

Martinka CJ. Rationale and options for management in grizzly bear sanctuaries. *N Am Wildl Conf* 1982;47:470–475.

Mattson DJ. Human impacts on bear habitat use. *Int Conf Bear Res Manage* 1990;8:33–56.

Mattson DJ, Blanchard BM, Knight RR. Food habits of Yellowstone grizzly bears, 1977–1987. *Can J Zool* 1991a;69:1619–1629.

Mattson DJ, Blanchard BM, Knight RR. Yellowstone grizzly bear mortality, human habituation, and whitebark pine seed crops. *J Wildl Manage* 1992;56:432–442.

Mattson DJ, Craighead JJ. The Yellowstone Grizzly Bear Recovery Program: Uncertain Information, Uncertain Policy. In TW Clark, RP Reading, AL Clarke (eds), *Endangered Species Recovery: Finding the Lessons, Improving the Process.* Washington, DC: Island Press, 1994:101–130.

Mattson DJ, Gillin CM, Benson SA, Knight RR. Bear feeding activity at alpine insect aggregation sites in the Yellowstone ecosystem. *Can J Zool* 1991b;69:2430–2435.

Mattson DJ, Herrero S, Wright RG, Pease CM. Science and management of Rocky Mountain grizzly bears: A model and some considerations. *Cons Biol* 1996;10 (in press).

Mattson DJ, Knight RR. Spring Bear Use of Ungulates in the Firehole River Drainage of Yellowstone National Park. In JD Varley, WG Brewster (eds), *Wolves for Yellowstone? A Report to the United States Congress. IV. Research and Analysis.* Mammoth, WY: U.S. National Park Service, 1992:5/94–5/120.

Mattson DJ, Knight RR, Blanchard BM. The effects of developments and primary roads on grizzly bear habitat use in Yellowstone National Park, Wyoming. *Int Conf Bear Res Manage* 1987;7:259–273.

Mattson DJ, Reid MM. Conservation of the Yellowstone grizzly bear. *Cons Biol* 1991;5: 364–372.

Mattson DJ, Reinhart DM. Bear Use of Whitebark Pine Seeds in North America. In WC Schmidt, F-K Holtmeier (compilers). *Proceedings—International Workshop on Subalpine Stone Pines and Their Environment: The Status of Our Knowledge.* Ogden, UT: U.S. Department of Agriculture Forest Service General Technical Report (INT-GTR-309), 1994:212–220.

McCrory W, Herrero S, Whitfield P. Using Grizzly Bear Habitat Information to Reduce Human-Grizzly Bear Conflicts in Kokonee Glacier and Valhalla Provincial Parks, BC. In GP Contreras, KE Evans (compilers). *Proceedings—Grizzly Bear Habitat Symposium.* Ogden, UT: U.S. Department of Agriculture Forest Service General Technical Report (INT-207), 1986:24–30.

McCrory WP, Herrero SM, Jones GW, Mallam ED. The role of the B.C. provincial park system in grizzly bear preservation. *Int Conf Bear Res Manage* 1990;8:11–16.

McCullough DR. Behavior, bears, and humans. *Wildl Soc Bull* 1982;10:27–32.

McLellan BN. Dynamics of a grizzly bear population during a period of industrial resource extraction. III. Natality and rate of increase. *Can J Zool* 1989a;67:1865–1868.

McLellan BN. Dynamics of a grizzly bear population during a period of industrial resource extraction. II. Mortality rates and causes of death. *Can J Zool* 1989b;67:1861–1864.

McLellan BN. Relationships between human industrial activity and grizzly bears. *Int Conf Bear Res Manage* 1990;8:57–64.

McLellan BN. Population regulation of grizzly bears. In M Taylor (ed), Density-dependent population regulation of black, brown, and polar bears. *Int Conf Bear Res Manage Monogr* 1994;3:15–24.

McLellan BN, Shackleton DM. Grizzly bears and resource extraction industries: Effects of roads on behaviour, habitat use and demography. *J Appl Ecol* 1988;25:451–460.

McNab J. Carrying capacity and related slippery shibboleths. *Wildl Soc Bull* 1985;13: 403–410.

Meagher M, Fowler S. The consequences of protecting problem grizzly bears. In Bromley (ed), *Bear–people conflicts: Proceedings of a Symposium on Management Strategies.* Yellowknife: Northwest Territories Department of Renewable Resources, 1989:141–144.

Nadeau MS. 1989. Movements of Grizzly Bears near a Campground in Glacier National Park. In M Bromley (ed), *Bear–people conflicts: Proceedings of a Symposium on Management Strategies.* Yellowknife: Northwest Territories Department of Renewable Resources, 1989:27–33.

Nagy JA, Gunson JR. Management plan for grizzly bears in Alberta. *Wildlife Management Planning Series No. 2.* Edmonton: Alberta Forestry, Lands and Wildlife, Fish and Wildlife Division, 1990.

Nagy JA, Hawley AWL, Barrett MW, Nolan JW. Population characteristics of grizzly and black bears in west central Alberta (Report AEC V88-R1). Vegreville, Alberta: Alberta Environmental Centre, 1989.

Nowak R. Uganda enlists locals in the battle to save the gorillas. *Science* 1995;267: 1761–1762.

Paradiso JL, Nowak RM. Wolves—*Canis lupus* and Allies. In JA Chapmon, GA Feldhamer (eds), *Wild Mammals of North America: Biology, Management, and Economics.* Baltimore, MD: The Johns Hopkins University Press, 1982:460–474.

Powell RA. *The Fisher—Life History, Ecology, and Behavior* (2nd ed). Minneapolis: University of Minnesota Press, 1993.

Primm SA. Grizzly conservation in Greater Yellowstone. Boulder, CO: University of Colorado M.Sc. Thesis, 1993.

Purves HD, White CA, Paquet PC. Wolf and grizzly bear habitat use and displacement by human use in Banff, Yoho, and Kootenay National Parks: A preliminary analysis. Banff, Alberta: Canadian Parks Service, 1992.

Raine RM. Grizzly Bear Mortality in Yoho and Kootenay National Parks, British Columbia. In RK McCann, (ed), *Grizzly Bear Management Workshop.* Revelstoke, British Columbia: Mt. Revelstoke and Glacier National Parks, and Friends of Mt. Revelstoke and Glacier, 1991:31–38.

Reading RP, Kellert SR. Attitudes towards a proposed reintroduction of black-footed ferrets (*Mustela nigripes*). *Cons Biol* 1993;7:569–580.

Reinhart DP, Mattson DJ. Bear use of cutthroat trout spawning streams in Yellowstone National Park. *Int Conf Bear Res Manage* 1990;8:343–350.

Ruggiero LF, Aubry KB, Buskirk SW, Lyon LJ, Zielinski WJ (eds). *The Scientific Basis for Conserving Forest Carnivores: American Marten, Fisher, Lynx, and Wolverine in the Western United States.* Fort Collins, CO: U.S. Department of Agriculture Forest Service General Technical Report (RM-254), 1994.

Russell RH, Nolan JW, Woody NA, Anderson G. A study of the grizzly bear (*Ursus arctos* L.) in Jasper National Park, 1975–1978: Final report. Edmonton, Alberta: Canadian Wildlife Service, 1979.

Schleyer BO. Activity patterns of grizzly bears in the Yellowstone ecosystem and their reproductive behavior, predation, and the use of carrion. Bozeman, MT: Montana State University M.Sc. Thesis, 1983.

Schonewald-Cox C, Bayless JW. The boundary model: A geographic analysis of design and conservation of nature reserves. *Biol Cons* 1986;38:305–322.

Schonewald-Cox C, Buechner M. Housing Viable Populations in Protected Habitats: The Value of Coarse-Grained Geographic Analysis of Density Patterns and Available Habitat. In A Seitz, V Loeschcke (eds), *Species Conservation: A Population-Biological Approach.* Basel: Birkhäuser, 1991.

Schullery P. The fire and fire policy. *BioScience* 1989;39:686–694.

Servheen C. The status and conservation of the bears of the world. *Int Conf Bear Res Manage Monogr 2,* 1990.

Shaffer ML. Keeping the grizzly bear in the American west: A strategy for real recovery. Washington, DC: The Wilderness Society, 1992.

Shrader-Frechette KS. *Risk and Rationality: Philosophical Foundations of Populist Reforms.* Berkeley, CA: University of California Press, 1991.

Simberloff D, Cox J. Consequences and costs of conservation corridors. *Cons Biol* 1987;1:63–71.

Smith RB, Barnes VG, Jr, Van Daele LJ. Brown Bear–Human Conflicts in the Kodiak Archipelago, Alaska. In M Bromley (ed), *Bear–People Conflicts: Proceedings of a Symposium on Management Strategies.* Yellowknife: Northwest Territories Department of Renewable Resources, 1989:111–120.

Stirling I, Derocher AE. Factors affecting the evolution and behavioral ecology of the modern bears. *Int Conf Bear Res Manage* 1990;8:189–204.

Storer TI, Tevis LP, Jr. *California Grizzly.* Berkeley, CA: University of California Press, 1955.

Stringham SF. Grizzly bear reproductive rate relative to body size. *Int Conf Bear Res Manage* 1990;8:433–443.

Thomas CD. Extinction, colonization, and metapopulations: Environmental tracking by rare species. *Cons Biol* 1994;8:373–378.

Thomas JW, Gill JD, Pack JC, Healy WM, Sanderson HR. Influence of forestland characteristics on spatial distribution of hunters. *J Wildl Manage* 1976;40:500–506.

Titus K, Beier L. Population and habitat ecology of brown bears on Admiralty and Chichagof Islands. *Federal Aid in Restoration Research Project Report* (W-23-4). Juneau: Alaska Department of Fish and Game, Division of Wildlife Conservation, 1992.

Tracy DM. Reactions of wildlife to human activity along Mount McKinley National Park road. Fairbanks, AK: University of Alaska M.Sc. Thesis, 1977.

U.S. Fish and Wildlife Service. *Grizzly Bear Recovery Plan.* Missoula, MT: 1993.

Wakkinen WL, Zager P. Selkirk grizzly bear ecology project. *Threatened and Endangered Species Project Report* (E-3-6). Boise, ID: Idaho Department of Fish and Game, 1991.

Warner SH. Visitor impact on brown bears, Admiralty Island, Alaska. *Int Conf Bear Res Manage* 1987;7:377–382.

Weaver J, Escano R, Mattson D, Puchlerz T, Despain D. A Cumulative Effects Model for Grizzly Bear Management in the Yellowstone Ecosystem. In GP Contreras, KE Evans (eds), *Proceedings—Grizzly Bear Habitat Symposium.* Ogden, UT: U.S. Department of Agriculture Forest Service General Technical Report (INT-207), 1986:234–246.

Wielgus RB, Bunnell FL. Dynamics of a small, hunted brown bear *Ursus arctos* population in southwestern Alberta, Canada. *Biol Cons* 1994;67:161–166.

Wilcove DS, McLellan CH, Dobson AP. Habitat Fragmentation in the Temperate Zone. In ME Soulé (ed), *Conservation Biology: The Science of Scarcity and Diversity.* Sunderland, MA: Sinauer Associates, 1986:237–256.

Wilcox BA, Murphy DO. Conservation strategy: The effects of fragmentation on extinction. *Am Nat* 1985;125:879–887.

Wondolleck JM, Yaffee SL, Crowfoot JE. A Conflict Management Perspective: Applying the Principles of Alternative Dispute Resolution. In TW Clark, RP Reading, AL Clarke, (eds), *Endangered Species Recovery: Finding the Lessons, Improving the Process.* Washington, DC: Island Press, 1994:305–326.

Young TP. Natural die-offs of large mammals: Implications for conservation. *Cons Biol* 1994;8:410–418.

9

Expansion of the U.S. National Park System in Alaska

R. Gerald Wright

Factors Influencing Park Establishment

As described in earlier chapters, national parks and associated areas have been one of the major mechanisms used in the United States to preserve the nation's natural, scenic, and historic heritage. The process of identifying and preserving the remaining unique and unprotected resources in the United States continues today, although the effort is subject to much scrutiny and controversy given the contemporary demands to use lands for consumptive resource production. Also, as discussed in Chapter 7, today there is probably as much attention given to screening new park proposals, because of the added burden they place on the limited resources of the National Park Service (NPS), as there is to identifying unique areas that are unprotected.

Most park areas in the United States have been created from lands already in the public domain, such as former state parks and lands administered by other federal agencies, or the lands have been donated or sold willingly by private citizens, the latter being more common in the early years of the twentieth century. Public sentiment and politics generally prevent the condemnation of use of private lands for parks in all but the most exceptional cases. Political considerations have also played an important role in dictating the extent and area of lands to be included in parks.

Even at times when land availability was not a problem (e.g., in 1872 when Yellowstone National Park was established), the parks created were often inadequate, in part because of an incomplete understanding of the resources contained in the selected areas (see Chap. 7) but also because of economic reluctance to set aside any more land than was absolutely necessary. Because of these constraints and problems, parks in the conterminous United States lack land areas sufficient to protect important components of their ecological systems. The management problems resulting from these deficiencies have been cataloged repeatedly (NPS, 1980; Keiter, 1985; Freemuth, 1991; see also Chapters 6, 7, 8).

It can be argued in principle that parks or any other landscape can never be regarded as a completely self-contained ecological unit. All are dependent on or influenced by, to some degree, factors external to their boundaries. The degree to which this occurs, however, is relative, and it could be argued that it is just as important to develop park boundaries that will mitigate external influences (Polunin & Eidsvik, 1979) as it is to include lands in parks that optimize resource protection.

The Establishment of the New Alaskan Parks

This chapter discusses the planning process used by the NPS to select the lands included in the 13 parks established or expanded in Alaska as a result of the Alaska National Interest Lands Conservation Act (ANILCA) in 1980. The objective of this process was the creation of new natural-area parks that would protect the scenic beauty and biodiversity of the state, while freeing the parks from the management problems that have beset so many existing protected areas. It is still too early to judge to what extent this objective has been achieved, but some insights are now possible. The original planning process is reviewed by focusing on one of the new parks, the 4.98-million-hectare Wrangell–St. Elias National Park and Preserve, hereafter referred to as Wrangell.

Prior to the passage of ANILCA, one of the largest gaps in representative natural regions within the national park system occurred in Alaska (see Table 1.1). The passage of the Alaska Native Claims Settle Act (ANCSA) in 1971 provided an opportunity to correct this situation. This act authorized the Secretary of the Interior to withdraw up to 32 million hectares of unreserved public lands to study their suitability for inclusion in a system of national parks, forests, wildlife refuges, and scenic rivers. As the great majority of these lands already were under federal ownership and largely in a wilderness condition, at the beginning of the process there was an opportunity to identify viable landscapes without the usual constraints imposed by existing land uses and ownerships.

Land selections and management decisions for the new parks were made under ANCSA between 1972 and 1980. During this time, several bills that attempted to set aside varying amounts of land under different management strategies were drafted and introduced into the U.S. Congress. The final resolution, culminating in the passage of ANILCA, established ten new parks and preserves, totaling 15.3 million hectares, and added 2.3 million hectares to three existing parks in Alaska.

The Planning and Legislative Process

The Wrangell area initially was chosen for study as a potential park primarily because of its scenic beauty. However, in a state where awesome scenic beauty is commonplace, aesthetic factors alone were not considered sufficient to qualify a region for park status. Therefore, significance was keyed to the fact that the Wrangells offered a unique opportunity to view the action of geological, hydrological, and ecological processes that have dynamically altered the landscape of the region and continue to do so. The region contains active or recently active volcanoes; more than 100 named glaciers, including the 123-kilometer long Nebesna Glacier; and numerous large, wild, glacially derived rivers, including the Copper River, which, in terms of water volume, is among the largest in the United States (Wright, 1981).

The huge land area encompassed by the Wrangell–St. Elias Mountains and associated lowlands and the virtual lack of human development offered a unique opportunity to design a park of such size and configuration that it could contain all of the habitat attributes needed by the faunal communities while buffering these species from external influences. To accomplish this, strong reliance was placed on the use of remote-sensing techniques to map vegetal communities and define habitat types (Wright, 1985).

Sport hunting, as one of the major recreational uses of the Wrangells, posed a major conflict. It was, in fact, a major issue of contention in the planning process for all of the new Alaskan parks, because, traditionally, U.S. national parks are closed to sport hunting. In the Wrangells, trophy hunting for Dall sheep was the most important component of the sport-hunting issue. At the time, an average of 29 percent of the annual state harvest of 1300 animals occurred in this region, and about 60 guides worked in the area at least part of the time. The magnitude of these numbers, combined with local and national pressures, made it necessary politically to allow sport hunting to occur on some of the lands under study, or otherwise be forced to settle for a much smaller park, with diminished ecological integrity. To resolve the sport-hunting conflict, a compromise was reached that required a certain proportion of the lands under study in the Wrangells be classified as a preserve. In turn, preserve lands would be subject to all NPS policies, except that sport hunting would be permitted under state regulation. The preserve-lands classification was also used to resolve sport-hunting conflicts in seven other Alaskan park proposals, and two areas, Bering Land Bridge and Yukon Charley Rivers, were established solely as national preserves.

Preserve Establishment

Decisions on which lands to include within the Wrangells preserve were based on species-distribution and sport-harvest data. Aerial wildlife surveys conducted by the author from 1976 through 1979, combined with records from the Alaska Department of Fish and Game, were used to help define the external boundaries of the park and preserve units. Information compiled yearly from sport hunters by the Alaska Department of Fish and Game, containing the general location of each hunt and harvest, was used to determine hunting intensity and success in defined zones within the study area (Murphy & Dean, 1978). The harvest of mountain goats (*Oreamnos americanus*), Dall sheep (*Ovis dalli*), moose (*Alces americana*), and caribou (*Rangifer tarandus*) was considered initially. However, because the harvest and survey units for each species differed, only the alternatives based on Dall sheep harvest—the most important species—were used (Wright, 1984). The goal of the compromise was to configure the park-preserve boundaries such that approximately 50 percent of the former sport-hunting activity would be able to continue. Table 9.1 shows the average yearly harvest of the four big-game species that occurred on lands placed under park or preserve management in the final proposal. These figures indicate that we were relatively successful in achieving the 50 percent goal.

Evaluation of the Park Planning Process

The intent of the planning process was to create a park in the Wrangells that encompassed a complete ecological unit. This goal was achieved to a certain ex-

Table 9.1 Past annual harvest of species on lands proposed for park or preserve status in the Wrangells

Species	Total state harvest occurring in region (%)	Wrangells (%)	
		Park	Preserve
Dall sheep	29	46	54
Mountain goat	7	50	50
Moose	2	49	51
Caribou	7	53	47

Source: Modified from RG Wright, Wildlife resources in creating the new Alaskan parks and preserves. *Environ Manage* 1984:8:121–124.

tent. The Wrangell–St. Elias National Park and Preserve is bordered on the north and south by natural physiographic divides that, thus far, have sheltered it from adverse land-use activities external to the park. The international boundary with Canada forms the eastern limit of the park. Here, the park shares a common border and management philosophy with Canada's Kluane National Park, a 2.2 million-hectare area of similar physiography.

While there are occasional poaching and hunting infractions, and there clearly is movement of animals between the park and preserve units, particularly in northern and midsections of the park, the preserve designation has thus far proved to have been a workable compromise. The most significant land-management conflicts to date have arisen from the limited number of private and large expanse of corporate (native land selections) inholdings in the park and from disagreements about subsistence use. While the potential for both of these factors to impair the integrity of the park was recognized during the park-planning process, planners and scientists had virtually no latitude with respect to these issues.

With respect to native corporation land selections, it should be understood that the original purpose of ANCSA was to settle aboriginal native land claims. The settlement embodied in ANILCA was very complex and, along with the new parks, wildlife refuges, and scenic rivers, involved the establishment of 12 native corporations—all with extensive land holdings and within the individual corporate boundaries of extensive native village land selections (Arnold, 1976). In many cases, these selections fell within the boundaries of the new parks and preserves and, because of the selection criteria, tended to form a checkerboard of selected townships bounded by park lands. The pattern was similar to that created by railroad land selection in the American West in the nineteenth century. As privately owned lands held by for-profit corporations, these areas may be subjected to timber harvests, mining, oil and gas drilling, and other forms of development. It is, however, still too early to tell whether some of these events will occur in the Wrangells and to what extent the park will be able to mitigate their effects.

ANILCA contains detailed provisions for maintaining the opportunities for traditional hunting and gathering (i.e., subsistence activities), by certain rural residents on all of the lands covered in the settlement. Since the passage of ANILCA, state legislation and state court decisions have greatly expanded subsistence opportunities to include a broad spectrum of state residents, even those living in urban areas. These actions have tended to blur distinctions between sport and subsistence hunting and have the potential to seriously impact wildlife populations in the park, as well as to complicate park management and limit options. Again, it is too early to say how this situation will work out over the long term.

It has been 124 years since the creation of Yellowstone National Park. Even though, at 866,000 hectares, Yellowstone is the largest park in the continental United States, after its creation management learned (and continues to learn) that its insufficient size, illogical configuration, and lack of integrated ecosystem management with the surrounding lands are detriments to the ecological and biophysical resources of the park (see Chap. 8).

Conversely, it has been only 15 years since the Wrangell–St. Elias National Park and Preserve was established. In that brief period of time, the park/preserve has thus far been spared some of the external transboundary problems that have beset most parks in the United States. However, the potential impacts resulting from the internal transboundaries associated with inholdings and the management restraints imposed by subsistence activities have created problems, which, while not unforeseen 15 years ago, are probably of greater magnitude and potential severity than originally anticipated. The problems facing the Wrangells are not insurmountable, neither are those faced by the greater Yellowstone ecosystem. Whether they can be overcome will depend as much on political good fortune as it does on ecological science.

The process of creating and establishing the new parks in Alaska was proclaimed the "last chance to do it right the first time" (Frankfourth, 1982). Like the Western parks 100 years ago, the new Alaskan parks were carved out of broad expanses of undeveloped wilderness. Time will judge whether the biopolitical process used to establish them was a success. As with Yellowstone and many other parks today, 100 years from now there will probably be little opportunity to rectify any mistakes made in Alaska. The last chance to do it right was also likely the last chance to do it wrong.

Literature Cited

Arnold R. *Alaska Native Land Claims.* Anchorage: Alaska Native Foundation. 1976.

Frankfourth D. National Park Ideals. In E. Connally (ed), *National Parks in Crisis.* Washington, DC: National Parks and Conservation Association, 1982:335–341.

Freemuth JC. *Island Under Siege: National Parks and the Politics of External Threats.* Lawrence, KS: University of Kansas Press, 1991.

Keiter RB. On protecting the national parks from the external threats dilemma. *Land and Water Law Rev.* 1985;20:355–420.

Murphy E, Dean F. *Hunting Activity and Harvest in the Wrangell St. Elias Region, Alaska* (Final Report, CX-9000-6-0154). Fairbanks, AK: National Park Service Cooperative Parks Studies Unit, University of Alaska, 1978.

National Park Service. *State of the Parks 1980: A Report to the Congress.* Washington, DC: National Park Service, 1980.

Polunin N, Eidsvik H. Ecological principles for the establishment and management of national parks and equivalent reserves. *Environ Cons* 1979;6:21–26.

Wright RG. Wrangell St. Elias: International Mountain Wilderness. *Alaska Geographic* 1981;8(1):1–144.

Wright RG. Wildlife resources in creating the new Alaskan parks and preserves. *Environ Manage* 1984;8:121–124.

Wright RG. Principles of new park-area planning as applied to the Wrangell–St. Elias region of Alaska. *Environ Cons* 1985;12:59–66.

10

The Role of Networks and Corridors in Enhancing the Value and Protection of Parks and Equivalent Areas

Larry D. Harris
Thomas Hoctor
Dave Maehr
Jim Sanderson

Parks and Preserves as Self-Contained Units

Parks and protected areas have become part of the traditional human land use and land fragmentation paradigm: They are isolates increasingly contained in a disarticulated, dismembered landscape. Despite their importance, parks and reserves are necessary but not sufficient to protect biological diversity and ecological processes and functions. The comments of Sax (1991), referring to Yellowstone National Park, support our proposition:

> Perhaps I can best introduce the subject with a slightly shocking statement: A fundamental purpose of the traditional system of property law has been to destroy the functioning of natural resource systems. What I mean is this. Under our legal system we cut up the land into arbitrary pieces (such as square 160-acre tracts) and then endow the owner with the right, indeed with every encouragement, to enclose the land and make it exclusive. . . . What fencing accomplishes is the severance of wildlife lifelines and the destruction of

> wildlife habitat. That is what the law, as the handmaiden of the new settlers' incipient economy, sought to accomplish, and it succeeded.

Parks and preserves are not in themselves fragmenting forces nor do they destroy the functioning of natural systems. However, they are in danger of becoming habitat isolates surrounded by increasingly hostile land uses. By setting aside isolated parks and reserves and assuming that they can maintain their native biological diversity, we are gambling that the context surrounding these areas either will not change or does not matter. The parks and preserves paradigm, though necessary, represents a limited approach where isolated areas are set aside for people to enjoy "nature." The geological and evolutionary record indicates that the long-term viability of these set-asides depends on maintaining the integrity of the surrounding landscape.

The inadequacy of parks and preserves was suggested by naturalists and ecologists who were quick to comment and hypothesize about the ecological consequences of fragmentation and isolation. Over 100 years ago, a prominent European ecologist observed that "[t]he breakup of a large landmass into smaller units would necessarily lead to the extinction or local extermination of one or more species and the differential preservation of others" (de Candolle, 1855, quoted in Browne, 1983).

The first national parks in the United States presented a new opportunity to protect significant examples of the natural heritage of this country, and they have served as a conservation paradigm that has been emulated across the globe. However, within decades of the establishment of the first U.S. national parks, ecologists were already recognizing the inadequacy of these areas. In a series of landmark reports, Wright et al. (1933) and Wright and Thomson (1934) recognized that even the largest national parks were not sufficient to protect the full complement of native biological diversity: "The preponderance of unfavorable wild-life conditions confronting superintendents is traceable to the insufficiency of park areas as self-contained units. At present, not one park is large enough to provide year-round sanctuary for adequate populations of all resident species" (Wright et al., 1933). Likewise, Preston (1962) concluded more than 30 years ago that "[i]f what we have said is correct, it is not possible to preserve in a State or National Park, a complete replica on a small scale of the fauna and flora of a much larger area. If the major part of the state, for instance, is given over to a complete disclimax, whether urbanization or mining or agriculture, the preserved area becomes an isolate or an approximation thereto, and the number of species that can be accommodated must apparently fall to some much lower level."

Wright et al. (1933), Wright and Thompson (1934), and Shelford (1936) recognized that the national parks in the United States were not sufficient in size to support the needs of large, wide-ranging animals. They understood that these parks were failing to conserve species because of the inability of parks to function as "self-contained units" and because they were affected by the land uses and management practices that surrounded them. These practices included predator control and uncontrolled harvest of game animals and fur bearers: "This matter of external influence incessantly acting upon the faunal resources of a national park cannot be overestimated. The fate of the carnivores and fur bearers is too well known. The ungulates are robbed of their winter range and held within park during winter, by virtue of hunting and civilization just outside the boundaries" (Wright & Thompson, 1934). Cahalane (1948) supported these conclusions in an article on the status of native mammals in the U.S. national parks: "Most of the park areas are too small to protect adequate numbers of the species that are widely ranging. Even in most of the largest parks, wildlife is influenced by hunting and trapping or by agricultural or other practices on exterior lands."

Wright and Thompson (1934) and Shelford (1936) suggested buffer zones as potential mitigation for these problems, though they offered no specific proposals on how large such zones might need to be. Advances in conservation science since then, though helpful, have done little to elaborate on what these ecologists knew six decades ago.

Reserves and Florida's Large Carnivores

The focus of many of the examples presented in this chapter will be the state of Florida. There, problems for wide-ranging carnivores are even more challenging. Though national parks in the western United States are often surrounded by larger multiple-use landscape matrices that offer potentially suitable habitat, reserves in Florida are increasingly threatened by urban and agricultural development. Therefore, species such as Florida panthers (*Felis concolor coryi*) and black bears (*Ursus americanus*), which require very large areas, are threatened by the pace and magnitude of overdevelopment. Indeed, it has been estimated that a population of 50 to 70 panthers or 200 black bears (that may or may not be viable populations) would need at least 800,000 to 1.6 million hectares and 200,000 to 400,000 hectares, respectively (Cox et al., 1994). The only potentially secure population, in the short term, may be the black bear population centered in the Big Cypress National Preserve and Everglades National Park reserve complex (Cox et al., 1994), although the more productive portions of both the bear and panther populations in this area are located on private lands that are under development pressure to the north of the BCNP/ENP complex (Maehr, 1990).

Table 10.1 Documented nonvolant mammalian faunal collapse on several Caribbean Islands

Island	Area (km^2)	Mammals lost (%)	Mammals remaining (#)
Cuba	114	81	5
Hispaniola	76	92	2
Jamaica	11	80	1
Puerto Rico	9	100	0

The islands of Cuba and Puerto Rico are only slightly more insular than is the peninsula of southern Florida. The depletion of flora and fauna in the West Indies since human colonization represents a powerful demonstration of the fate of biodiversity resources in the presence of nonvigilant humans. The four westernmost large Caribbean islands are biogeographically very similar to Florida. Paleontological research offers a clear record of the levels of vertebrate diversity of the islands both prior and subsequent to human colonization approximately 7000 years ago (Morgan & Woods, 1986). At least for the nonvolant mammals, the loss of species since human colonization is directly inverse to island size. As shown in Table 10.1, even an island as large as Puerto Rico has no remaining native terrestrial mammals.

The point to be made from these examples is that it seems likely that the vast majority of national parks and similar reserves are not large enough to satisfy conservation goals completely. Rather, these areas are heavily influenced by the surrounding landscape mosaic. Effective conservation therefore requires management of the mosaic itself rather than of selected habitat units within the mosaic (Harris, 1984; Hobbs, 1993; Wiens, 1994). As Rodgers et al. (1993) note, "In the absence of significant consideration of large scale spatial processes and disruptions at the level of the landscape, conservationists are prone to set isolated ecological vignettes such as state parks and small reserves aside as though they could persist through time despite being severed from their former surroundings. Even the largest of such 'set-aside' preserves fail to protect and maintain their intrinsic biodiversity when the surrounding landscape matrix becomes sufficiently modified."

Understanding Contextual Management

For more than 10 years (see Harris, 1984), it has been argued that the context within which a biodiversity resource area occurs is at least as important for biodiversity maintenance as is the "content" of the habitats and management that occur within the boundaries of the area. For small areas, such as state parks in increasingly urban settings (e.g., San Felasco State Preserve, Florida), the context is probably much more important and determinant than anything that can be done within the confines of the area per se. This dichotomy between content and context is now the topic of many works in conservation theory and, increasingly, in applied biological conservation.

Consider New York City's Central Park. When established, the park was surrounded by many quasi-natural tracts of land that served as connections to the regional biophysical system. A century later, the park is totally isolated and maintains little native biological diversity. Some would argue that this is because it is small: this is folly. Some would argue that this is because it is an urban park: this is true. Either way, the fact is that the same phenomenon is occurring to greater or lesser degrees with all biodiversity preserves on earth.

Consider the following simple analogy: The element hydrogen totally dominates our universe. On earth, two atoms of hydrogen conjoin with an atom of oxygen to form a molecule of water or H_2O. Knowing the chemical makeup of this molecule tells us nothing about its characteristics. Only by knowing the context can we predict and/or know the nature and behavior of the molecule. In other words, the content (i.e., the substance) is controlled by the context. If the pressure is "normal" and ambient temperature is below 0° Celsius (C), the water molecule will be solid. If the pressure is "normal" and ambient temperature is above 0°C but lower than 100°C, the water molecule will be liquid. If the pressure is "normal" and ambient temperature is greater than 100°C, the molecule will change yet again to a gaseous state. This example varies only a single variable, temperature, and not the full suite of potential variables (e.g., pressure). Clearly, few things can be said about the characteristics of H_2O unless the context is known. Likewise, a small stand of trees surrounded by a universe of pasture is not a forest and, for that matter, neither is an arboretum or a botanical garden. When and if the same stand of trees is surrounded by a variety of native trees, it most certainly does constitute a forest. A small patch cannot and will not function through time as a biodiversity preserve unless it is surrounded by a native landscape.

Biodiversity Erosion in the Absence of Content or Context Management

The national park and nature preserve movement has, so far, succeeded for nearly 100 years in forestalling erosion of much of the earth's biological diversity. This is principally because the more famous parks and preserves were created at a time when their contextual settings were not significantly different from their content. However, the overwhelming majority of biodiversity parks and preserves on earth (approximately 75%) have been established in the last 25 years. They have been created at a time when their contextual surroundings have already been sufficiently impacted by human dominance, clearly signaling their need. These parks and preserves generally did not have the advantage of beginning with a full complement of native fauna and flora nor do they have naturally dominant ecological functions. Moreover, no authoritative biodiversity scientist or conservationist believes that any of these set-aside areas can remain viable in the future without massive levels of contextual management.

The integrity of one of the largest national parks in the lower United States, Yellowstone National Park, is dependent on effective management of approximately 2.4 million hectares of surrounding national forest lands (Newmark, 1985). Similarly, Big Cypress–Everglades is the second largest national park and preserve complex in the contiguous United States. It has already lost numerous species of vertebrates, perhaps countless species of invertebrates, and demands billions of dollars for restoration to maintain any semblance of its pre-European character. Remarkably, Everglades National Park is less than 50 years old.

At least five prominent principles derive from the above-mentioned phenomena:

1. The loss of species from the preserves that have their contextual settings benignly managed is inverse to the size of the preserve area.
2. The loss of species from preserves that have their contextual settings benignly managed is directly related to the time since their establishment.
3. The loss of species from biodiversity parks and preserves is directly related to the degree of alteration of the contextual setting from its natural state and/or the state of the biodiversity preserve.
4. The intensity and cost of management per unit area of preserves set aside for protection of biodiversity will be inverse to their size and time since establishment, and directly proportional to the degree of alteration to surrounding contextual setting.
5. Given the objections of United States citizens to active management within the boundaries of biodiversity parks and preserves, the degree of security of biodiversity within the areas will depend almost entirely on the degree of management of the surrounding contextual landscape.

The Loss of Biodiversity from Isolates

Few areas of public land occurring in the contiguous United States can be shown to possess their native complement of vertebrate species. To the best of our knowledge, no national park or biodiversity preserve on earth has been shown to be maintaining its native complement of biological diversity. In the absence of contextual considerations, we contend that no amount of management of the areas themselves will maintain the native complement of biological diversity over many decades.

For migratory species, such as the monarch butterfly (*Danaus plexippus*), no single or small collection of set-aside preserves will secure their futures as free-ranging species on the North American continent (Brower & Malcolm, 1991). Might one reasonably expect to create biodiversity reserves that are larger than the island of Cuba, which is thirteen times as large as Yellowstone National Park? It should be obvious that factors more than size and management are called for to maintain native biological diversity well into the future.

Land Management and Conservation Bias Toward Weedy Species

Present land management and conservation efforts are biased in favor of "weedy" wildlife species and against most rare and endangered species. The ongoing loss of biological diversity from earth is sufficiently startling to most authorities, and many alternative approaches have been proposed. The most stark and bold statement of how biodiversity conservation must proceed in the absence of context is asserted by the distinction between in situ and ex situ conservation. *In situ conservation* refers to that which occurs in its place, as opposed to *ex situ conservation,* which occurs outside of its place. In other words, conservationists must concede that several species, such as Pe're David's deer (*Elaphurus davidianus*), can and do occur only in human-created environments that are detached from any native contextual settings whatsoever. This is labeled explicitly as ex situ conservation.

Without massive changes in the way that land resources are managed, most professional conservationists concede that ex situ conservation will have to play an ever-increasing role in the maintenance of biodiversity. Without these changes, the only species that will be secure in their own native settings are the small, and/or sedentary, and/or anthrophilic, easily managed species, such as the gray and red fox (*Urocyon cinereoargenteus, Vulpes vulpes*), raccoon (*Procyon lotor*), coyote (*Canis latrans*), and opossum (*Didelphis virginiana*), which are generally referred to as "weedy" species. Without altered conservation programming, very large, wide-ranging, and/or human-incompatible species, such as the black bear, Florida panther, and red wolves, will of necessity be relegated to zoological gardens and other human-dominated contexts. Because these artificial alternatives totally change the mix and intensity of evolutionary forces that led to the creation and maintenance of the species in the first place, much of the cause for conservation will be lost.

Landscape Ecology and Ecological Processes

The discipline of landscape ecology teaches that a distinction must be drawn between the concept of ecosystem and that of the landscape. Landscapes are not the same as large ecosystems; this argument is much more than semantics. Landscapes generally consist of more than one ecosystem that interacts in time and space. Spatial heterogeneity, edges, patches, and expansiveness are critical aspects of landscape ecology. None of them is implicit or explicit in the concept of ecosystem.

Landscapes, which incorporate the patterns and processes of ecological interaction of different patches (communities or ecosystems), are an appropriate scale for conservation efforts (Turner, 1989). Landscape ecology emphasizes broad spatial scales and the interactions between spatial patterning and ecological processes; however, an emphasis on landscapes does not preclude the importance of other scales. Nonetheless, landscape ecology is an essential tool for ameliorating the effects of ongoing habitat destruction and fragmentation.

Adjacency and connectivity are important considerations for many of the landscape processes critical to biodiversity formation and maintenance. We define a *landscape process* as one that occurs in a spatial domain, is propagated essentially parallel to the land surface, serves to either help structure and/or is influenced by the spatial structure or the boundaries between two or more ecosystems in a landscape, and whose strength of impact is dependent on landscape structure.

One aspect of a landscape approach to conservation is the consideration of gamma, or landscape-level, species. Florida panthers and black bears can be considered gamma species because individuals have large home ranges (up to 500 km^2) and utilize a variety of habitat types (Maehr et al., 1991). Florida panthers are top-level carnivores that require large areas to meet their prey needs for individual maintenance and to raise young. Black bears are opportunistic carnivores that depend on plant food resources and need to move seasonally in response to plant phenology.

Functional Landscape Juxtaposition

If conservation efforts are to be successful into the future, there must be a shift toward a landscape-management emphasis that recognizes natural patterns of heterogeneity and the ecological processes responsible for these patterns. Managed landscapes must incorporate sufficient natural levels of spatial and temporal heterogeneity, while minimizing the negative effects of artificial edges and barriers. Natural processes, including dispersal, are easily disrupted in culturally fragmented landscapes (Hansson & Angelstam, 1991). These disruptions can lead to changes in plant-community composition that create process barriers. As a result, sensitive fauna are impacted by reduced or precluded dispersal, demographic stochasticity, and increased predation, parasitism, and competition from "weedy" edge species (Harris & Silva-Lopez, 1992).

One of the key requirements for maintaining functional landscapes is the maintenance of natural juxtapositions of communities. Landscape managers must strive to protect full gradients of ecosystem types, from aquatic channel/basin, to wetlands, to mesic uplands, to xeric uplands, in an effort to maintain critical processes and species interactions that are integrated across such landscapes. For example, it is the nature of bottomland riverine systems to "flip" from one trophic system to another depending on the season. When water is low, the bottomlands are dominated by terrestrial food chains. When the water is high, the same area is dominated by aquatic food webs. Furthermore, the pulsing nature of the system is responsible for the great productivity of both the terrestrial and aquatic energy-transfer systems.

Slow, but highly recurrent riverine flooding over a landscape that has remarkably little topographic relief is a prominent feature of the southeastern coastal plain. This gradient of inundation creates a gradient of niches, from the consistently wet and flood-prone sites that support species such as cypress (*Taxodium distichum*), to the consistently dry and fire-prone sites that favor longleaf pine. But even more important, for our purposes, is the migratory ecotone that moves both seasonally and spatially across the landscape, sometimes several kilometers and with imperceptibly varying depths that further amplify the niches. When integrated with the niches of sedentary species such as trees and shrubs, the migratory ecotone creates an even more powerfully manifested gradient of foraging niches for mobile species, such as shellfish and finfish;

amphibious species, such as wading birds, otters, and raccoons; and terrestrial species, such as white-tailed deer (*Cervus virginianus*). Only when the temporally varying subsystems are allowed to function integratively in a spatial context can species such as Shermans' fox squirrel (*Sciurus niger shermani*) find ideal habitat conditions.

In contrast, fire has a dominant role in determining the nature of upland systems in Florida. Somewhere in the middle of this gradient, within the wetland to mesic portions, fire and flood meet to determine the nature of ecotones, such as pitcher plant bogs and seepage slopes. Florida's native fauna is adapted to this zonation of patterns and processes across landscapes. Many of Florida's amphibians live as adults in upland areas but require wetlands to reproduce. Furthermore, riparian corridors require the protection of adjacent uplands to satisfy the life-history requirements of upland-dependent native fauna.

Integrated Regional Landscape Management

The endangered red-cockaded woodpecker (*Picoides borealis*) of the southeastern United States illustrates the necessity of integrated landscape management. Mature pine-forest corridors might serve the dispersal needs of the woodpeckers (Walters et al., 1988; McFarlane, 1992). The native longleaf pine (*Pinus palustris*) and wiregrass (*Aristida stricta*) community type is the breeding habitat so that linear corridors of longleaf pine might constitute an interconnection between metapopulations. However, this scheme puts the emphasis on the characters (i.e., woodpeckers) and the stage setting (old-growth islands and corridors) and does not focus on the larger ecological picture.

Noss (1988) suggested that frequent, low-intensity fire is a critical ecological process necessary for the maintenance of the ecosystem. Other frequent disturbances, such as floods, freezes, lethal fungi, hurricanes, and tornadoes, created one of the greatest forests on earth (Wells & Strunk, 1931). Fires started by frequent lightning strikes spread across the interconnected landscape on fuel-loads of wiregrass and accumulated longleaf pine needles, maintaining an open understory of diverse plant species. As Harris et al. (1995) suggest:

> Therefore, simply saving some islands of old-growth longleaf pine is not the issue; saving the red-cockaded woodpecker is not the issue; and providing an interconnected habitat system that protects both the longleaf pine and the red-cockaded woodpecker should not be the issue. Rather, we see the issue as

> being that of restoring and maintaining a spatially integrated longleaf pine ecosystem that can and will maintain the full suite of landscape ecological processes including fire that is ignited in one place but allowed to disperse across the system; a system that can withstand the effects of major hurricanes and still remain viable and resilient because of its extensive nature; a system that is capable of sustaining natural outbreaks of beetles and fungi; and a system that is interdigitated with other community types that provide seasonally important services for the longleaf pine community and vice versa.

Protecting biological diversity is a minimum first step. Additionally, ecological processes and functions must be protected and/or restored. For instance, natural selection must be emphasized over artificial selection. Therefore, any approach to conservation that promises the possibility of less management and less manipulation, while protecting biological diversity and ecological and landscape processes, and facilitates future natural selection and adaptation is considered to be superior over more manipulative, less comprehensive approaches.

Connectivity and Comprehensive Landscape Conservation

The concept of connectivity must be included with the other conservation objectives such as buffering core areas and protecting isolated "hotspots" of biological diversity. The connectivity approach is based on the thesis that an interconnected conservation system of reserves would be greater in function than the sum of its parts (see Chap. 6). Connectivity, in essence, is the opposite of fragmentation (Noss, 1992). A connectivity approach to conservation is simply an attempt to protect landscapes from further fragmentation and effects and, where possible, to restore connectivity to culturally fragmented landscapes. It does not consist of attempts to somehow link populations that were naturally isolated.

We argue that the majority of isolated reserves will not maintain diversity and that to effectively conserve both biological diversity and the ecological interactions responsible for creating and maintaining this diversity, single reserves need to be at least on the order of several million acres (Pickett & Thompson, 1978; Frankel & Soulé, 1981). However, this does not take into account long-term environmental change. Even if a reserve is large enough to maintain internal patch dynamics and viable populations over a short time frame (say, 100 years or less), climate change or other large-scale environmental changes may eventually make the reserve unsuitable for the systems it was designed to protect. Without the capability of each ecosystem component to make large-scale geographic changes in response to such events, even extremely large, isolated reserves are doomed to fail.

Several papers critical of connectivity approaches to nature conservation have appeared in the last decade (Simberloff & Cox, 1987; Simberloff et al., 1992). The arguments presented in these papers are based on examples of small populations (usually found on islands) in an attempt to argue that inbreeding and other problems associated with the survivability of small, artificially isolated populations may be inconsequential. Other arguments are based on claims that not enough evidence exists regarding the functioning of corridors and whether certain species use them, or how wide corridors need to be to function. This is why for so many years the popular approaches to conservation were species oriented instead of ecosystem- or landscaped-based, despite the fact that many case studies have suggested the futility of a purely species approach to conservation (Hunter et al., 1988; Noss, 1991; Scott et al., 1993).

The most common argument against the utility of a landscape-connectivity approach is that there is little evidence that corridors "work," especially in regards to animals actually using them. In fact, Beier and Loe (1992) and Simberloff et al. (1992) have argued that until well-designed, replicated hypothesis-testing experiments prove that corridors have utility, conservation money should be invested only in isolated reserves. Admittedly, there have been few experiments showing that corridors function for their target purposes. However, observation and study of how animals move across both naturally and culturally fragmented landscapes provide some evidence for their utility (Beier, 1995). Lacking definitive evidence, most conservation scientists therefore agree that connected reserve systems will be more viable for maintaining biological diversity than a disjointed collection of reserves increasingly isolated through habitat fragmentation. Corridors may also serve other conservation purposes such as the maintenance or facilitation of key ecological processes and services. This could include corridors that benefit keystone species, which provide essential services for other species. Corridors may have water conservation functions that could include groundwater recharge and runoff filtration for adjacent water bodies. Large corridors such as landscape linkages also might, in some instances, facilitate the movement of fire in fire-dependent ecosystems. As Harris and Atkins (1991) explain:

> The crucial questions are not whether corridors are all good or all bad, not whether they are the sole answer to the present biodiversity crisis, or even

whether there are costs and liabilities. Rather, the key question is whether an integrated system of protected and managed natural habitats will be less discriminatory against our rare and endangered native species and also less supportive of alien, exotic, and pest species. Will such an interconnected system of habitats be superior for more natural assemblages of native species and more natural levels of ecological processes, such as competition, predation, and parasitism, than a disjunct system of isolated preserves? The consensus of leading conservationists is a resounding yes.

Components of Connectivity

Greenways, corridors, and landscape linkages that link larger core reserves are the most commonly considered applications of a connectivity approach. *Greenways* are linear parks of varying sizes that serve both conservation and recreational objectives. Greenways primarily meant to serve ecological functions are also called environmental corridors or landscape linkages. Depending on the size and other characteristics of a greenway system, these linkages could perform vital ecological functions by providing habitat for sensitive species, movement corridors linking conservation lands, and by buffering conservation lands from negative impacts associated with nearby intensive land development.

Corridors are considered to be smaller linear features that facilitate the movement of animals and other ecological processes. Landscape linkages are larger areas that can be narrowly defined as connectors capable of meeting the entire habitat needs of at least one member of a target species. Landscape linkages are preferable because there is a much higher probability that functional connectivity will be achieved, although this has not been demonstrated. The best regional example of landscape linkage is the Pinhook Swamp, which connects the Osceola National Forest to the Okefenokee National Wildlife Refuge. It should be obvious that landscape linkages can or could serve as important habitat areas in their own right as well as connectors to facilitate movement through an "inter-refuge" system. (See Fig. 6.2, a map of existing and potential conservation reserves and connecting greenways for Florida.)

According to Noss (1992), connectivity offers the following species-specific benefits:

1. Facilitates daily or seasonal movements.
2. Allows dispersal that might facilitate gene flow between populations and buffer small populations.

3. Allows range shifts in response to catastrophic events or long-term environmental change.

Ecological Connectivity and the Florida Greenways Initiative

The state of Florida has initiated a statewide greenways program in an attempt to help comprehensively manage natural resources and conservation lands in a coherent, integrated fashion. The establishment of integrated conservation systems is necessary to achieve effective, comprehensive land management. These systems could be variable in size but would be most effective if they incorporated entire landscapes apportioned into functional networks of reserves, multiple-use buffer zones, and landscape linkages (Harris, 1984; Noss, 1987). Although actual percentages will vary with the resource utilization capabilities of particular taxa, research suggests that connectivity reaches a critical threshold when random habitat connectedness is reduced to below 60 percent (Pearson et al., 1994). However, integrated conservation systems could lower these critical thresholds for functional landscapes by emphasizing connectivity and integrated resource management. In theory, comprehensive landscape management would maximize connectivity by linking reserves and multiple-use conservation lands into an optimized, functioning system. The goal is to protect and manage a landscape to preserve and/or mimic natural processes and evolutionary forces.

Riparian Connectivity

Riparian ecosystems or riparian conservation areas are another important form of linkage. Properly designed riparian corridors facilitate connectivity and landscape function. Rivers and streams serve as natural landscape features that help to guide animal movement. Furthermore, sufficiently wide riparian corridors include the gradient of ecosystems, ranging from aquatic habitats, to wetlands, to mesic habitats, and xeric uplands. These linear bands of habitat serve both the habitat needs of most species and the movement needs of all species, including specialists that restrict their movement to only one particular habitat type. Although ridgeline corridors have important functions, properly designed riparian corridors are the natural "skeleton" of an effective integrated conservation system.

Riparian corridors also have other important values (Schaefer & Brown, 1992). Riparian zones are often diverse in plants and animals, are highly productive, and serve as buffer zones to protect hydrological processes and quality. If wide enough, these areas also serve as habitat for forest-interior bird species and cavity-nesting birds (Harris, 1989). Riparian strips act as habitat and corridors for otters and mink (*Mustela vision*), and other wetland-dependent mammals sensitive to fragmentation (Harris, 1989; Noss & Harris, 1986; Schaefer & Brown, 1992).

Riparian buffers have a positive effect on water quality, by reducing the impact of nonpoint pollution, and provide critical habitat for some important species. However, these buffer zones often are not large enough to support viable populations of many species (Schaefer & Brown, 1992). Therefore, even a comprehensive network of riparian buffers will not take the place of larger reserves. Instead, integrated conservation design, reserves, and riparian buffers would be mutually beneficial. Riparian buffer zones will not likely support viable populations of wide-ranging species such as black bear, bobcat, otter, mink, panther, or many raptors. However, if linked to larger reserves that do support such species, riparian buffer zones provide additional, potentially critical habitat while benefiting from the ecological services these species provide. For example, black bears use the Wekiva River basin in Florida, but the established Wekiva River buffer zone cannot support a viable population of bears. Only linkage to the Ocala National Forest might ensure into the future the existence of bears in the Wekiva basin. Riparian buffers could also provide important primary, secondary, or seasonal habitat for many other species inhabiting linked reserves.

Connectivity and Black Bears

Several studies suggest that black bears use corridors both for seasonal movements and dispersal. Two studies found that bears used riparian strips to move within fragmented landscapes (Kellyhouse, 1977; Weaver et al., 1990). It has also been noted that in agricultural landscapes, bears restricted movements to wooded areas, such as ravines, shelterbelts, and riparian zones (Klenner, 1987; Weaver et al., 1990). In another study, when suitable cover was limited in a naturally fragmented landscape, bears used north-south aligned canyons with enough cover to make seasonal foraging trips and would return using the same corridors (Mollohan & Lecount, 1989). Beecham (1983) concluded that migration corridors connecting his study area with other bear habitat were critical in maintaining black bear numbers in the mountains of Idaho.

Florida's Biological Diversity Challenge

Florida is blessed with a rich biological heritage that includes over 600 vertebrate species, of which 115 (19%) are endemic, and approximately 3500 species of vascular plants, of which approximately 280 (8%) are endemic (Muller et al., 1989). Florida is also home to a complex array of community types (81 total, with 13 [16%] endemic to Florida), ranging from scrub and xeric sandhill ridges millions of years old, to the flatwoods of the coastal plain, to the fresh and saltwater marsh and swamp wetlands (Cox et al., 1994). However, Florida is in danger of losing much of this heritage to urbanization.

Agricultural activities also have had a major impact on natural systems in Florida. Florida's landscape is now apportioned as approximately 30 percent agricultural, 13 percent urban, and 57 percent forested or "semi-natural," with much of this 57 percent subjected to some form of human use or disturbance, including intensive forestry operations and ranching (Kautz, 1993). The product of such massive landscape alteration has been a general erosion of landscape function and habitat quality. Once common biological communities such as scrub, sandhill, and pine rockland have largely disappeared. Sandhill, which once covered over 20 percent of the state, has declined from approximately 2.8 million hectares to 325,000 hectares, with much of the remainder found in small, fragmented patches and in highly degraded condition (Davis, 1967; Cox et al., 1994). As a result of this massive loss of habitat and landscape function, Florida now has more federally threatened and endangered species than any other state, except California and Hawaii.

Habitat fragmentation has played a key role in the erosion of biological diversity in Florida. *Habitat fragmentation* is the combination of habitat loss with the increasing isolation of remaining habitat patches. As fragmentation progresses, habitat patches tend to become smaller and more isolated. Florida has experienced an increasingly intense level of fragmentation through various development activities: urban development, residential development, agricultural activities (e.g., citrus groves), roads, mining, and intensive silvicultural practices. Both forest clearance and development of forested lands for intensive human use have reduced Florida's forest land at a rapid rate—more than twice as rapidly as the loss rate for developing countries such as Brazil—and Florida continues to lose approximately 60,000 hectares of forest per year. In the last 50 years, Florida has lost over 3.6 million hectares of forest and wetland communities through conversion to intensive agricultural and urbanization (Cox et al., 1994).

At the same time, human-dominated landscapes favor a host of mid-sized generalist species that often proliferate in fragmented landscapes. With the reduction or extirpation of large predators and the increase in human-altered habitat, ecological relations and processes change drastically. The increase of mesomammals, such as raccoons and opossums, and opportunistic bird species, like crows, grackles, and blue jays, results in much greater predation pressure on ground-nesting birds, forest songbirds, reptiles, and amphibians.

These trends have important implications for natural resource conservation. However, there is a growing consensus that many existing and protected habitat remnants, including what are now considered to be "expansive" reserves, will be too small to provide suitable conditions to maintain viable populations of whole suites of sensitive species. These species include both *forest-interior species* that require interior conditions free from the negative effects of edges, and *area-sensitive species* that require large expanses of suitable habitat, such as black bear, panthers, and mink. Biologists, planners, and managers will need to consider designing and managing landscape systems that will meet the sustainable needs of humans while effectively protecting biological diversity.

Challenging Conservation: The Everglades System and Big Cypress National Preserve

Everglades National Park illustrates many of the typical problems of a conservation reserve that is not large enough, lacks connectivity, and has a highly altered context. It also illustrates why emphasis on the interactions of different ecosystem types in time and space is necessary for effective biodiversity conservation. "The Everglades ecosystem" implies a homogeneous unit; nothing could be farther from the truth. Consider only two of the score of endangered species in Everglades National Park, the wood stork (*Mycteria americana*) and the snail kite (*Rostrahamus sociabilis*): Both are dependent on wide expanses of shallow water throughout the Everglades landscape, but occurrence and reproductive success of the two species involve virtually opposite hydrologic regimes (Harris & Frederick, 1990; Kushlan, 1979).

The wood stork depends on tactile detection of fish for successful feeding. In the dry season when water levels are low, isolated ponds that concentrate fish and other aquatic organisms become feeding pools in an otherwise desiccated landscape. Not only are concentrations of fish necessary for wood stork reproduction, a relatively rapid drawdown is also necessary to trigger successful reproductive behaviors.

The snail kite now breeds only sporadically in Everglades National Park. This bird depends almost exclusively on the air-breathing apple snail (*Pomacea paludosus*) as a food source. The snail is substantially dependent on a shallow layer of water that covers large areas. Although most of the snail's life cycle is spent below the water, it must occasionally ascend to breathe, at which time it becomes exposed to predation by the kite. Generally, ideal snail-kite habitat is synonymous with ideal apple-snail habitat—large expanses of permanently, but shallow, inundated marshes that support populations year-round. Such a hydro-period contrasts starkly with that necessary for wood storks. As such, viable populations of the two species cannot coexist indefinitely without significantly different water regimes (Harris & Frederick, 1990), which are easily accommodated in a time- and space-varying landscape, but which are almost impossible to achieve in an "ecosystem" that is either consciously or unconsciously managed as a homogeneous entity.

At the scale of the landscape, isolated water holes with concentrations of aquatic organisms and hardwood-tree islands with concentrations of terrestrial creatures play commensurate ecological roles depending on the season. During the wet season, the hardwood islands serve as lifeboats for numerous terrestrial species that are sequestered in a water-covered landscape. Thus, depending on season, the predominance of strong ecological interactions shifts between the pools of water that serve as lifeblood, and hardwood islands that serve as lifeboats.

For more than 10,000 years, an array of carnivorous Florida land mammals have tolerated fluctuating climate, sea level, and prey opportunities across an uninterrupted, though constantly changing, peninsula. Modern south Florida supports three of the five large (>10 kg) carnivores that were present before European settlement of the New World, and it is the only region in eastern North America where these three species still share a common landscape. Ironically, the two largest preserves in Florida, Everglades National Park and Big Cypress National Preserve, are probably insufficient to ensure a future for this trio. While it appears that bobcat and black bear populations south of Lake Okeechobee are reasonably secure, the Florida panther, with its even larger spatial requirements (Maehr et al., 1991), is more immediately threatened by further habitat loss on private forest lands in southwest Florida. Furthermore, the large (1.5 million acres) national park and preserve complex only provides marginally suitable to inadequate habitat for panthers. Maehr and Cox (1995) conclude that the naturally occurring fragmented forests and predominantly wet herbaceous landscapes of Big Cypress National Preserve and Everglades National Park are incapable of supporting a stable panther population. Everglades National Park was home, until very recently, to the only subpopulation of panthers that is known to have gone extinct in the last two decades (Bass & Maehr, 1991), and Big Cypress National Preserve supports fewer panthers than preserves 5 percent of its size (i.e., the Florida Panther National Wildlife Refuge is 25,000 acres and is used by as many as 10 panthers annually).

Maehr (1990) observed that the loss of high-quality panther habitat on private lands to the north of the Big Cypress National Preserve and Everglades National Park reserve complex would result in the loss of more than half of the existing panther population. Conversely, the loss of all U.S. National Park Service land in south Florida would result in only a minimal (<10%) population reduction—primarily the loss of individuals that contribute irregularly to reproduction. Again, large preserve size is not, in and of itself, sufficient. For much of south Florida to become the good panther habitat that many managers imagine their efforts can create, sea level must drop and upland forest vegetation must replace most of the present-day Everglades. Currently, however, sea levels are rising (Henry et al., 1994).

The coastal preserves of south Florida will be early victims of rising sea levels that have already caused reductions in freshwater forest systems (Craighead, 1971). Before the advent of human landscape alterations such as canals, highways, and deforestation, wide-ranging mammals in south Florida were capable of following the gradual movements of plant communities as the climate changed. Black bears are known to be capable of negotiating barriers such as the dredged, channeled Caloosahatchee River and relatively large unforested areas (Maehr et al., 1988), but neither panthers nor bobcats have been observed doing the same. For bobcats, the consequences of sea-level rise with no escape route are less onerous because of their wide distribution and less extensive spatial requirements, but for panthers the impacts could be lethal. Within this century, all continuous upland connections between south Florida and points to the north have been dredged, cleared, impounded, farmed, or paved. However, corridors between south Florida preserves do appear to serve a local connecting function (Fig. 10.1).

Black bears and panthers are landscape creatures. Dispersal movements of individual black bears link disjunct populations as well as distant drainage basins (Maehr et al., 1988). Resident panthers routinely negotiate unrelated forested upland and wetland systems. In some cases, the connections between larger habitat blocks are no more than 100-meter-wide corridors of cypress that are bordered by farm fields and bisected by highways (Maehr, 1990). These artificially narrowed travelways vary in their use by mammalian carnivores: Some serve as portions of home ranges, while others act only as movement conduits (see Fig. 10.1). The maintenance of these linkages will become increasingly important as changing environmental conditions compel populations to move.

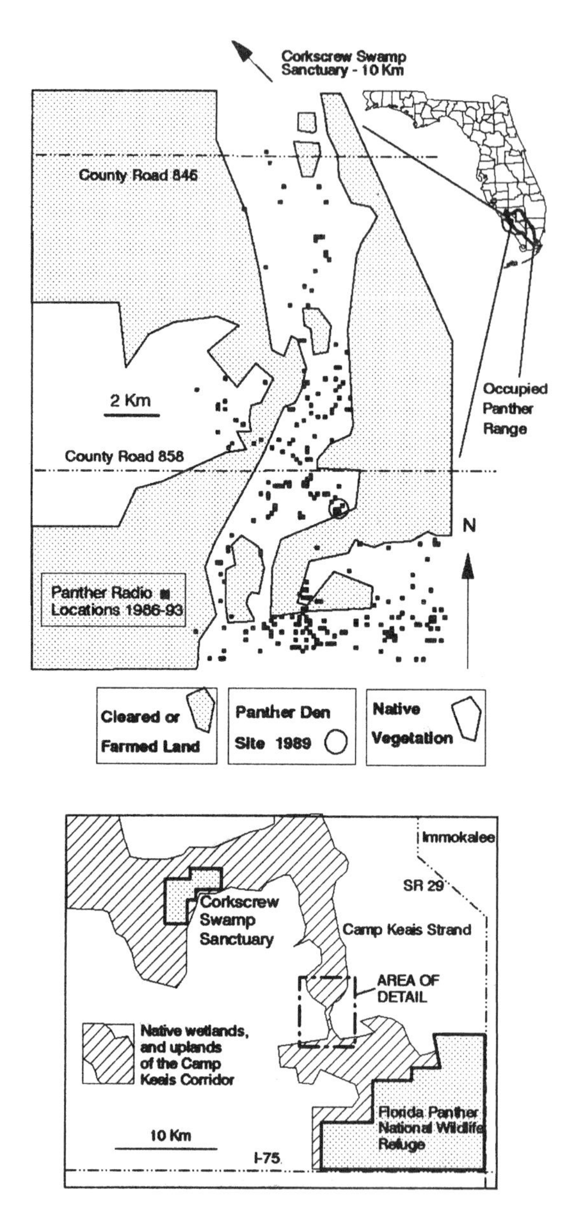

Figure 10.1 Panther radio-collar locations and the south Florida context. Note that most locations are found in native vegetation.

Clearly, a large area is required for evolution to occur among large mammalian predators and their prey. However, without a way for panthers to evade an encroaching seascape, it is unlikely that evolution will continue in south Florida. Given that two very large preserves contribute insignificantly to the future of the Florida panther and large terrestrial mammals in general, planning for these wide-ranging species must go well beyond the traditional approach of static-boundary nature preserves.

Conclusion

Throughout the past there has been a tendency to treat ecological systems as discrete units: Land was subdivided, property lines drawn, and management prescriptions developed based only on the intrinsic characteristics of the site, and often on the narrow objectives of the manager. This resulted in increasingly fragmented and otherwise degraded landscapes with greatly reduced ability to support native biological diversity as well as sustain the needs of humans. Developments in ecological sciences and a growing environmental awareness have led to important changes in both how we perceive our natural heritage and how we utilize natural resources.

Increasing acceptance of Aldo Leopold's land ethic signifies a major change in our relationship to the land (and waters), that is, we as humans are members of a greater community that includes soils, waters, plants, and animals (Leopold, 1949). Because our well-being is based on the health of the land and because we have an ever-growing power to affect it, we also have an impending ethical responsibility to protect those relationships between components that are necessary to ecosystem maintenance. This imperative led Forman (1995) to formulate his ethics of isolation: "Simply stated, in land use decisions and actions, it is unethical to evaluate an area in isolation from its surroundings or from its development over time. Ethics impel us to consider an area in its broadest spatial and temporal perspectives."

Only a comprehensive conservation approach applied at the scale of the regional landscape can embody this philosophy. It effectively protects biological diversity while managing natural resources, and provides for economic sustainability by planning land use in an integrated and synergistic manner. The regional landscape focus bridges the gap between our tendency to manage ecological systems as discrete units and the need for effective planning and management that protects the functional interrelationships between these ecosystems. A comprehensive regional approach is critical to the creation of an effective conservation program such as the statewide Florida Greenway Network and conservation system. Such an approach is necessary to ensure that connectivity is enhanced.

Environmental corridors and landscape linkages are the most commonly considered applications of a connectivity approach. Corridors and landscape linkages are therefore critical components of an integrated land conservation program. The ecological value of greenways will only be realized when they serve as effective linkages within a larger system composed of biological reserves, multiple-use conservation areas, buffer zones, and agricultural lands. Such integrated ecological systems will become increasingly necessary for effective conservation of native biological diversity as human population and human-built environments continue to grow unabated. This regional landscape conservation framework will facilitate the protection of our natural heritage while providing compatible recreational and sustainable economic activities that will also be protected from continued intensive land use changes.

Literature Cited

Bass OL, Maehr DS. Do recent panther deaths in Everglades National Park suggest an ephemeral population? *National Geograph Res Explor* 1991;7:427.

Beecham JJ. Population characteristics of black bears in west central Idaho. *J Wildl Manage* 1983;47:405–412.

Beier P. Dispersal of juvenile cougars in fragmented habitat. *J Wildl Manage* 1995;59: 228–237.

Brower LP, Malcolm SB. Animal migrations: Endangered phenomena. *Am Zool* 1991;31: 265–276.

Browne J. *The Secular Ark: Studies in the History of Biogeography.* New Haven, CT: Yale University Press, 1983:44.

Cahalane VH. The status of mammals in the U.S. national park system, 1947. *J Mamm* 1948;29:247–259.

Cox J, Kautz R, MacLaughlin M, Gilbert T. *Closing the Gaps in Florida's Wildlife Habitat Conservation System* (Draft report). Tallahassee, FL: Office of Environmental Services, Florida Game and Fresh Water Fish Commission, 1994.

Craighead FC. *The Trees of South Florida* (Vol. 1). Coral Gables, FL: University of Miami Press, 1971.

Davis JH. *General Map of Natural Vegetation of Florida.* Gainesville, FL: University of Florida, Agricultural Experiment Station, Institute of Food and Agricultural Science, 1967.

Forman RTT. *Land Mosaics: The Ecology of Landscapes and Regions.* New York: Cambridge University Press, 1995.

Frankel OH, Soulé M. *Conservation and Evolution.* New York: Cambridge University Press, 1981.

Hansson L, Angelstam P. Landscape ecology as a theoretical basis for nature conservation. *Landscape Ecol* 1991;5:191–201.

Harris LD. *The Fragmented Forest: Island Biogeography Theory and the Preservation of Biotic Diversity.* Chicago: University of Chicago Press, 1984.

Harris LD. The Faunal Significance of Fragmentation of Southeastern Bottomland Forests. In DD Hook, R Lea (eds), *Proceedings of the Symposium: The Forested Wetlands of the Southern United States,* General Technical Report SE-50. USDA Forest Service, Southeastern Forest Exper. Sta., Asheville, NC, 1989:126–134.

Harris LD, Atkins K. Faunal Movement Corridors in Florida. In W Hudson (ed), *Landscape Linkages and Biodiversity.* Washington, DC: Island Press, 1991:117–134.

Harris LD, Frederick P. The Role of the Endangered Species Act in the Conservation of Biological Diversity: An Assessment. In J Cairns Jr, T Crawford (eds), *Integrated Environmental Management.* Chelsea, MI: Lewis, 1990:99–117.

Harris LD, Hoctor T, Gergel S. Landscape Processes and Their Significance to Biodiversity Conservation. In O Rhodes, R Chesser, M Smith (eds), *Spatial and Temporal Aspects of Population Processes.* Chicago: University of Chicago Press, in press.

Harris L, Silva-Lopez G. Forest Fragmentation and the Conservation of Biological Diversity. In P Fielder, S Jain (eds), *Conservation Biology: The Theory and Practice of Nature Conservation, Preservation, and Management.* New York: Chapman and Hall, 1992:197–237.

Henry JA, Portier KM, Coyne J. *The Climate and Weather of Florida.* Sarasota, FL: Pineapple Press, 1994.

Hunter ML, Jacobson GL, Webb T. Paleoecology and the coarse-filter approach to maintaining biodiversity. *Cons Biol* 1988;2:1988.

Hobbs RJ. Reintegrating Fragmented Landscapes: Towards Sustainable Production and Nature Conservation. New York: Springer-Verlag, 1993.

Kautz R. Trends in Florida wildlife habitat 1936–1987. *Florida Scient* 1993;56:7–24.

Kellyhouse DG. Habitat utilization by black bears in northern California. *Int Conf Bear Res Manage* 1977;4:221–227.

Klenner W. Seasonal movements and home range utilization patterns of the black bear, *Ursus americanus,* in western Manitoba. *Can Field Nat* 1987;101:558–568.

Kushlan JA. Design and management of continental reserves: Lessons from the Everglades. *Biol Cons* 1979;15:281–290.

Leopold AA. *Sand County Almanac.* New York: Oxford University Press, 1949.

Maehr DS. The Florida panther and private lands. *Cons Biol* 1990;4:167–170.

Maehr DS, Cox JA. Landscape features and panthers in Florida. *Cons Biol* 1995;9: 1008–1019.

Maehr DS, Belden RC, Land ED, Wilkins L. Food habits of panthers in southwest Florida. *J Wild Manage* 1990;54:420–423.

Maehr DS, Land ED, McCown JW. Social ecology of Florida panthers. *National Geograph Res Explor* 1991;7:414–431.

Maehr DS, Layne JN, Land ED, McCown JW, Roof JC. Long distance movements of a Florida black bear. *Florida Field Nat* 1988;16:1–6.

McFarlane RW. *A Stillness in the Pines, the Ecology of the Red-cockaded Woodpecker.* New York: W. W. Norton, 1992.

Mollohan CM, Lecount AL. Problems of Maintaining a Viable Black Bear Population in a Fragmented Forest. Conference on the Multiresource Management of Ponderosa Pine Forests, Northern Arizona University, November 1–16, 1989.

Morgan G, Woods C. Extinctions and the zoogeography of the West Indian land mammals. *Biol J Linn Soc* 1986;28:167–203.

Muller JW, Hardin ED, Jackson DR, Gatewood SE, Caire N. Summary report on the vascular plants, animals, and plant communities endemic to Florida. *Nongame Wildlife Program Technical Report No. 7.* Tallahassee, FL, Florida Game and Fresh Water Fish Commission, 1989.

Newmark WD. Legal and biotic boundaries of western North American national parks: A problem of congruence. *Biol Cons* 1985;33:197–208.

Noss RF. From plant communities to landscapes in conservation inventories: A look at The Nature Conservatory (USA). *Biol Cons* 1987;41:11–37.

Noss RF. The longleaf pine landscape of the Southeast: Almost gone and almost forgotten. *Endang Spec Up* 1988;5(5):8.

Noss RF. Landscape Connectivity: Different Functions at Different Scales. In WE Hudson (ed), *Landscape Linkages and Biodiversity.* Washington, DC: Defenders of Wildlife and Island Press, 1991:27–39.

Noss RF. Issues of Scale in Conservation Biology. In PL Fiedler, SK Jain (eds), *Conservation Biology: The Theory and Practice of Nature Conservation, Preservation, and Management.* New York: Chapman and Hall, 1992:239–250.

Noss RF, Harris LD. Nodes, networks, and MUMs: Preserving diversity at all scales. *Environ Manage* 1986;10:299–309.

Pearson SM, Turner MG, Gardner RH, O'Neill RV. Scaling Issues of Biodiversity Protection. In RC Szaro (ed), *Biodiversity in Managed Landscapes: Theory and Practice.* London: Oxford University Press, 1995.

Pickett STA, Thompson JN. Patch dynamics and the design of nature reserves. *Biol Cons* 1978;13:27–37.

Preston FW. The canonical distribution of commonness and rarity. Part II. *Ecol* 1962;43:410–432.

Rodgers WH, et al. *Setting Priorities for Land Conservation: Committee on Scientific and Technical Criteria for Federal Acquisition of Lands for Conservation.* Washington, DC: National Academy of Sciences Press, 1993.

Sax JL. Ecosystems and Property Rights in Greater Yellowstone: The Legal System in Transition. In RB Keiter, MS Boyce (eds), *The Greater Yellowstone Ecosystem.* New Haven, CT: Yale University Press, 1991:77–84.

Schaefer JM, Brown, MT. Designing and protecting river corridors for wildlife. *Rivers* 1992;3(1):14–26.

Scott JM, et al. Gap analysis: A geographical approach to protection of biological diversity. *Wildl Monogr* 1993;123:1–41.

Shelford VE. Conservation of Wildlife. In AE Parkins, JR Whitaker (eds), *Our Natural Resources and Their Conservation.* New York: John Wiley and Sons, 1936: 485–526.

Simberloff D, Cox J. Consequences and costs of conservation corridors. *Cons Biol* 1987;1:63–71.

Simberloff D, Farr JA, Cox J, Mehlman, DW. Movement Corridors: Conservation bargains or poor investments? *Cons Biol* 1992;6:493–504.

Turner MG. Landscape ecology: The effect of pattern on process. *Ann Rev Ecol System* 1989;20:171–197.

Walters JR, Hansen SK, Carter JH III, Manor PD, Blue RJ. Long-distance dispersal of an adult red-cockaded woodpecker. *Wilson Bull* 1988;100:494–496.

Weaver KM, et al. Bottomland hardwood forest management for black bears in Louisiana. *Proc Ann Conf Southeast Assoc Fish Wildl Agen* 1990;44:342–350.

Weins JA. Habitat fragmentation: Island v landscape perspectives on bird conservation. *Ibis* 1994;137:97–104.

Wells B, Strunk I. The vegetation and habitat factors of thc coarser sands of the North Carolina plain: An ecological study. *Ecol Monogr* 1931;1:465–521.
Wright GM, Dixon JD, Thompson BH. Fauna of the National Parks of the United States (Series 1). Washington, DC: Government Printing Office, 1933.
Wright GM, Thompson BH. Fauna of the National Parks of the United States (Series 2). Washington, DC: Government Printing Office: 1934.

11

Fragmentation of a Natural Area: Dynamics of Isolation for Small Mammals on Forest Remnants

L. Scott Mills

A casual airplane flight over nearly any human-modified landscape might bring to mind thoughts of oceanic islands. In tropical and temperate forests, for example, formerly continuous tracts have rapidly come to resemble "islands" of natural areas in a "sea" of modified habitat. But, is this analogy a good one? Are species that remain on fragmented natural areas actually isolated? And, if so, how does isolation occur? Also, are there edge effects that modify the functional size of a persisting remnant for wildlife species? Clearly, the crux of contemporary natural-area management has become the prediction of change as an area is progressively modified by humans.

To learn more about wildlife responses to coniferous forest fragmentation in the Pacific Northwest, I studied small mammals on small remnants of forest surrounded entirely by clearcuts. At the same time, I considered small mammals in the clearcuts, as well as those on large unfragmented (control) areas. I focused on the California red-backed vole (*Clethrionomys californicus*), a species whose food habits made them especially likely to be negatively impacted by forest fragmentation. The primary food of this species is mycorrhizal fungi sporocarps (i.e., truffles; Maser et al., 1978; Ure & Maser, 1982), and previous studies had linked these voles to coarse woody debris (e.g., Doyle, 1987; Hayes & Cross, 1987). Thus, it seemed that a close examination of the distribution of red-backed voles with respect to truffles, logs, and abiotic factors would reveal something about particular mechanisms leading to isolation or edge effects.

While trapping California red-backed voles, I also captured a number of other species in the remnants, clearcuts, and controls. In total, the detailed data on voles and subsidiary information on other species allow me to consider several issues related to the patterns and process of fragmented natural areas. First, I synthesize and extend previous papers describing mechanisms of isolation for California red-backed voles on these remnants (Mills, 1993; Tallmon & Mills, 1994; Clarkson & Mills, 1994; Mills, 1995). Second, I ask whether it is appropriate to extend an assumption of isolation to other small-mammal species on the remnants. Island biogeography models that assume extinction and/or recolonization dynamics like those present on oceanic islands are sometimes used in the study and management of fragmented natural areas. I assess both the validity of the isolation analogy for species on my remnants, and the consequences of applying island models. Finally, I end with recommendations about how to approach the study and conservation of species on natural areas fragmented by human modification.

Study Area

All study sites were in southwest Oregon, within a 100-kilometer radius of Grants Pass (Fig. 11.1). The sixteen forest remnants were unlogged mature to old-growth forest (primarily Douglas fir [*Pseudotsuga menziesii*] communities) surrounded entirely by land clearcut 1 to 30 years previously. The surrounding clearcuts included primarily *Rubus, Ribes, Ceonothus,* and *Epilobium* plant associations. Remnants varied in elevation, slope, and aspect (Table 11.1). They also varied in biogeographic descriptors such as size (0.3–7.2 ha), width of the surrounding clearcut, and time elapsed since harvest of the last link with nearby forest (see Table 11.1). Three remnants (K, C, and A; see Table 11.1) are included here in the consideration of biogeographic effects, though they were deleted from the edge-effect analysis of Mills (1995) because they were too large to accurately measure the distance from edge to trap (sites K and C) or because a bisecting road confounded the edge measurement (site A).

Control sites were defined as the interior of unlogged mature to old-growth coniferous forest more than 250 hectares in size. Although such sites were difficult to find due to extensive forest fragmentation in the region, I used five controls, with at least one within 13 kilometers of each remnant (see Fig. 11.1, Table 11.1).

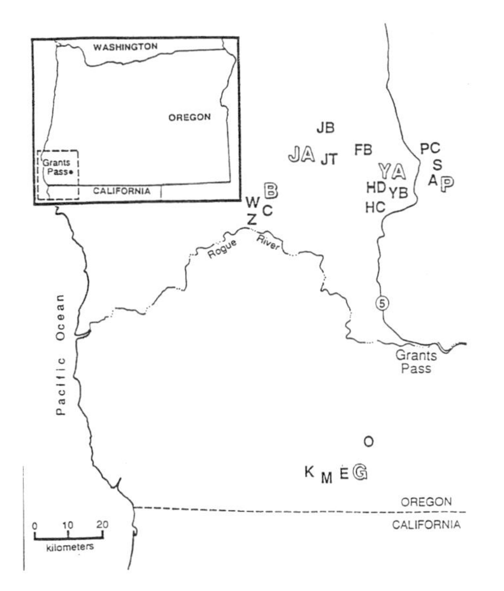

Figure 11.1 Location of remnant and control sites analyzed for small-mammal responses to forest fragmentation in southwestern Oregon, 1990 and 1991. Control sites (250 ha) are indicated by outlined letters, remnants by solid type. (From LS Mills, Edge effects and isolation: Red-backed voles in forest remnants. *Cons Biol* 1995;9:395–403.)

Methods

Small-Mammal Trapping

Trapping was conducted from June 25 to September 14, 1990, and from June 7 to September 6, 1991. For each site and each field season, up to eight-five Sherman live traps were set for four nights in a measured, regularly spaced trapping grid; the trap protocol involved 15-meter spacing, with adjustment of up to 4 meters allowed if a trap could be placed under or near fallen logs. Additional details on trapping protocol may be found in Mills (1995).

The group of remnants (and associated clearcuts) nearest each control was trapped within a month of when the control was trapped. Simultaneous with sampling a remnant, the surrounding clearcuts were sampled with sixteen Sherman traps placed in four lines of four traps each. As in the forest, traps were set for four nights, located under logs whenever possible, and spaced 15 meters apart within lines.

Because the primary study animal was the California red-backed vole, only this species was individually marked on capture.

Remnant Edges and Edge Classes

Although the edge, or clearcut/forest interface, was generally obvious, I ensured objective delineation of the edge by applying a consistent definition

Table 11.1 Biogeographic characteristics of forest remnants and controls (>250 ha) used to study small-mammal responses to forest fragmentation.[a]

Site	Size (ha)	Number of traps[b]	Last clearcut[c]	Distance to forest (m)[d]	Aspect	Elevation(m)[e]	Slope (degrees)
Remnants							
K	7.2	85	1989	200	N	915	99
C	6.0	85	1980	55	S	793	99
O	3.6	85	1981	150	W	1403	31
HC[f]	2.5	84	1974	95	W	976	24
E[f]	2.5	85	1987	100	W	1342	32
JB[f]	2.5	69	1976	140	SE	686	13
A	2.3	84	1962	60	SE	702	99
FB	2.0	81	1973	120	W	671	25
W[f]	1.4	6.7	1983	60	SE	732	14
S[f]	1.3	63	1977	150	E	991	33
YB	1.3	68	1989	75	N	915	35
HD[f]	1.1	42	1965	110	S	976	22
Z[f]	1.0	52	1983	60	S	732	22
PC[f]	0.9	52	1986	170	N	640	35
JT	0.6	30	1988	50	E	991	2
M	0.3	20	1978	100	W	1220	10
Controls							
G	N/A	85	N/A	N/A	S	1586	19
B	N/A	85	N/A	N/A	SE	1083	9
P	N/A	85	N/A	N/A	E	884	31
YA	N/A	85	N/A	N/A	E	817	37
JA	N/A	85	N/A	N/A	—	991	2

[a]Sites with single letter designations were sampled in both 1990 and 1991; all others were sampled only in 1991.
[b]Number of traps set for four nights each on measured grid with 15 m spacing (see text). Lack of perfect concordance between trap number and remnant size reflects the influence of remnant shape on trapping grid layout.
[c]Year in which the remnant became completely surrounded by clearcuts.
[d]Distance to nearest forest >7 ha in size (the area of the largest remnant).
[e]Elevation at approximate center of site.
[f]Remnant site sampled for abiotic edge effects.

(Mills, 1995) across remnants: The edge was operationally defined as the minimum convex polygon comprised of the outermost conifer trees that were at least two-thirds the mean diameter of conifers in the remnant center. The edge was allowed to be concave if the distance from one edge tree to the next was more than 30 meters. To eliminate the influence of occasional unharvested trees outside the remnant, a conifer was not considered an edge tree if the line of travel along the remnant perimeter and out to the tree formed an angle of less than 135 degrees, and if the tree was farther than 20 meters from any two trees inside the edge.

By measuring in from the edge to the interior, four edge classes were described: 0 to 15 meters, 16 to 30 meters, 31 to 45 meters, and 46 to 90 meters. Each small-mammal trap could be assigned to one edge class.

Sampling of Truffles, Logs, and Abiotic Factors

To sample these factors we first established two 20-meter line transects randomly in each edge class of each remnant. Log volume (m^3/ha) and log number (number/ha) for each edge class on each remnant were estimated using line-intercept sampling (DeVries, 1974; see also Mills, 1995). Truffle sampling involved 136 plots on four remnants and 80 plots in clearcuts surrounding two of the forest remnants (Clarkson & Mills, 1994; Mills, 1995).

We sampled abiotic factors (Mills & Reinertson, unpublished data) during the day on a subset of eight of the remnants (see Table 11.1). On each of the transects, we subsampled soil temperature and soil moisture at 5-meter intervals. Thus, means for each edge class of each remnant were based on a total of ten samples (five subsamples on each of two transects). Soil temperatures were obtained with a Reotemp thermometer placed 10 centimeters beneath the cleared litter layer for 30 seconds or until the temperature reading stabilized. The weights of collected and dehydrated soil samples provided the estimates of percent soil moisture. No samples were collected within a week of rain on a site, and the sampling scheme avoided bias due to time of day.

Red-Backed Vole Responses and Potential Proximate Mechanisms

California red-backed voles are strongly and negatively affected by forest fragmentation (Mills, 1995). Voles on remnants are virtually isolated, with only three voles captured in the clearcuts (Fig. 11.2); preliminary indications from DNA fingerprinting show that voles on remnants have less genetic variation than those on control sites (Mills, 1993). This species also shows a strong negative edge effect on the remnants studied, with four times the density in the most interior compared to the most peripheral edge class (see Fig. 11.2).

My search for a mechanism of the isolation and edge effect began with coarse woody debris, as downed logs have been closely linked to the presence of California red-backed voles (Doyle, 1987; Hayes & Cross, 1987). Using radiotelemetry to closely monitor the use of logs by red-backed voles on one

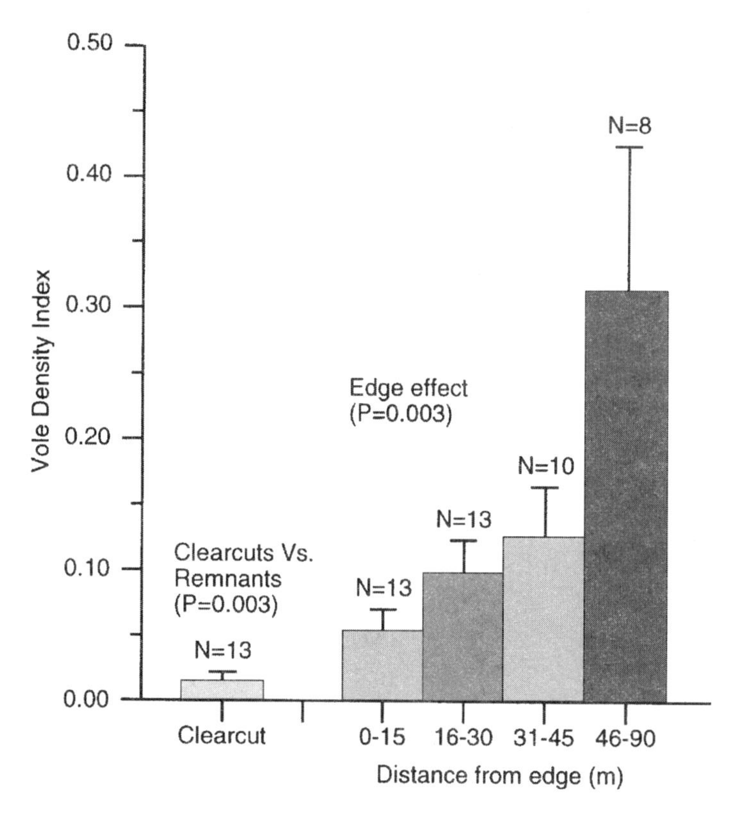

Figure 11.2 Mean and standard errors of the California red-backed vole density index (number of different voles captured per trap over four nights of trapping) in southwestern Oregon, 1990 and 1991. Forest edge classes represent number of meters from trap to the nearest forest-clearcut interface. The N refers to the number of sites trapped. Small remnants often did not have any area farther than 30 meters from an edge. Statistical tests are described in Mills (1995). (Modified from LS Mills, Edge effects and isolation: Red-backed voles in forest remnants. *Cons Biol* 1995;9:395–403.)

remnant (site E), we found that voles were found under logs 98 percent of the time, even though logs covered only 7 percent of the collective home range area (Tallmon & Mills, 1994). In concert, these studies pointed toward an expectation that log and vole edge effects would coincide.

Surprisingly, however, I found that both volume and number of logs tended to increase toward the remnant edges—probably due to cut trees falling into the forest and from death and blow-down of trees on the edge (Mills, 1995; see also Laurance, 1991; Chen et al., 1992)—indicating that it is not coarse woody debris per se that drives vole distribution. Rather, the radiotelemetry work indicated that these voles have a significant preference for decayed logs (Tallmon & Mills, 1994), which are often moister and cooler than the recently fallen logs that dominate at remnant edges.

My investigation of abiotic factors indicated that cooler and moister conditions also predominate in remnants versus clearcuts, and in the interior of remnants relative to edges. Soil temperature decreased significantly from the edge to the interior of remnants and was higher in clearcuts (Fig. 11.3A). Increased light penetration due to vegetation removal plays an obvious role here (Ranney et al., 1981; Williams-Linera, 1990), and such trends have been observed elsewhere (e.g., Kapos, 1989; Laurance & Yensen, 1991; Matlack, 1993). The response of soil moisture to forest fragmentation was more complex: In our study it generally increased from remnant edge to interior, and from clearcuts to remnants (Fig. 11.3B), although the trends were not statistically significant. The lack of strong trends may have been because our relatively small sample size was overwhelmed by variation arising from differences in aspect, slope, depth of humus layer, overstory characteristics, and soil texture (see Ranney et al., 1981). The trend in soil moisture is also complicated by factors such as wind, which affects humidity and temperature in clearcuts, edges, and forest interior (Saunders et al., 1991; Chen et al., 1992).

I did not conduct experiments to determine if abiotic conditions affected red-backed vole physiology or fitness either directly or indirectly through factors such as changed susceptibility to predation or competitive interactions (Grant, 1970; Reese & Ratti, 1988; Laurance, 1995). The vole's major food source—hypogeous sporocarps of mycorrhizal fungi (truffles)—was, however, differentially distributed across the edge gradient and between clearcut and remnants. This result had not previously been demonstrated but could be anticipated given the sensitivity of truffle production to temperature and moisture (Fogel, 1976; 1981). In particular, truffles were virtually absent from the clearcuts, as well as in the first edge class (0–15 m) of the remnants (Mills, 1995). Furthermore, truffles were more likely to be found under logs than elsewhere, and voles were more likely to be caught at traps having truffles nearby (Clarkson & Mills, 1994). A related factor of likely importance to both truffles and voles is the soil organic layer, which decreases following clearcutting and burning (see Rosenberg et al., 1994).

In sum, our work and that of others begins to paint a coherent picture of proximate mechanisms driving California red-backed vole response to forest fragmentation. Both voles and truffles are rarely found in clearcuts, where the soil organic layer tends to be shallower and the daytime microclimate is hotter and potentially drier. Within remnants, both voles and truffles are more likely to be found in the interior of the remnants and under logs—particularly more decayed logs—again favoring the more cool and moist microclimates. Importantly, there is a feedback loop that adds another dimension to the effect of clearcutting on truffles and voles (see Clarkson & Mills, 1994). Mycorrhizae

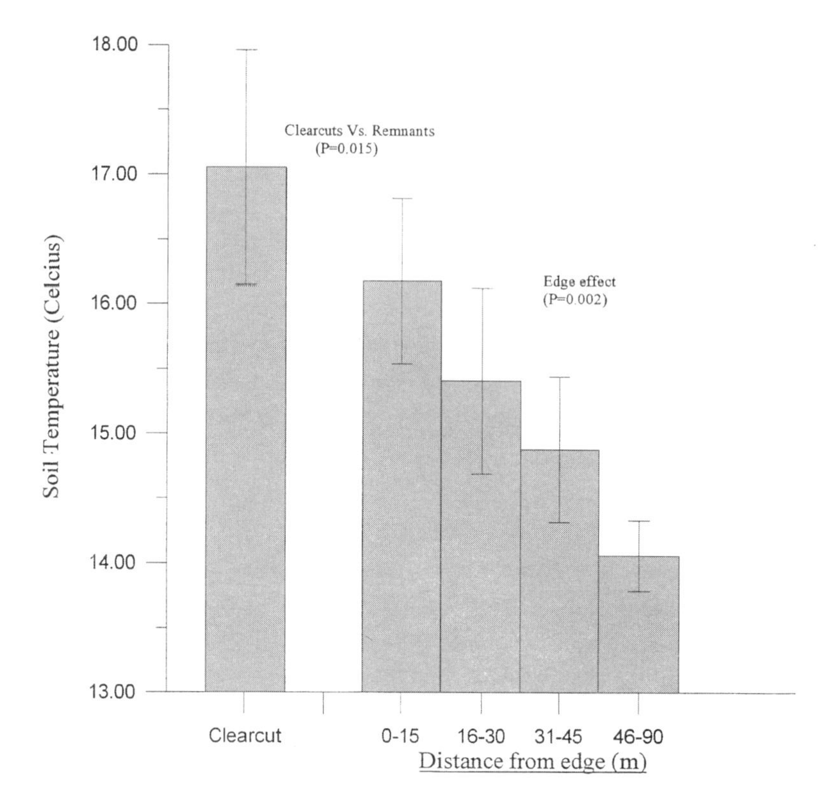

Figure 11.3 Abiotic factors measured in eight remnants and surrounding clearcuts in southwest Oregon in the summer of 1991. Forest edge classes represent the number of meters to the nearest forest-clearcut interface. **A.** Mean (and standard error) soil temperature (°C). **B.** Mean (and standard error) soil moisture content. The P-value for clearcuts versus remnants was based on a paired T-test of the mean of each remnant and its surrounding clearcut. The edge effect P-value was calculated using an ordered heterogeneity test (described in Mills [1995]).

("fungus roots") are critical to the water and nutrient functioning of nearly all vascular plants, and one of the primary dispersers of sporocarps for hypogeous taxa are red-backed voles. Thus, small remnants of forest can serve as vital refuges for both truffles and the small-mammal mycophagists that eat and disperse them. In turn, the dispersal of these mycorrhizae help benefit young trees as the clearcut forest regenerates.

Biogeography of Small Mammals

Are the Remnants "Islands" for All Species?

The human observer viewing these remnant natural areas in a stark landscape of clearcuts finds it difficult to resist labeling these patches as "islands." Indeed,

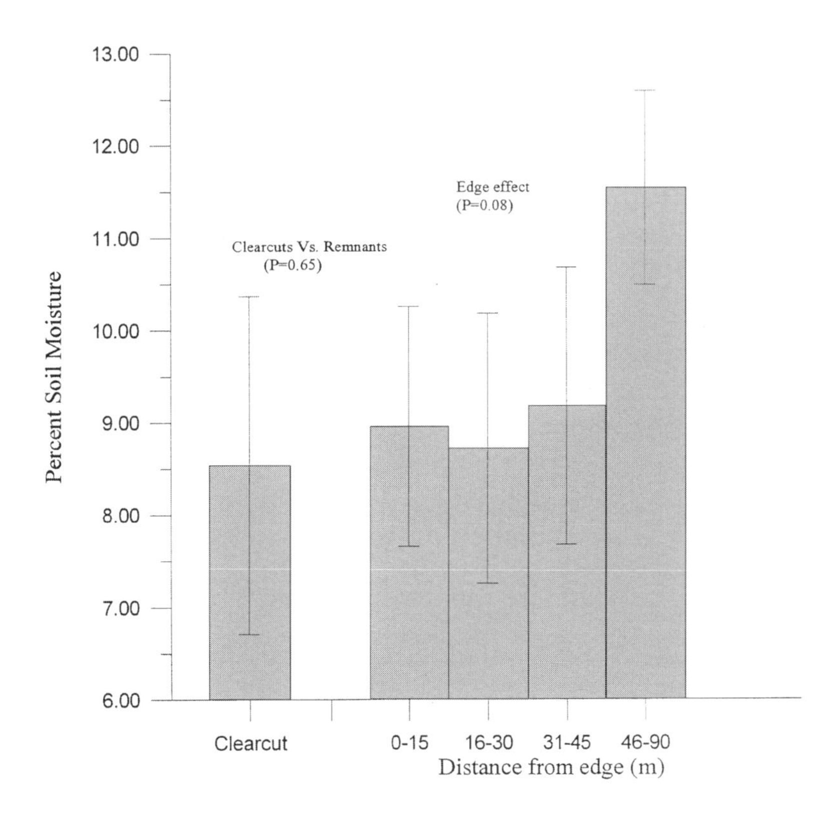

B

Figure 11.3 *(cont.)*

it seems that the island analogy holds for red-backed voles, as abiotic and biotic mechanisms isolate this species on remnants surrounded by clearcuts 3 to 30 years old. However, can we make this analogy in a general way for species occupying these forest remnants?

Although my study was designed primarily to examine the distribution of one species—California red-backed voles—I used capture data as a rough index of responses for three other species commonly captured on remnants. Based on trapping success on clearcuts and remnants for the four most commonly captured species (Fig. 11.4), it is obvious that not all species respond as do red-backed voles. For example, likelihood of capture for Trowbridge shrews (*Sorex trowbridgii*) did not differ between forest remnants and surrounding clearcuts, or among different edge classes on the twelve remnants where they were caught (Fig. 11.4A). Townsend chipmunks (*Eutamius townsendii*) also exhibited no edge effect but used clearcuts significantly less than remnants (Fig. 11.4B). Thus, the clearcut may impede movement of chipmunks between remnants, although they do not appear to differentiate between areas within a

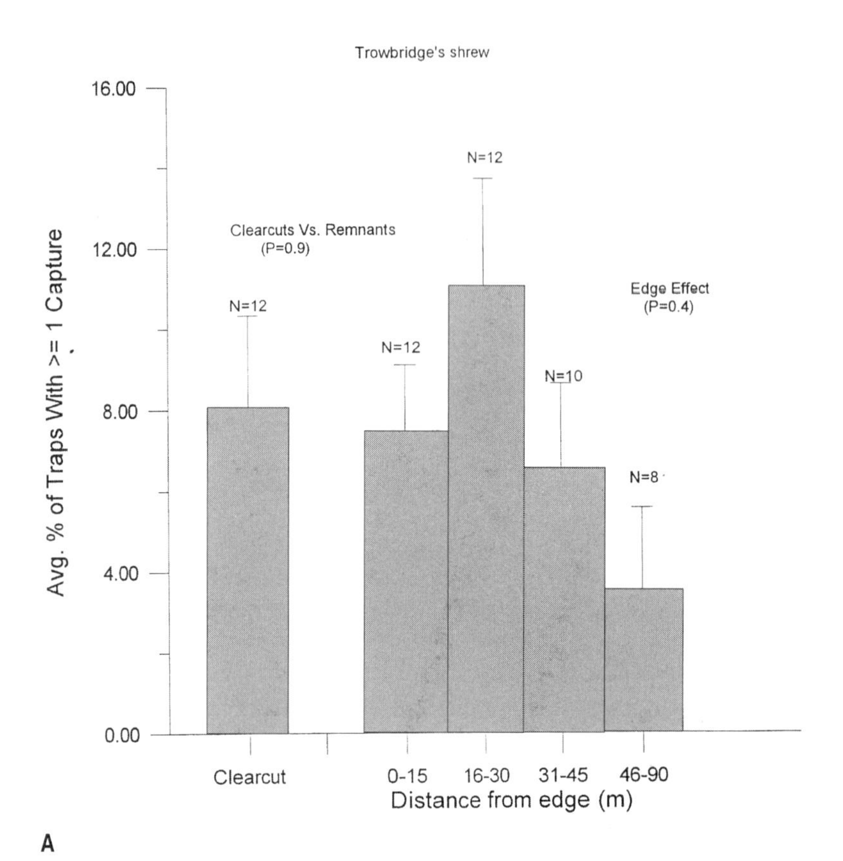

A

Figure 11.4 Mean and standard errors of trapping-success indices for all remnants where at least one animal of that species was captured. The index is the percent of all traps in that edge class that captured at least one animal of that species in the four nights of trapping. Although changes in the index could include both increased numbers or increased captures of the same individuals, I use it only to examine relative differences in space-use across edge and clearcut classes. Note that, using this index, the trend for red-backed voles is identical to the direct density index used in Fig. 11.2 (the density index could not be used for the other three species because only red-backed voles were individually marked at capture). Forest-edge classes represent number of meters from trap to forest-clearcut interface; numbers above the bars are the number of remnants analyzed with traps in that edge class. **A.** Trowbridge shrew. **B.** Townsend chipmunk. **C.** Deer mouse (capture data for *Peromyscus maniculatus* and *Peromyscus truei* were combined). **D.** California red-backed vole. P-values for the clearcut-remnant comparison were based on a Wilcoxon signed-rank test; for the edge-effect analysis an ordered heterogeneity test was used (described in Mills 1993, 1995).

remnant. Edge effects for deer mice (*Peromyscus* spp.) were positive, with traps on the edge seven times more likely to capture a mouse than traps in the interior of remnants (Fig. 11.4C); the mice used clearcuts significantly more than remnants. In contrast to deer mice, California red-backed voles, as shown earlier, were nearly isolated and negatively affected by edge (Fig. 11.4D).

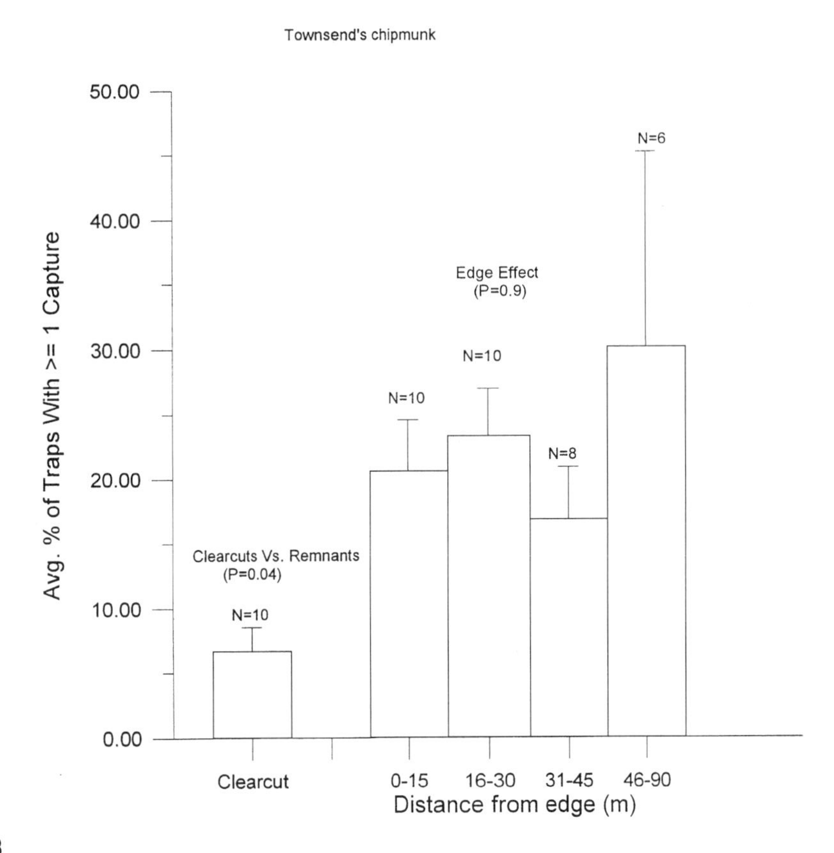

B

Figure 11.4 *(cont.)*

Clearly, in this system of forest remnants, the four most commonly captured small mammals vary greatly in the extent to which they appear isolated by the "sea" of clearcuts (see also Laurance, 1995). Furthermore, at least one species (dusky-footed woodrat) was rarely caught in control sites but often captured in clearcuts and remnants (Mills, 1993), implying that some species are colonizing remnants not from the control "mainland" but, rather, from the modified "sea." In short, the twin criteria for a remnant to be an "island" in the strict sense—isolation of species and "emptiness" of the surrounding matrix—do not seem to be met for even this subset of species occupying the remnants.

Testing Biogeographic Expectations Assuming an Island Analogy

We have just seen that for some species captured on remnants the island analogy is not a good one. And yet, quite often managers and researchers concerned with

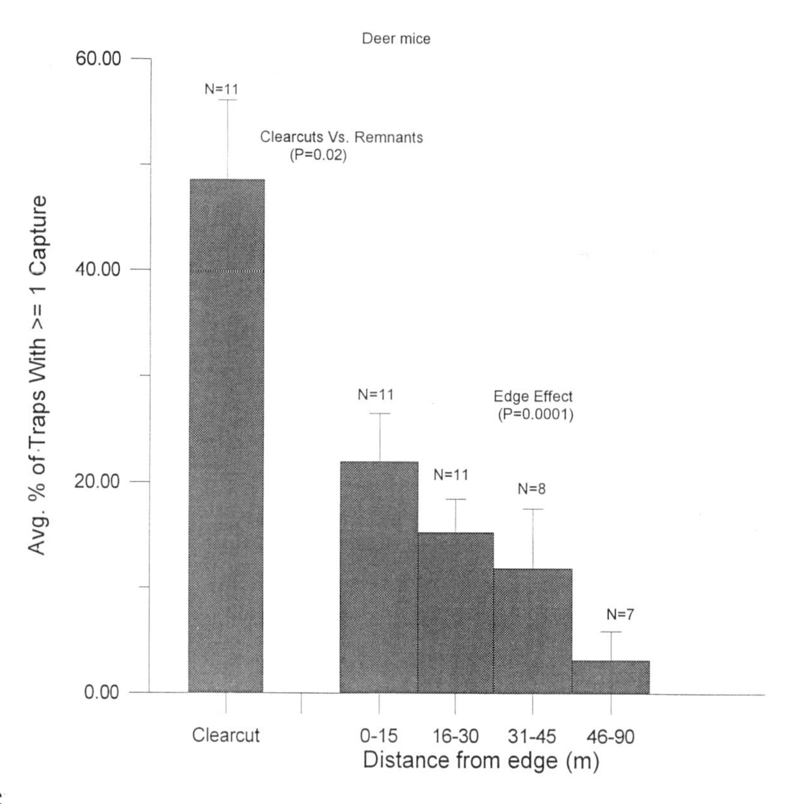

Figure 11.4 *(cont.)*

consequences of natural area fragmentation embrace the assumption that human-caused habitat remnants are analogues of islands (for review, see Doak & Mills, 1994). Indeed, during the course of my field work I was often asked questions that implicitly assumed an island analogy. For example, "How strong is your species-area curve?" And, "What is the order of extinction of species on the remnants?"

Therefore, it is useful to consider the outcome of applying a traditional island biogeographic analysis to the capture data. That is, what might we have concluded if we had not analyzed individual species responses, but instead taken a biogeographic approach that assumed that species on these remnants were, in fact, experiencing the same dynamics as "landbridge islands"—islands created by factors such as rising sea level (e.g., Diamond, 1972; Terborgh, 1974; Soulé et al., 1979; Patterson, 1987)?

Traditional island biogeography theory (Macarthur & Wilson, 1967) leads to several predictions involving the characteristics of remnants and composi-

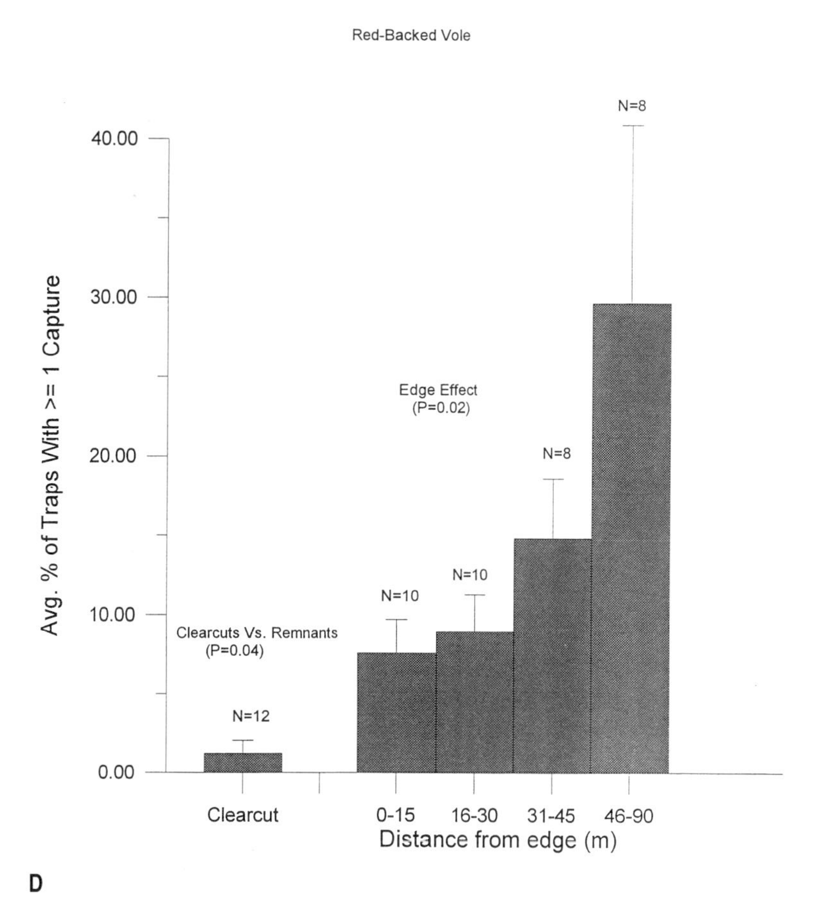

Figure 11.4 (*cont.*)

tion of mammal species. The remnant size, or area prediction, follows directly from the fundamental tenet of island biogeography; that is, an archipelago of islands should exhibit a positive relationship between number of species and island area (Diamond, 1975; Newmark, 1987; Soulé et al. 1988). The other predictions follow from specific characteristics of landbridge islands. In particular, there should be a negative correlation between the time of isolation (remnant age) and number of species in the remnant, because the isolated remnants should lose species at a greater rate than they are colonized (Diamond, 1972; Wilcox, 1978; Newmark, 1987); to the extent that colonization does affect species number, greater distances from colonizing sources should lead to lower species number (Brown, 1971, 1978; for more complex discussion, see Lomolino et al., 1989). Finally, because the species most vulnerable to extinction should persist only on the largest areas, the species present on smaller remnants

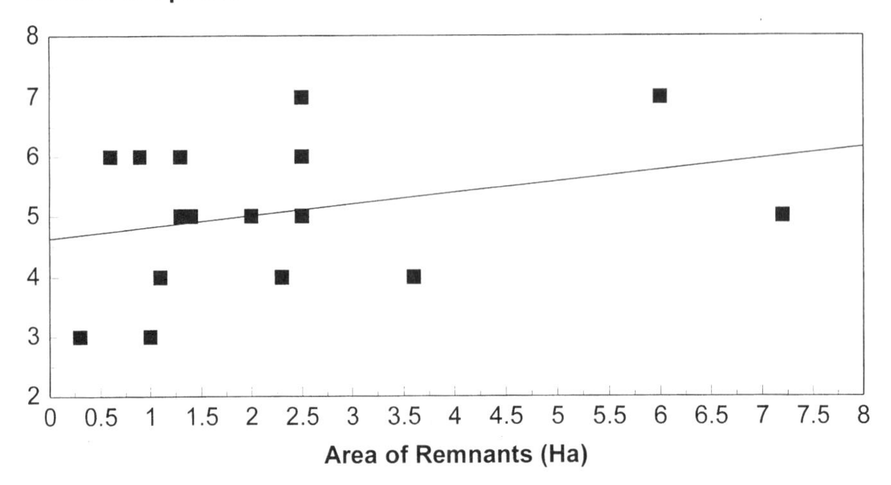

Figure 11.5 Regression of number of species on remnants versus area of remnants ($r^2 = 0.087$, $P = .27$). Because of potential concerns due to sampling effects caused by area (Bolger et al., 1991) or species detectability (Grayson & Livingston, 1993; Laurance, 1994), I also analyzed data based only on species captured commonly on controls (i.e., on at least four out of five control sites); the species-area curve only improves slightly ($r^2 = 0.155$, $P = .13$). The multiple regression (see text) for only the commonly caught species is $r^2 = 0.16$, $P = .52$.

are expected to be a nested subset of those occupying larger remnants (Patterson & Atmar, 1986; Patterson, 1987; Cutler, 1991).

In my study, the number of species on remnants did not change predictably with the size of the remnant (Fig. 11.5); the other simple biogeographic factors (isolation time and distance) were even poorer predictors of species numbers. In fact, even considering all three variables in a multiple regression accounts for only 10 percent of the variance in species numbers (logarithmic transformation does not do much to improve r^2).

The poor explanatory power of these factors can be explained by the fact that, as mentioned earlier, the island analogy is violated for many species, and remnants are colonized by matrix-dwelling species. As a result, changes in species composition obscure any faunal relaxation (extinction) that occurred for isolated species on small remnants (see Kirkland, 1990). Similarly, the simple measures of "isolation time and distance" are not appropriate given the violation of the island analogy; that is, the response to clearcutting varies not only across species but also across the time elapsed since the forest was clearcut, planted, and regenerated.

I used two techniques to determine whether a predictable order of extinction caused species occurring on species-poor remnants (islands) to be subsets

Table 11.2 Species distribution matrix for small mammals captured in Oregon on forest remnants.

	Species codes*										
Site	A	B	C	D	E	F	G	H	I	J	Species richness
JB	X	X	X	X	X	X	X				7
C	X	X	X	X	X		X	X			7
JT	X	X	X	X	X			X			6
PC	X	X	X	X	X				X		6
S	X	X	X	X		X				X	6
HC	X	X		X	X	X			X		6
YB	X	X	X	X		X					5
FB	X	X	X	X	X						5
W	X	X	X	X	X						5
E	X	X	X	X			X				5
K	X	X	X	X	X						5
O	X	X	X	X							4
A	X	X	X	X							4
HD	X	X	X		X						4
M	X	X		X							3
Z	X		X		X						3
Totals	16	15	14	14	10	4	3	2	2	1	

*A = *Peromyscus* spp.; B = *Sorex trowbridgii;* C = *Eutamias townsendii;* D = *Clethrionomys californicus;* E = *Neotoma fuscipes;* F = *Neurotrichus gibbsii;* G = *Microtus oregoni;* H = *Neotoma cinerea;* I = *Scapanus orarius;* J = *Sorex pacificus.*
Note: The species distribution is significantly nested ($P = .01$) based on the Monte Carlo simulation program developed by Cutler (1991; see also Patterson & Atmar, 1986). The P-value represents the probability that the observed deviation from nestedness could be drawn from the distribution of values generated by simulations.

of the species on biotically richer remnants (islands). Nested-subset analysis of the remnants indicated that the distribution of species is nested (Table 11.2), suggesting that there was a rank order to the extinction of species on remnants (Patterson, 1987; Cutler, 1991; McDonald & Brown, 1992). However, the second technique, cumulative species-area analysis (Fig. 11.6), contrasts with the nested-subset analysis because it implies that there is not a predictable order of extinction as remnants become smaller. In particular, cumulative species-area analysis indicates that a collection of small remnants contained more species than an equal cumulative area of large remnants across nearly the entire range of cumulative area. The implication that a collection of small remnants more often contained more species than a collection of large remnants of equivalent cumulative area is an unexpected result if species are nested in composition from large to small islands.

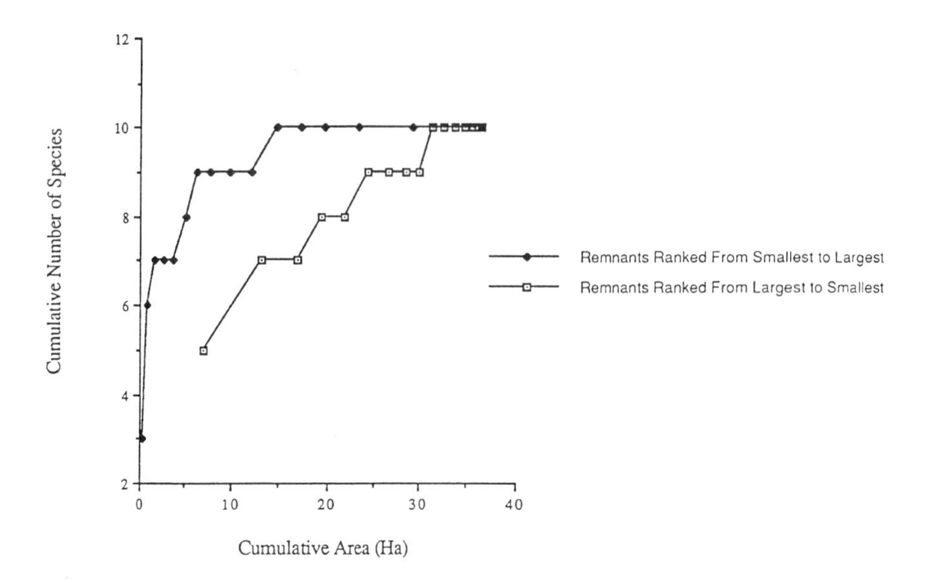

Figure 11.6 Cumulative species-area curves for small mammals captured on forest remnants in Oregon, 1990 and 1991. Remnants are ranked by size, and then two species-area curves are drawn; the first begins with the largest remnant and adds successively smaller ones, while the second begins with the smallest remnant and adds successively larger ones (Quinn & Harrison, 1988). The consistent dominance of small remnants over large remnants indicates that as remnants become smaller they do not lose the same species.

These conflicting results can be explained by the fact that nested-subset analysis, per se, is independent of the species-area relationship; it merely quantifies the order of extinction from remnants with progressively more to fewer species (Doak & Mills, 1994). The order of extinction for individual species relates to island size only if the number of species on a remnant (faunal richness) correlates closely with island size. In this study there is no such correlation (see Fig. 11.5), so that even if the order of extinction is relatively predictable (i.e., significantly nested), these losses are not necessarily associated with smaller remnants. In sum, a collection of remnants, or "islands," must exhibit near-perfect nestedness and correlation between area and species number before cumulative species area curves (see Quinn & Harrison, 1988) will support the premise that the order of extinction can be predicted as remnant size decreases.

Overall, the lack of correspondence with expectations of biogeographic theory, and the contradictory results of nested-subset and cumulative species-area curves are readily explained by recognizing that the studied remnants were not analogous to landbridge islands: (1) the size of suitable habitat for the suite of small mammals was not the same as the measured size of remnants; (2) isolation of all species did not begin in the year of clearcutting; and (3) species colonizing a remnant were likely different than the species present at the time of fragmentation. Indeed, the surrounding matrix appears to play a pivotal role in explaining deviations from standard biogeographic predictions (see Bierregaard et al., 1992, for similar conclusions related to small mammals, butterflies, and primates).

Conclusions and Management Recommendations

Two broad lessons emerge from this work. The first is that human-caused habitat fragmentation and isolation is a real and unquestionably negative phenomenon faced by mammals such as the California red-backed vole (see Newmark [1987] for consequences of negative isolation-effects on a range of mammals). For the vole species studied, negative edge effects translated into a functional reduction in remnant size (Mills, 1995). Furthermore, if the detailed genetic assay I am currently conducting supports my preliminary findings of decreased genetic variation for voles on remnants (Mills, 1993), it is possible that remnant populations are more susceptible to extinction due to inbreeding depression (Mills & Smouse, 1994).

However, the second important lesson from this study is that different species respond differently to forest fragmentation. Whether an island analogy is appropriate depends entirely on how the natural habitat has been modified and the response of individual species to that modification (see McCoy, 1982; Zimmerman & Bierregaard, 1986; Janzen, 1983; Margules et al., 1982). In the case of species on Pacific Northwest forested areas surrounded by clearcuts, an island model is appropriate for some, but not all, species.

These lessons are consistent with important recent work on mammals and forest structure by Laurance in the tropical forests of Queensland, Australia (Laurance, 1990, 1991, 1994, 1995). Laurance found a variety of responses to forest fragmentation, including vegetation changes and a range of isolation and edge effects among the small mammals. Competitive interactions among small mammals were more intensively structured in remnants than in control sites, and changes in species composition led to shifts in predation pressures. The ability of a species to tolerate the habitat matrix surrounding the remnants was identified as the most important ecological predictor of vulnerability (see Bierregaard et al., 1992; Franklin, 1993). Species that used the matrix remained stable or increased in remnants, probably because they were better able to move between forest remnants and were preadapted to subsequent changes in remnants. These results led Laurance to a conclusion relevant to this book: "Land managers must not restrict their attention to parks and corridors, but must also work to conserve and rehabilitate key elements in the surrounding landscape" (1995).

Undoubtedly, then, as natural areas break up, decision-making and prioritization of species protection must include evaluation of species dynamics in unfragmented (control) areas and the modified matrix, as well as the remnants themselves. Species commonly caught in controls and rarely in the matrix are most likely to be vulnerable to isolation-driven processes. Note that low numbers in remnants, without control or matrix data, do not help to identify vulnerability because these animals may be absent from controls and simply colonizing from the modified matrix to the remnants (see Bolger et al., 1991). Similarly, abundance in controls (e.g., "natural rarity") without data from the modified landscape can be misleading because it does not account for the likelihood of being isolated or extinction-prone once isolated (Laurance, 1995).

Certainly, a proactive approach mandates that we not limit our focus in modified landscapes to only the species that are most isolated or most prone to extinction following disturbances. Rather, as McIntyre and Barrett (1992) suggest, there also is benefit in emphasizing species that are vulnerable, but that show tolerance to the habitat modification.

There is no doubt that large-scale, intensive forest fragmentation can have deleterious effects on wildlife species native to the forests. Our detection of these effects can be blurred if we apply simplistic island models that treat all species as responding in the same way. By evaluating which species are most likely to be isolated and which are likely to suffer under future fragmentation, we can focus efforts on maximizing connectivity and persistence for those species. Such an approach would best direct scarce resources to preserve biodiversity in the wake of fractured natural areas.

Acknowledgments

This work absolutely would not have been possible without the dedicated enthusiasm of "Team Vole": David Clarkson, Richard Nauman, Chris O'Connor, Tara Reinertson, and David Tallmon; funding was provided by National Audubon Society and the U.S. Forest Service (Pacific Northwest Research Station). This research was completed as part of a Ph.D. dissertation at University of California, Santa Cruz, under the advisement and good humor of Michael Soulé. Preparation of this book chapter began as Visiting Assistant Professor in the Fisheries and Wildlife Department, University of Idaho. David Mattson, David Tallman, and Bill Laurance provided insightful editorial comments.

Literature Cited

Bierregaard RO, et al. The biological dynamics of tropical rainforest fragments. *BioScience* 1992;42:859–866.

Bolger DT, Alberts AC, Soulé ME. Occurrence patterns of bird species in habitat fragments: Sampling, extinction, and nested species subsets. *Am Nat* 1991;137:155–156.
Brown JH. Mammals on mountaintops: Nonequilibrium insular biogeography. *Am Nat* 1971;105:467–478.
Brown JH. The theory of island biogeography and the distribution of boreal birds and mammals. *Great Basin Naturalist Memoirs,* 1978;2:209–227.
Chen J, Franklin JF, Spies TA. Vegetation responses to edge environments in old-growth Douglas-fir forests. *Ecol Appl* 1992;2:387–396.
Clarkson DA, Mills LS. Hypogeous sporocorps in forest remnants and clearcuts in Southwest Oregon. *Northwest Sci* 1994;68:259–265.
Cutler A. Nested faunas and extinction in fragmented habitats. *Cons Biol* 1991;5:496–505.
Devries PG. Multi-stage line intersect sampling. *For Sci* 1974;20:129–133.
Diamond JM. Biogeographic kinetics: Estimation of relaxation times for avifaunas of Southwest Pacific Islands. *Proc Nat Acad Sci* 1972;69:3199–3203.
Diamond JM. The island dilemma: Lessons of modern biogeographic studies for the design of natural reserves. *Biol Cons* 1975;7:129–146.
Doak DF, Mills LS. A useful role for theory in conservation. *Ecol* 1994;75:615–626.
Doyle AT. Microhabitat separation among sympatric microtines, *Clethrionomys californicus, Microtus oregoni* and *M. richardsoni. Am Midl Natur* 1987;118:258–265.
Fogel R. Ecological studies of hypogeous fungi. II. Sporocarp phenology in a western Oregon Douglas-fir stand. *Can J Bot* 1976;54:1152–1162.
Fogel R. Quantification of Sporocarps Produced by Hypogeous Fungi. In DT Wicklow, CG Carroll (eds), *The Fungal Community, Its Organization and Role in the Ecosystem.* New York: Marcel Dekker, 1981:553–568.
Franklin JF. Preserving biodiversity: Species, ecosystems, or landscapes? *Ecol Appl* 1993;3:202–205.
Grant PR. Experimental studies of competitive interaction in a two-species system. II. The behaviour of *Microtus, Peromyscus,* and *Clethrionomys* species. *Ani Behav* 1970;18: 411–426.
Grayson DK, Livingston SD. Missing mammals on Great Basin mountains: Holocene extinctions and inadequate knowledge. *Cons Biol* 1993;7:527–532.
Hayes JP, Cross SP. Characteristics of logs used by western red-backed voles, *Clethrionomys californicus,* and deer mice, *Peromyscus maniculatus. Can Field Nat* 1987;101:543–546.
Janzen DH. No park is an island: Increase in interference from outside as park size decreases. *Oikos* 1983;41:402–410.
Kapos V. Effects of isolation on the water status of forest patches in the Brazilian Amazon. *J Trop Ecol* 1989;5:173–185.
Kirkland GL. Patterns of initial small mammal community change after clearcutting of temperate North American forests. *Oikos* 1990;59:313–320.
Laurance WF. Comparative responses of five arboreal marsupials to tropical forest fragmentation. *J Mamm* 1990;71:641–653.
Laurance WF. Edge effects in tropical forest fragments: Application of a model for the design of nature reserves. *Biol Cons* 1991;57:205–219.
Laurance WF. Extinction and Survival of Rainforest Mammals in a Fragmented Tropical Landscape. In WZ Lidicker (ed), *Landscape Approaches in Mammalian Ecology and Conservation.* Minneapolis: University of Minnesota Press, 1995:46–63.

Laurance WF, Yensen E. Predicting the impacts of edge effects in fragmented habitats. *Biol Cons* 1991;55:77–92.
Lomolino MV, Brown JH, Davis R. Island biogeography of montane forest mammals in the American Southwest. *Ecol* 1989;70:180–194.
Macarthur RH, Wilson EO. The theory of island biogeography. Princeton, NJ: Princeton University Press, 1967.
Margules C, Higgs AJ, Rafe RW. Modern biogeographic theory: Are there any lessons for nature reserve design? *Biol Cons* 1982;24:115–128.
Maser C, Trappe JM, Nussbaum RA. Fungal small mammal interrelationships with emphasis on Oregon coniferous forests. *Ecol* 1978;59:799–809.
Matlack GR. Microenvironment variation within and among forest edge sites in the eastern United States. *Biol Cons* 1993;66:185–194.
McCoy ED. The application of island-biogeographic theory to forest tracts: Problems in the determination of turnover rates. *Biol Cons* 1982;22:217–227.
McDonald KA, Brown JH. Using montane mammals to model extinctions due to global climate change. *Cons Biol* 1992;6:409–415.
McIntyre S, Barrett GW. Habitat variegation, an alternative to fragmentation. *Cons Biol* 1992;6:146–147.
Mills LS. Extinction in habitat remnants: Proximate mechanisms and biogeographic consequences. Ph.D. Dissertation, Santa Cruz, CA: University of California, 1993.
Mills LS. Edge effects and isolation: Red-backed voles in forest remnants. *Cons Biol* 1995;9:395–403.
Mills LS, Smouse PE. Demographic consequences of inbreeding in remnant populations. *Am Nat* 1994;144:412–431.
Newmark WD. A land-bridge island perspective on mammalian extinctions in western North American parks. *Nature* 1987;325:430–432.
Patterson BD. The principle of nested subsets and its implications for biological conservation. *Cons Biol* 1987;1:323–333.
Patterson BD, Atmar W. Nested subsets and the structure of insular mammalian faunas and archipelagos. *Biol J Linn Soc* 1986;28:65–82.
Quinn JF, Harrison SP. Effects of habitat fragmentation and isolation on species richness: Evidence from biogeographic patterns. *Oecologia* 1988;75:132–140.
Ranney JW, Bruner MC, Levenson JB. The Importance of Edge in the Structure and Dynamics of Forest Islands. In Burgess RL, Sharpe DM (eds), *Forest Island Dynamics in Man-Dominated Landscapes.* New York: Springer-Verlag, 1981:67–95.
Reese KP, Ratti JT. Edge effect: A concept under scrutiny. *Trans N Am Wildl Nat Res Conf* 1988;53:127–136.
Rosenberg DK, Swindle KA, Anthony RG. Habitat associations of California red-backed voles in young and old-growth forests in Western Oregon. *Northwest Sci* 1994;68:266–272.
Saunders DA, Hobbs RJ, Margules CR. Biological consequences of ecosystem fragmentation: A review. *Cons Biol* 1991;5:18–32.
Soulé ME, Wilcox BA, Holtby C. Benign neglect: A model of faunal collapse in the game reserves of East Africa. *Biol Cons* 1979;15:259–272.
Soulé ME, et al. Reconstructed dynamics of rapid extinctions of chaparral-requiring birds in urban habitat islands. *Cons Biol* 1988;2:75–92.

Tallmon DA, Mills LS. Use of logs within homeranges of California red-backed voles on a remnant forest. *J Mamm* 1994;75:97–101.
Terborgh J. Preservation of natural diversity: The problem of extinction prone species. *BioScience* 1974;24:715–722.
Ure DC, Maser C. Mycophagy of red-backed voles in Oregon and Washington. *Can J Zool* 1982;60:3307–3315.
Wilcox BA. Supersaturated island faunas: A species-age relationship for lizards on post-Pleistocene land-bridge islands. *Science* 1978;199:996–998.
Williams-Linera G. Vegetation structure and environmental conditions of forest edges in Panama. *J Ecol* 1990;78:356–373.
Zimmerman BL, Bierregaard RO. Relevance of the equilibrium theory of island biogeography and species-area relations to conservation with a case from Amazonia. *J Biogeo* 1986;13:133–143.

Management Opportunities and Conflicts in Parks

Ecological processes ultimately govern the character of the landscapes occurring in protected areas, as well as the composition of their flora and fauna. The degree to which natural processes (e.g., fire, flood, disease, erosion, succession, and population growth and decline) are allowed to operate unhindered by anthropogenic controls has lacked definition since the first parks were established. Indeed, even after 125 years, there are still no clear guidelines as to when, where, and the extent to which natural processes can operate unimpeded in protected areas. Questions are numerous, solutions few. The next five chapters focus on a few of the more contentious issues relative to natural process management. I have been selective in picking these issues and probably biased—most concern the management of animal species.

Chapter 12 begins the discussion by examining the problems caused by overabundant animals in parks. The focus is on the dramatic increases of white-tailed deer populations in historic parks in the eastern United States—areas that usually are small and lack natural controls to regulate populations. Chapter 13 takes a different look at a related issue—how to manage large ungulates in large parks that still contain functional ecological processes.

One of the most important and controversial natural processes in parks and protected areas is fire. Fire has shaped the composition and structure of nearly every terrestrial vegetation type on earth. In many "natural areas," fire is now excluded or tightly controlled. To what extent has this control been detrimental to natural systems, and what should be the role and use of fire in natural areas? These and other issues are explored in Chapter 14.

The restoration of missing or degraded ecological processes and the reestablishment of previously extirpated species are issues of acknowledged importance and controversy in protected area management. Stream and aquatic systems are perhaps the most altered of the natural systems, particularly in the western United States. Dams, channelization, diversion, and drainage have taken a tremendous toll on aquatic systems. Protected areas have not been immune from these changes. Now, possibly for the first time at a meaningful scale, a national park is seriously contemplating the removal of a dam. Chapter 15 examines the ecological and economic considerations involved in this effort.

Finally, we turn to what is ostensibly the first step in the successful reestablishment of an ecological process missing or compromised in most parks in the continental United States—predation. Chapter 16 examines the role of the wolf as the preeminent ungulate predator and describes the efforts to establish wolves in Yellowstone National Park and the Idaho wilderness.

12

Management of Overabundant Species in Protected Areas: The White-Tailed Deer as a Case Example

William F. Porter

The dilemmas associated with the management of overabundant wildlife populations are in stark contrast to the management challenges posed by endangered species. National parks and related areas preserve valuable remnants of natural ecological systems. When native species in such areas reach unusually high levels of abundance, managers are faced with a paradox: Not controlling these populations may result in dramatic changes in the natural ecosystem, whereas initiating management action means intervening in natural processes that parks seek to preserve. The dilemma is compounded by political volatility when the issues involve charismatic vertebrates.

Serious debates over wildlife management in national parks have been ongoing since the 1950s. Precipitated by eruptions of elk populations (*Cervus elaphus*), debates in the western parks generated a philosophical shift in National Park Service (NPS) policy from active management to "hands off" (Leopold et al., 1963; Despain, 1986). In the East, white-tailed deer (*Odocoileus virginianus*) provoked similar debates in dozens of national parks during the 1980s and 1990s. While much has been written about the management of overabundant species in western parks, little is available describing deer problems in the East. Thus, this chapter focuses on white-tailed deer as a case example of the challenges posed by overabundant species in protected areas.

White-tailed deer, which were uncommon in most national parks in the 1950s, erupted in population during the 1980s. Populations numbering as many as 60 deer per square kilometer occurred in some parks (Storm et al., 1989; Underwood et al., 1995). Such densities were extraordinary because they were beyond the experience of most biologists and the general public. High densities had been reported in the past, but most of these cases were isolated and associated with installations where public access was limited (Hesselton et al., 1965; Woolf & Harder, 1979). What made the eruptions of the 1980s extraordinary was the pervasive nature of high densities. By the early 1990s, nearly 50 national parks, historical sites, and recreation areas were reporting problems with deer. Many biologists and managers perceived these populations to be *overabundant* based on a value system whose origins were tied to experiences in the "north woods," where densities rarely achieved 10 deer per square kilometer and winter kill was common. Historical paradigms such as the mule deer (*Odocoileus hemionus*) on Kaibab Plateau caused biologists to be concerned that white-tailed deer populations were no longer constrained by natural forces and were destined to crash (Caughley, 1970; Flader, 1974).

Why did deer populations erupt in so many eastern parks? Have the natural processes been so altered that deer populations grow unconstrained and require intervention by park managers? If intervention is warranted, how should it be done? Why did the NPS hesitate to undertake reductions of deer populations? How do we proceed in view of the enormous ecological complexity and the political volatility of this issue? These questions are at the heart of the dilemma posed by overabundant deer in parks. The intent of this chapter is to address these questions.

It appears that the interplay of science and politics is responsible for the conservative response of the NPS. Thus, the chapter begins with an overview of the basic ecology of the species, examining habitat requirements and key aspects of behavior and population biology. Basic ecology is used to examine the reasons for the eruption of deer populations during the 1980s and the management alternatives available to the NPS. In conclusion, the actions taken by the NPS to deal with deer are explored, and the roles of science and politics in management decisions evaluated.

Autecology of Deer

Habitat Requirements

The foundation of a deer population is habitat. White-tailed deer consistently reach the highest densities in fragmented environments, areas composed of forest and field (Halls, 1984). The best habitats are those which provide as a food resource an abundance of green, rapidly growing plants with a low stature, and larger shrubs and trees for essential cover. Important food items include leaves and twigs of woody vegetation and fruits and nuts, but deer prefer to feed in areas with grass or herbaceous vegetation because the quality and quantity of food is higher. Agricultural fields constitute the richest food resource for deer (Harlow, 1984; Short, 1986). Deer find protection from weather, predators, and humans in areas of woody vegetation, ranging from shrublands to mature forests.

Deer are tolerant of a broad range of environmental conditions. They can survive in areas that are 100 percent forested and in areas with less than 10 percent forest (Halls, 1984). Deer also show high tolerance to human development. For example, northern suburbs of Minneapolis–Saint Paul and the parks of the Chicago metropolitan area contain deer populations exceeding 40 deer per square kilometer (i.e., 100 deer/mi^2) (Sillings, 1987; Witham & Jones, 1987). Low-to-medium density housing development generally creates suitable habitat and often eliminates major predators and precludes hunting. Optimal conditions appear to be about one house per 60 hectares. Deer densities decrease as housing density increases because vegetation tends toward individual trees and linear arrangements, and numbers of dogs and automobiles increase (Vogel, 1989).

Movement Behavior

Movement behavior for deer in eastern forest and/or field ecosystems is characterized by a high degree of fidelity to a specific geographic area (Thomas et al., 1964; Mathews, 1989). The area traversed by an individual, or *home range,* varies by locale and season. Depending on the juxtaposition of food and cover, summer (i.e., growing season) home ranges include 50 to 500 hectares. Home ranges tend to be larger in relatively open environments and smaller in forested areas. Agricultural and grassland environments may require greater movement because food and cover resources are not always in close proximity (Marchinton & Hirth, 1984).

In many regions, the food and cover conditions preferred by deer in summer and winter are not immediately juxtaposed. Although food is most important during summer, cover is critical during winter. Where adequate summer food and winter cover are separated, deer migrate between the areas. Distances of 10 to 20 kilometers between summer and winter ranges are common; 50 kilometers appears to be the extreme (Marchinton & Hirth, 1984). During summer, energy requirements associated with raising young are high, and deer move to sites where they give birth to young and are able to feed extensively on rich food resources (Verme & Ullrey, 1984). In early winter, deer shift their movements to areas that allow them to optimize the thermal conditions of their environment. In northern latitudes, deer use the thick overhead canopy of conifer stands as a blanket during the coldest periods of winter. Deer also congregate on south- and west-facing slopes because the increased solar radiation melts the snow, reducing its depth and providing a more favorable thermal condition for the deer (Moen, 1973). Food supplies are less important during the winter because the combination of fat reserves accumulated during late summer and fall and a much lower metabolic rate enables deer to survive with relatively little energy intake (Verme & Ullrey, 1984).

Philopatry or love of homeland is an important quality of movement behavior in deer. Studies have shown that most females are philopatric, remaining in the area where they were born for life, whereas most males disperse (Hawkins et al., 1971; Kammermeyer & Marchinton, 1976). In many areas, 90 percent of females establish their summer home range adjacent to and overlapping that of their mother. When the first female offspring is born, it usually establishes a home range overlapping that of its mother; consequently, the home ranges of subsequent generations of offspring overlap those of their mothers and relatives (Mathews, 1989). Those females that disperse from their natal areas establish a new home range.

There appears to be an upper limit to both the total area occupied by a family group and the number of individuals in the group. In the Adirondack Mountains of northern New York, once the total area occupied by a family group reaches about 4 square kilometers, the cumulative size of the home range stops increasing. Adding new members to the group requires that more individuals be packed into the same area (Porter et al., 1991). As density increases, the per-capita resources available in the group's home range necessarily must decrease. Eventually, the resources per deer become insufficient to support recruitment of any additional members to the group.

Population Biology

Abundance of deer is determined primarily by three factors: reproduction, survival, and carrying capacity. Two measures are commonly used to assess reproductive performance in deer. *Natality* is the average number of fawns born per year. Although white-tailed deer can produce up to three fawns per year, most females 2 years and older produce two fawns each year. *Recruitment* is the number of fawns that survive to sexual maturity; 40 to 60 percent recruitment of young born each year is common.

Age of first reproduction has a greater effect on population growth than litter size (Cole, 1954), and variation in age of first estrus appears to be related to food resources. The estrous cycle in deer is regulated by accumulation of fat reserves and day length (Verme & Ullrey, 1984). Northern forests provide lower quality food resources and a shorter growing season in comparison to southern forests and agricultural environments (Harder, 1980). As a result, deer accumulate fat reserves more slowly, and age of first estrus is generally delayed (Underwood, 1990). A similar delay occurs in areas where the deer are malnourished as a result of intense competition for food resources (McCullough, 1979).

Reproductive performance is often indexed by measuring antler development in yearling males (Severinghaus & Moen, 1983). Males face the same challenges from growth as females, and antler development is directly related to nutrition. Measurement of the antler-beam diameter 2.5 centimeters above the pedicel on 18-month-old males is a good index to nutritional quality of the environment and, consequently, to reproductive performance of females in the same population. This is a relatively easy statistic to obtain from deer harvested during fall hunting seasons and is frequently used to predict reproduction in a local or regional population (Taber 1958; Severinghaus & Moen, 1983).

Although age of first reproduction is the best indicator of a population's potential to grow, realized growth is determined by recruitment. Fawns can constitute 50 percent of the deer population in early summer and, if their survival is high, total abundance in the populations can increase rapidly.

Survival of adults in most areas is largely dependent on vulnerability to sport hunting and automobiles. Where sport hunting occurs, it is the dominant mortality factor for adult deer. Many states regulate the harvest to achieve an annual harvest of 30 to 40 percent of the females and up to 60 percent of the males in the fall population. Under these conditions, deer generally do not live beyond 2.5 years. Where sport hunting is not a factor, females often live to 12 years and can live to 16 years (Masters & Mathews, 1990). Automobiles are also an important cause of mortality, killing 30 to 50 percent as many deer each year as hunters (Cypher et al., 1985; Storm et al., 1989; Decker et al., 1990). Overwinter mortality is often thought to be a key determinant in deer population dynamics, but it actually is an infrequent factor in all except the most northern regions of the United States and southern Canada.

Carrying capacity is the third major component of population abundance and growth. In the absence of significant mortality among adults, as is the case in most eastern parks, populations grow until competition for food resources becomes intense. There are three demographic consequences of intense competition in deer: (1) delayed age of first reproduction, (2) declining rates of recruitment, and (3) increasing mortality among adults (McCullough, 1979). Carrying capacity is commonly conceived in two ways: *economic carrying capacity* and *ecological carrying capacity.* The conceptions differ in their interpretations of the consequences of competition. Economic carrying capacity corresponds to the density at which (1) the three demographic consequences of competition first become manifest and (2) the rate of population growth begins to diminish. Ecological carrying capacity is reached when competition causes the survival of fawns to equal the mortality of adults and the population ceases to grow (Caughley, 1979; Macnab, 1985).

A relatively simple mathematical formulation, the logistic model, captures this pattern of growth:

$$dN/dt = rN(1 - N/K)$$

> where dN/dt is an expression for the *change in numbers of deer per change in time,* or the number of deer added to the population every year after subtracting the losses; r represents the *intrinsic growth rate,* or the potential number of offspring a deer can produce in a year; N is the *number of deer in the population;* K is the *maximum number* of deer the environment will support; and $1 - N/K$ causes the growth pattern to curve in an S-shaped, symmetrical fashion when population abundance versus time is plotted.

There are three important simplifying assumptions often associated with the logistic curve that constrain its ability to represent growth in deer populations. First, deer populations do not grow in a perfectly symmetrical fashion as suggested by the logistic curve. The right side of the recruitment parabola has not been well defined by empirical studies but appears to be a skew in the curve such that the maximum recruitment is closer to the upper limit of abundance. Nevertheless, the symmetrical representation is generally close enough for predictive purposes (McCullough, 1979; Downing & Guynn, 1985) (Fig. 12.1).

A more serious problem is the assumption that K is constant. If we think of K as the maximum number of deer that can be supported by the quantity and nutritional quality of the vegetation on the landscape, we realize that K fluctuates. For instance, in years of drought or unusually long winters, K is lower than average. The more widely and unpredictably K fluctuates, the more difficult it is to predict population growth (Fig. 12.2).

The population is always responding to K, increasing or decreasing in a centripetal fashion (Caughley et al., 1987). The logistic model assumes deer respond instantaneously to changes in K; however, in reality, deer populations do not. Following a hurricane or a fire in a forested environment, regeneration results in a more than 100-fold increase in biomass of seedlings and saplings. Deer populations are able to grow at a maximum rate of less than two times per year.

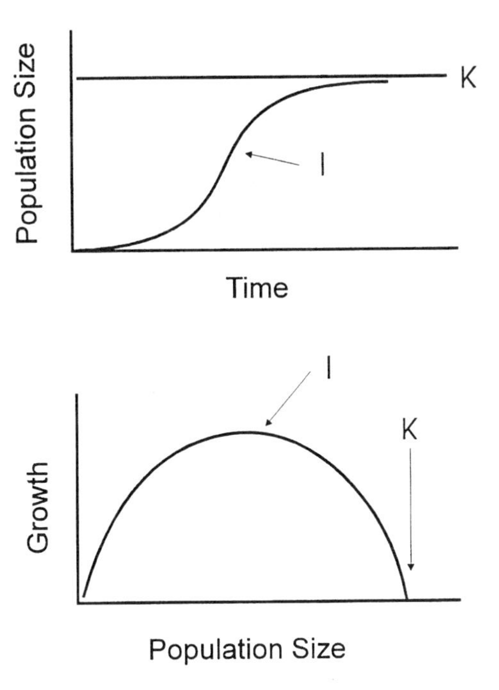

Figure 12.1 Standard logistic curve (A) and recruitment curve **(B)** provide approximate representations of growth in white-tailed deer populations.

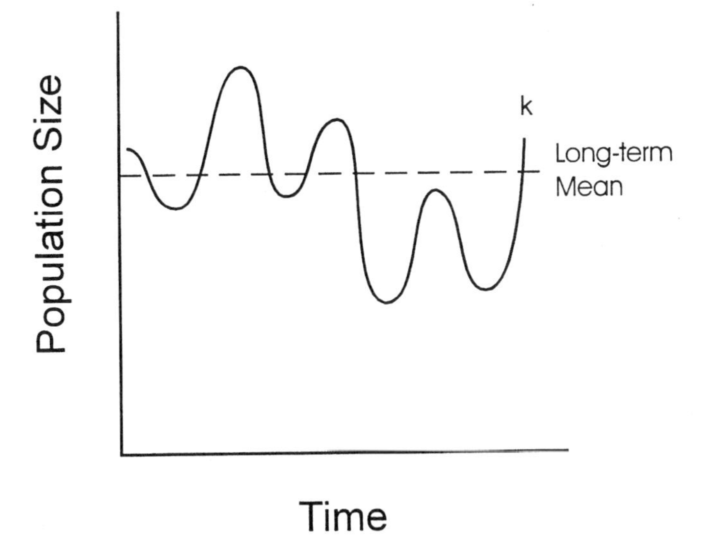

Figure 12.2 Ecological carrying capacity is not constant but fluctuates around a long-term mean.

Thus, while nutritional conditions would support a larger population, the numerical response of the population will require years to reach this level.

Time lags are also the cause for populations "overshooting" *K*. Recruitment is much more sensitive to changes in *K* than is survival of adults. For instance, *K* declines rapidly when forest regeneration reaches the tall sapling stage and most food resources are out of reach of deer. Recruitment ceases, but adults are adapted to surviving intense competition by supplementing food intake with use of internal fat and muscle tissue. It may take months or years for population abundance to drop to *K*.

Why Deer Erupted in Parks

The background in population biology helps us understand why deer populations erupted in parks during the 1980s. First, fragmentation of the eastern forests with agriculture created a combination of grass fields and crop areas interspersed with deciduous and coniferous forest, which provided ideal habitat conditions for deer. While the fragmented environment has been common to the East for decades, until recently deer were not able to take advantage of these conditions. Deer populations were so low throughout the 1940s and 1950s that even a small amount of illegal harvest of females was sufficient to keep the population low on the growth curve (see Fig. 12.3). Once enforcement of game laws improved in the 1950s and conservation programs restored previously extirpated deer populations, deer numbers began to increase.

Through the 1960s and 1970s, harvest regulations were generally conservative. Hunting traditions from the 1940s and 1950s emphasized removal of only males (Flader, 1974). Because deer are polygamous, removal of males had little effect on the growth potential of the population. When harvest of females was allowed, the number of females removed by hunters was tightly controlled by a license-permit system. The harvest of females kept deer populations in check so long as there was sufficient hunter pressure to remove the quota of females each year.

As suburban development began to encroach into rural areas, habitat remained good for deer, but hunting was restricted because of concerns for human safety. As hunting diminished, deer populations grew. Deer were rare in most eastern suburbs during the 1950s but were seen with greater frequency during the 1960s. Populations grew slowly through the 1960s and early 1970s and more rapidly in the late 1970s. In the 1980s, suburban greenbelts and parks began to experience deer populations exceeding economic carrying capacity and approaching ecological carrying capacity (Fig. 12.3). The apparent syn-

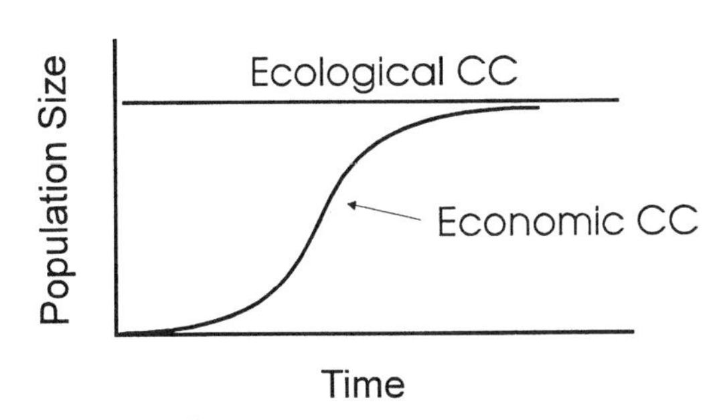

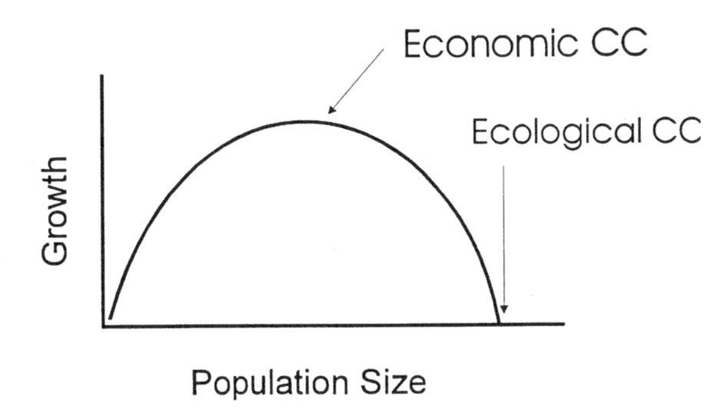

Figure 12.3 Two common conceptions of carrying capacity (CC) are ecological and economic CC. Note that economic CC is at the inflection point on the sigmoid curve and is the maximum value for the parabola describing recruitment.

chrony in the timing of this growth across broad regions of the East was probably the result of the widespread, rapid movement of humans from cities to suburbs in the 1950s and 1960s.

The pattern of growth of deer populations in national parks throughout the East was pronounced for the same reasons. Most eastern parks provided extensive areas of grass, and, in some cases, agricultural crops, interspersed with wooded areas. Many parks were established during the 1950s and 1960s, and at the same time enforcement of no-hunting regulations began in earnest. Anecdotal reports suggest growth followed the sigmoid curve. By the mid-1980s, deer on many parks reached density levels well-above economic carrying capacity.

Managing Deer on National Parks

In the face of burgeoning deer populations, there was considerable discussion in the NPS about direct reduction of deer populations, but no decisive action was taken. The lack of a firm NPS decision about deer management in most parks can be traced to the fact that the agency had not organized the necessary political and scientific support. Having experienced the controversy and media attention surrounding elk and bison management in the West, the NPS was likely loath to spark another controversy in the East. Deer management had the potential to erupt into an even greater controversy because many people had direct contact with deer, and the issues would literally have been debated in the backyard of major media markets. Thus, until the NPS was well prepared, it did not venture into active management.

Political Anchors and Legislative Framework

Political consensus is important to wildlife management in parks, but the issue of deer management was caught in a broader political battle over conservation ethics. Political consensus is an anchor-point because all management on national parks is done under the magnifying lens of public scrutiny and must reflect society's interests and aspirations (Keiter & Boyce, 1991; Wright, 1992). However, consensus was not possible on some fundamental issues. What qualities of natural ecosystems should we seek to preserve? Which species do we value most? What rights to existence do we ascribe to individuals of these species? Answers to these questions depend on human values (Caughley & Sinclair, 1994) and were the subject of broad societal debate during the 1980s and 1990s.

The only solid political foundation for deer management in national parks was in existing law. The legislative framework or park management policy includes the Organic Act of 1916, NPS *Management Policies* (NPS, 1988), *National Resources Management Guidelines* (NPS, 1991), and the legislation by which Congress establishes each park. In addition, the NPS must adhere to legislation such as the Endangered Species Act and the National Environmental Policy Act. NPS *Management Policies* and *National Resources Management Guidelines* (NPS, 1988) are built on this foundation of law.

Scientific Anchors and Litmus Tests

A key element of NPS management is the goal of protecting native animal populations against harvest, removal, destruction, or harm through human action. This policy provides some discretion, however, in that individual animals may be removed under two principal criteria: (1) to control native species when unnatural concentrations are caused by human activities, and (2) to prevent the loss of another species, preserve the integrity of natural and cultural resources, and protect human safety (NPS, 1988); the policy stipulates that there be scientific evidence to show that these criteria are met.

The scientific foundation necessary to support deer management decisions in parks was lacking. Consequently, the agency commissioned more than two dozen studies during the 1970s and 1980s. It then sought to design litmus tests to help managers discern when deer populations had reached unnatural densities and when these densities threatened cultural or natural resources (Porter et al., 1994).

Litmus tests proved more straightforward for assessing the impact of deer on cultural resources than on natural resources. For instance, at Gettysburg, Pennsylvania, during the Civil War battles on early July 1863, agricultural crops were an integral part of the landscape. An extensive photographic history provides a definitive basis for recreating the scene. However, in the 1980s, crops became increasingly difficult to grow, and exclosure studies showed that the cause was the high densities of deer in the park (Wright, 1990). Indeed, without deer-proof fences around the agricultural fields, many crops could not be grown at all. In this instance, science had provided a clear litmus test in the form of the exclosures. As a result, the NPS prepared an environmental impact statement with the intent of initiating direct manipulation of the deer population (Storm et al., 1989; U.S. Department of Interior, 1994).

Similar studies using exclosures at Saratoga, New York, showed that deer were having a significant impact on the native vegetation. Here again, the vegetation was considered important in portraying the scene of the Revolutionary War battles. In contrast to Gettysburg, information about the character of the landscape in Saratoga in 1777, the year when the battles were fought, was less precise. Thus, while science could show deer were having an impact, the importance of this impact on cultural resources was unclear and no action was taken.

The effect on natural resources was even less clear. Parks contain some of the best remnants of natural ecosystems in the East. As deer populations erupted, there was concern over the impact of deer on these natural ecosystems, and debate arose over the integrity of natural processes and the role of natural regulation in limiting abundance of deer. A paradigm evolved: Humans had so altered natural processes by removing predators and isolating deer in parks that processes regulating deer populations had been disrupted. Consequently, the ultimate fate of these populations involved destruction of the vegetation and death of large numbers due to malnutrition and disease. Public pressure began to mount for a "hands-on" approach to managing deer.

The most frequently cited lines of evidence proving a breakdown in natural regulatory processes were (1) the dramatic fluctuation in deer populations, and (2) the change in native vegetation associated with high deer populations. The premise to this inference was that deer had coevolved with their predators and the vegetation of eastern forest ecosystems over millennia and had reached a general equilibrium or state of constancy. If natural processes, and not anthropogenic forces, were still regulating population growth in deer, these forces should have constrained abundance at relatively constant levels. Further, these levels should have permitted native plant species to exist (Cole, 1971; Warren, 1991).

This logic was challenged for two reasons. First, the reality of ecosystems in eastern North America is fluctuation, not constancy. Hurricanes, fires, and severe winters frequently perturb the ecosystem. Empirical and modeling studies suggest that because of frequent perturbations and time lags in the responses of various species, plant-herbivore-predator systems are probably at equilibrium only rarely. The more likely pattern was centripetality (Caughley et al., 1987; Pimm, 1991).

Second, the argument implies that restoration of the predators and removal of barriers to dispersal would result in reestablishment of the historical limits to population size and a return of native vegetation. The argument that predators would damp fluctuations of deer populations to an equilibrium is doubtful but debatable. Empirical studies of predator-herbivore-plant interactions involving larger vertebrates provide a mixed assessment of this hypothesis (McLaren & Petersen, 1994). However, even the longest studies are just now beginning to provide the perspective of time that is essential to evaluating constancy. Modeling based on a wide array of taxa suggests that adding another trophic level to the ecosystem by restoring predators is likely to exacerbate fluctuations (Pimm, 1991). The argument assumes that dispersal is crucial to natural regulation and that because parks are isolated by suburban development, the potential for dispersal is limited. Such an assumption conflicts with the fact that deer inhabit suburban areas and the observation that, even when surrounded by fences, deer populations show regulated growth (McCullough, 1979).

The entire line of argument fails to recognize that the natural processes of regulation are functioning but within a different context. If, over long time intervals, population growth in deer is determined primarily by K (the maximum number of deer the environment will support), then regardless of the magnitude of K, the natural regulatory process is still functioning. Human actions have dramatically altered K, increasing it by perhaps as much as ten times. We should not be surprised by the effect this has on native vegetation, because it is likely that many plant species are not adapted to the intensity of herbivory that they face at modern deer densities. We could argue that this landscape is artificial and that, in the absence of human action, the landscape would return to an ecosystem dominated by forest: This misses the point. Abundances of deer in eastern parks are higher than they probably were historically, not because natural regulatory processes have been disrupted but because the limits (K) have been reset (Fig. 12.4). Whether we are willing to accept these changes is a question of value but not a test of natural regulation (Porter, 1992).

The paradigm of disrupted natural-regulatory processes was hard to dismiss because it corresponded closely with a common interpretation of NPS policy. In many ways, NPS goals are designed to maintain the parks in a static condi-

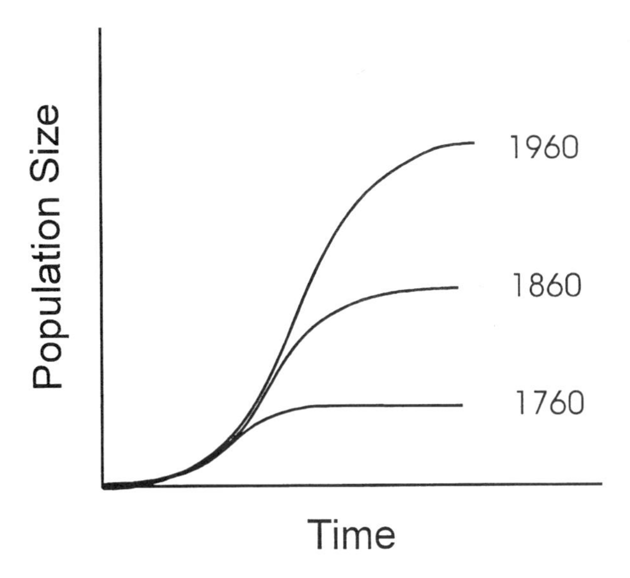

Figure 12.4 Ecological carrying capacity increased over the past two centuries as a result of the development of agriculture in the East and, later, by fragmentation of the extensive agricultural landscape by second growth forest when portions of the farmland were abandoned.

tion, as in a photograph. This policy seems in concert with public interest in preservation. For instance, demands for hands-on management by the public encouraged feeding of wildlife during difficult winters, preventing wildfires, and maintaining beaches against the forces of erosion. Yet, the park policy also dictates preserving the integrity of an ecosystem. Hard winters, large fires, and beach erosion are part of a normally functioning ecosystem. A goal that prohibits fluctuations that are inherent to natural processes is contrary to preserving ecosystem integrity. Policy creates a paradox of preserving a deer population at some constant level and preserving a process that allows deer populations to fluctuate in accord with normal ecological processes. Such mutually incompatible goals make for difficult management (Caughley & Sinclair, 1994).

Management Alternatives

If defining a litmus test for interventions limiting the growth of deer populations proved frustrating, determining a reasonable course of action proved equally difficult. Indeed, NPS hesitancy to take action, even in the face of clear deterioration of cultural resources, can be attributed to the lack of a politically and scientifically sound method for managing deer. A quick review of the common management approaches shows that all are plagued by controversy and untested assumptions.

No Action

In a strict sense, a decision of no action means that the status quo is maintained and existing management practices continue. As applied to most parks, this alternative is the basis for comparison. What would happen if there were no active efforts to intervene in growth of the deer population? Under this alternative, the deer population is allowed to fluctuate in response to environmental perturbations. An implicit assumption is that there is an upper limit to population growth and, that in the absence of perturbation, deer and vegetation would reach an equilibrium.

With no action, deer populations in most parks could be expected to continue to remain high relative to levels for the foreseeable future. As these populations approach ecological carrying capacity (K), average weights of individuals and reproductive rates would decline, especially among younger deer (McCullough, 1979). In some areas, diseases could become epidemic and cause substantial mortality (Wathen & New, 1980).

Species composition and structure characteristics of plant and animal communities respond to fluctuations in deer populations. Browsing by deer would alter the abundance and distribution of plant species (Tilghman, 1989). The effect on plant-species diversity and, especially, threatened or endangered species is of concern, but the impacts cannot be easily predicted (Miller et al., 1992; Underwood et al., 1995). Deer would maintain forest understories in a relatively open condition, and tree regeneration would be limited to species not preferred as food by deer (Storm et al., 1989). The mix of tree species dominant in the overstory would change and, as a consequence, the effects of high deer densities would persist for a century or more. The changes in plant community could also affect other animal species (Casey & Hein, 1983; Brooks & Healy, 1989; deCalesta, 1994).

Deer populations would be likely to fluctuate widely in response to severe winter weather, droughts, hurricanes, and other forms of disturbance. During the lows in the population, the plant community would be released from browsing pressure and would change rapidly in composition. If the deer population remained depressed for as few as five years, forest regeneration could attain sufficient height growth to escape the effects of browsing (Behrend et al., 1970; Underwood et al., 1995).

The chances of sustaining a program of no action would depend on the level of public debate. Substantial efforts would be required early in the program to

(1) dispel misconceptions about deer biology and NPS management policies, (2) measure true costs to cultural resources, and (3) study the long-term and larger scale consequences to ecological conditions.

Restoration of Predators

A decision to restore predators means in essence a program to actively translocate large predators to the parks in question and establish a free-ranging population. Under this alternative, the deer population would be allowed to fluctuate in response to predation and other environmental perturbations. Key assumptions are that a viable predator population could be established at a given park and that restoration would cause the long-term reduction in deer density.

If the underlying assumptions are valid, successful restoration of significant predator population would yield some change in the vegetation. Plant species that are widespread, but held in check by browsing, would respond quickly. The difficulty with this alternative is that the assumptions are untested, and available information suggests that they are questionable.

First, the assumption that a viable predator population can be established would depend on the size of the park and on surrounding land use. Predators are highly mobile, and long distance movements away from the release site are likely. Most large predators show some degree of territoriality (exclusive use of home ranges) and range over areas larger than 50 square kilometers. Establishing populations large enough to provide minimum genetic variability is probably not possible within the boundaries of most eastern parks (LaCava & Hughes, 1984). Individual animals would move off the park periodically and mortality rates would be high (Fritz et al., 1985).

Predator restoration would also face challenges stemming from ingrained public fear of predators. At the same time, restoration also would be challenged by those wishing to protect the predator population from control actions, even when problems with individuals occurred (Weise et al., 1975; Kellert,1985; Brocke et al., 1990; U.S. Department of Interior, 1990).

Finally, the species ecologically suited to most eastern parks, such as bobcat (*Felis rufus*), coyote (*Canis latrans*), and the red wolf (*C. niger*), would be unlikely to remove enough deer each year to produce significant population changes in most years. Deer populations would be driven by the same environmental factors that are important in the absence of predators. At best, we could expect that environmental conditions, in combination with predation, would cause periodic declines from which recovery by the deer population would be slow (Theberge & Gauthier, 1985; Fuller, 1990; Underwood, 1990).

This prolonged period of low deer densities would present a "window of opportunity" for changes in the plant community.

Fencing to Control Deer Movements

Fences may be appropriate for reducing access or eliminating deer from designated areas such as mall patches of rare plants, forest stands, agricultural fields, and highways. A variety of fencing designs is available (Hawthorne, 1980; Porter, 1983; Schafer & Penland, 1985). The key assumption is that conflicts between deer and cultural or natural resources are localized in areas of the park that can be fenced.

Fencing would immediately eliminate conflicts between deer and other management objectives on the protected areas. However, outside the restricted area, the deer population would fluctuate in response to environmental perturbations. Plant species held in check by browsing would respond quickly. Equitability among plant species would change dramatically on many sites (Underwood et al., 1995). Where the desired vegetation was woody, the fence could be removed after 5 to 10 years when trees had grown beyond the reach of deer; however, if the objective was to protect a low growing or herbaceous plant community, fencing would need to be permanent.

The probability of success would depend primarily on the size of the area to be fenced. Success would be highest where the areas were small. As exclosures get larger, there is increased disruption of deer behavior and increased visual impact to park visitors. Impacts of large exclosures on deer behavior are unknown. The fence itself and, later, the marked differences of vegetation inside and outside the fence could be obtrusive to the aesthetic or historic qualities of a landscape. From a natural resource perspective, the exclosure may create a condition that would be aberrant ecologically (Putman, 1986; Caughley, 1989).

Variations on the fencing alternative include "optical fences" and "deer repellents." Optical fences are created by devices that reflect automobile headlights and, in some circumstances, have been effective in reducing vehicle/deer accidents (Schafer & Penland, 1985). Chemical repellents are designed to discourage deer from browsing individual plants, but tests show repellents are ineffective (Hygnstrom & Craven, 1988; Swihart & Conover, 1990).

Hunting on the Periphery of Parks

Removing deer through legal hunter harvest on areas immediately surrounding a given park has long been considered a means of reducing park populations. This approach assumes that a significant proportion of the deer reside outside of the park part of the year and are vulnerable to hunting, and that increased removal of female deer can be accomplished with adjustments to the hunting season and bag limits. These assumptions are largely untested. Work at Saratoga National Historical Park shows that when 50 percent of the deer are outside of the park during hunting season, harvest can significantly reduce populations. Parks with large perimeter-to-area ratios would be more likely to experience an immediate impact of hunting on the periphery. However, no perimeter configuration is sufficient to allow absolute control of deer inhabiting the park (Schaberl, 1995). The utility of this alternative is questionable because movement behavior is learned; hunting on the periphery may serve only to eliminate those individuals whose movement patterns make them vulnerable. Ultimately, the portion of the population not vulnerable to hunting would grow, compensating for the loss of the long-distance migrants. The effectiveness of this approach to management predicated on the availability of lands outside a park that are open to sport hunting. In many cases, such lands are closed to hunting for various reasons.

Live-Trapping and Removal

Live-trapping and transfer of animals out of the park in order to reduce deer populations and maintain them at desired levels assumes that there are individuals or agencies willing to take large numbers of deer on a continuing basis and that translocating deer is humane. Successful reduction of deer populations that are near ecological carrying capacity (*K*) would yield a substantial change in plant communities and deer. In most eastern parks, forest regeneration would be noticeable within 5 to 10 years. However, as the deer population declined, the average size and reproductive output of individual animals would increase. Total recruitment of young would increase each year until the population was reduced to levels below half ecological carrying capacity (*K*/2). Consequently, the number of deer removed from the population must exceed the maximum annual recruitment of young possible for the population. A significant reduction and control of the deer population would require a removal of more than 60 percent of the female segment of the population annually.

The weakness of the implicit assumptions make success doubtful. First, few organizations are interested in having more deer. Donating or selling deer to game ranches that in turn sell the meat for profit would require changes in law in many states. Second, the assumption that this approach is more humane than killing deer is questionable. Survival rates of less than 30 percent within 1 year post-translocation are to be expected (O'Bryan & McCullough, 1985; Jones & Witham, 1990).

Reproductive Intervention

This alternative means delivering contraceptive drugs to female deer in order to reduce reproductive output in the population. This approach assumes that effective drugs could be delivered to deer in free-ranging populations, and that enough females could be treated to inhibit reproduction sufficiently so that population growth would be controlled. A key limitation to this approach is the long time interval before a decline would occur in the population. Given the life expectancy of deer in unhunted populations, little change would occur until 10 years after initiation of the treatment, when deer born prior to the treatment would reach senescence and die. It is likely that more than 60 percent of the females in a population would need to be treated and, with current drugs, treatments would need to be done annually. To be cost-effective, the delivery system must enable managers to treat deer without live-trapping and handling, and allow easy identification of animals once they have been treated. The system must also be suitable for continuing treatment, because if any reproduction is allowed, females six months of age will need to be treated (McCullough, 1979).

The potential for success of this approach depends on an effective method of delivering the proper dosages of the drug to target animals. Pharmacological studies show that effective contraception can be achieved by challenging the immune system of the deer, causing the system to produce antibodies to proteins found on ova or sperm. When the ova are released, or if sperm are present, the immune system destroys these cells and thus prevents pregnancy. Immunocontraception works well in captive deer herds, but there are concerns about its use in free-ranging populations. The logistics of delivering the proper doses to sufficient numbers of females in a free-ranging population are uncertain (Turner et al., 1992). Treating large numbers of females over areas the size of many eastern parks ($\geq$ 1,000 ha) appears daunting. Further, the effects on physiology and behavior of deer need to be explored. For instance, if pregnancy does not occur, estrous cycles continue in females and testosterone levels remain high in males. These are abnormal physiological states for deer. Prolonged periods of breeding behavior could disrupt annual cycles of fat acquisition that are necessary for survival.

Direct Reduction by Shooting

A decision to reduce the deer population by shooting means employing NPS personnel to kill a specified number of female deer throughout the park on an annual basis. This technique assumes that there are places in the park where shooting could be conducted safely or that deer could be effectively live-trapped and killed. Reduction of population densities to levels well below ecological carrying capacity (K) would probably produce effects similar to other population reduction techniques. The principal concern with direct reduction is the political furor it attracts. Protests would be heard from groups with widely divergent values, ranging from animal rights activists (opposed to killing) to sportsmen's organizations (opposed to the park being closed to public hunting). Unless all deer are removed, annual reductions would be necessary.

A modification of this approach is selective removal of family units of deer from areas where conflicts are most severe. Experiments to test this technique are now under way, and modeling suggests deer could be eliminated for 10 to 15 years, from areas as small as 400 hectares with one-time removal of 12 to 20 individuals (Porter et al., 1991). This modification may provide an important compromise solution that would be acceptable to the varied public interest groups.

Addressing Management Dilemmas in the Future

To date, the NPS has hesitated to invoke any of the population control alternatives described, primarily because the NPS likely believed it lacked the political consensus and scientific knowledge to be reasonably assured of success. There were no major legal suits filed against maintaining the status quo, but many were threatened if action to reduce populations was undertaken. The threat alone was sufficient to halt a special program to reduce deer populations on Fire Island (O'Connell & Sayre, 1989). Short-term scientific studies could help clarify cultural and human safety issues, but the long-term research needed to answer questions about the effects of deer on ecosystem integrity and biodiversity was not available.

The original dilemma remains: How does the agency proceed in view of the enormous ecological complexity and the political volatility of the issue? On one hand, if no action is taken, will deer populations so alter park ecosystems that they will jeopardize the very cultural and biological qualities that are held to be important? On the other hand, is knowledge about the system sufficient to intervene effectively? Clearly, decisions need to be made. The key question is not what alternative to choose, but how to turn the dynamic character of political views and scientific information into a sustained management action.

An approach to this dilemma is embodied in the concept of adaptive management (Walters, 1986) or experimental management (Caughley & Sinclair,

1994). The simplest definition of *adaptive management* is: try something; watch closely to see if it works; if it does not work, figure out why it does not quickly and try something else. The important elements include a specific objective, a monitoring program, and criteria by which we can judge failure (Caughley & Sinclair, 1994). (This process is outlined in Figure 20.2.) Adaptive management demands a team effort. The team includes resource managers, policymakers, scientists, and outside groups who have a stake in the decision (Decker et al., 1991). Each team member must be brought into the decision-making process from the outset in order to (1) express concerns and identify principal issues; (2) learn enough to be able to separate fact from dogma; and (3) reach agreements in which all parties have ownership. The intent of the team effort is to take advantage of the skills and knowledge of each specialist to design and evaluate creative solutions. There are four steps in the process: (1) bounding the management program, (2) constructing models of the system, (3) analyzing the uncertainty, and (4) designing management-science linkages.

Bounding the Management Program

The process of determining a course of action must begin with brain-storming sessions and open discussions of the possibilities. Basic premises from which various political positions arise must be identified and examined closely. Fact must be separated from dogma. Economic and ecological constraints must be recognized. Objective criteria for assessing program success or failure must be identified.

Public meetings on deer-management issues that include educational presentations and open discussions can pay large political dividends. The acrimony often dissolves when people are asked to listen to one another respectfully and when the perceptions about the situation are confronted with facts. Scientists can be especially helpful at the outset of this process because they are viewed as objective observers who carry the agenda of no interest group. In instances where the public understands NPS goals, deer ecology, and management alternatives, complaint levels have declined appreciably.

Constructing Models of the System

Working models in the form of computer simulations can avoid a lot of pitfalls and improve the design of a program. The purpose of these models is to define explicitly what is known about the system and to identify all of the assumptions underpinning a management plan. The behavior of the system under various management scenarios can be explored and the success or failure in light of assessment criteria can be evaluated.

An excellent example of the utility of modeling is the wolf recovery program for Yellowstone National Park. Here, the NPS commissioned models to explore the potential for successful restoration of wolves and the impacts such a program would have on ungulates in the park. Ecological data, assumptions, and ecosystem behavior were scrutinized closely by having several teams of scientists develop models independently of one another, and then comparing the results (U.S. Department of Interior, 1990). The result was a thorough analysis of what was and was not known about wolf-prey interactions. The similarity of ecosystem behavior predicted by the independent models helped politicians to understand and gain confidence in the plan.

Analyzing the Uncertainty

Frequently, the first two steps in the adaptive management process point out deficiencies in our knowledge. Lack of complete knowledge means that we will be moving forward with a degree of uncertainty about the actual outcome of the management action. We can quantify the degree of uncertainty using past experience, logic, and statistical analysis.

Perhaps the most common example is the uncertainty of the size of the deer population. Knowing how many deer are present in an area is generally the first question asked in discussions of management action. Yet, this is one of the most difficult pieces of information to obtain. We frequently must be satisfied with an estimate of the population because we are unable to conduct an absolute census. However, to say that we *think* there are about 600 deer is not as helpful as saying that we *know*, with 95 percent certainty, that there are 600 $\pm$ 100 deer. Knowledge that the actual population is somewhere between 500 and 700 communicates our level of uncertainty. Further, information on the possible range of values allows us to build contingencies into our management plan so that we can implement rapid adjustments as we learn more about the population size.

Designing Management-Science Linkages

The heart of adaptive management is a better linkage of politics and science. Most management programs improve with experience. The problems faced by many parks mean that they cannot afford to wait for absolute consensus and comprehensive knowledge. If the intent is to arrive at the optimal management solution as soon as possible, each management action must be considered as an experiment in a trial-and-error process. Management actions must be linked with solid scientific monitoring and analysis. Initiating a management action will allow an evaluation of the problem. Attention to experimental design will provide a better understanding of the problem, an objective evaluation of the solutions, and an adjustment in management.

The best management plan is one that will change or adapt as knowledge is acquired about costs, people's values, and the ecological system. If all deer management actions can be perceived as experiments and management actions are coupled with research efforts, both scientific knowledge and management capability will benefit.

Conclusions

Eastern North America is one of the most ecologically diverse regions in the world. The maintenance of this diversity is important as human development continues to expand in the region. Parks will serve a crucial role in achieving this goal.

The case of white-tailed deer illustrates that managing parks to maintain diversity will be contentious, messy, and filled with risk. At the same time, the experience with deer provides several lessons that help frame park management decisions:

1. The debates over deer management have demonstrated a tendency to rely on unsubstantiated ideas. For instance, it is argued that hands-on management of deer is necessary because the populations are no longer regulated by natural forces. Deer populations erupted because society created an ideal habitat; natural processes still set limits to deer abundance.
2. The NPS was wise to hesitate in taking definitive action with respect to deer control because management goals were politically and scientifically nonfunctional. While terms like ecosystem integrity, and deer-vegetation balance are superficially satisfying, they snare management programs in contradiction. There is a tendency to turn to science to reconcile the dilemmas, but management is a value-driven process. The role of science is to be an objective purveyor of factual information that aids in making more informed judgments.
3. Close linkage of management and science offers the most potential for resolving complex issues of deer in parks. Solutions untried will never work

and solutions tried are likely to fail at least once. It is better to cast management actions in the form of experiments from which to learn and adapt.

Literature Cited

Behrend DF, Mattfeld GF, Tierson WC, Wiley JE III. Deer density control for comprehensive forest management. *J For* 1970;68:695–700.

Brocke RH, Gustafson KA, Major AR. Restoration of the lynx in New York: Biopolitical lessons. *N Am Wildl Nat Resour Conf* 1990;55:590–598.

Brooks RT, Healy WM. Response of small mammal communities to silvicultural treatments in eastern hardwood forests of West Virginia and Massachusetts. In RZ Szaro, KE Severson, DR Patton (technical coordinators). Management of Amphibians, Reptiles and Small Mammals in North America. *USDA For Serv Gen Tech Rep* (RM-166). 1989:313–318.

Casey D, Hein D. Effects of heavy browsing on a bird community in deciduous forest. *J. Wildl Manage* 1983;47:829–836.

Caughley G. Eruption of ungulate populations, with emphasis on Himalayan thar in New Zealand. *Ecol* 1970;51:53–72.

Caughley G. What Is This Thing Called Carrying Capacity? In M S Boyce, L D Hayden-Wing (eds), *North American Elk: Ecology, Behavior, and Management.* Laramie, WY: University of Wyoming Press, 1979:2–8.

Caughley G. New Zealand plant-herbivore systems: past and present. *New Zealand J Ecol* 1989;12 (suppl.):3–10.

Caughley G, Shepherd N, Short J. *Kangaroos: Their Ecology and Management in the Sheep Rangelands of Australia.* New York: Cambridge University Press, 1987.

Caughley G, Sinclair ARE. *Wildlife Ecology and Management.* Boston: Blackwell Scientific, 1994.

Cole GF. An ecological rationale for the natural or artificial regulation of native ungulates in parks. *N Am Wildl Nat Resour Conf* 1971;36:417–425.

Cole LC. The population consequences of life history phenomena. *Q Rev Biol* 1954;29:103–137.

Cypher BL, Yahner RH, Cypher EA. Ecology and management of white-tailed deer at Valley Forge National Historical Park. Washington, DC: *Natl Park Serv Tech Rep* NPS/MAR-15, 1985.

deCalesta DS. Effect of white-tailed deer on songbirds within managed forests. *J Wildl Manage* 1994;58:711–718.

Decker DJ, Loconti Lee KM, Connelly NA. Incidence and costs of deer-related vehicular accidents in Tompkins County, New York. New York State College of Agriculture and Life Science Human Dimensions Research Unit, HDRU Series 89–7, 1990.

Decker DJ, Shanks RE, Nielson LA, Parsons GR. Ethical and scientific judgements in management: Beware of blurred distinctions. *Wildl Soc Bull* 1991;19:523–527.

Despain D, Houston D, Meagher M, Schullery P. *Wildlife in Transition.* Boulder, CO: Roberts Reinhart, 1986.

Downing RL, Guynn DC Jr. A Generalized Sustained Yield Table for White-tailed Deer. In SL Beasom, SF Roberson (eds), *Game Harvest Management.* Kingsville, : Ceasar Kleberg Wildlife Research Institute, 1985:95–103.

Flader SL. *Thinking Like a Mountain: Aldo Leopold and the Evolution of an Ecological Attitude Toward Deer, Wolves, and Forests.* Columbia, MO: University of Missouri Press, 1974.

Fritz SH, Paul WJ, Mech LD. Can relocated wolves survive? *Wildl Soc Bull* 1985;13:459–463.

Fuller TK. Dynamics of a declining white-tailed deer population in north-central Minnesota. *Wildl Monogr* 110, 1990.

Halls LK (ed.) *White-tailed Deer Ecology and Management.* Harrisburg, PA: Stackpole Books, 1984.

Harder JD. Reproduction of White-tailed Deer in the North Central United States. In RL Hine, S Nehls (eds), *White-tailed Deer Population Management in the North Central States.* Eau Claire, MN: North Central Section Wildlife Society, 1980:23–25.

Harlow RF. Habitat Evaluation. In LK Halls (ed), *White-tailed Deer Ecology and Management.* Harrisburg, PA: Stackpole Books, 1984:601–628.

Hawkins RE, Klimstra WD, Autry DC. Dispersal of deer from Crab Orchard National Wildlife Refuge. *J Wildl Manage* 1971;35:216–220.

Hawthorne DW. Wildlife Damage and Control Techniques. In SD Schemnitz (ed), *Wildlife Management Techniques Manual* (4th ed) Washington, DC: The Wildlife Society 1980:411–439.

Hesselton WT, Severinghaus CW, Tanck JE. Population dynamics of deer on the Seneca Army Depot. *NY Fish and Game J* 1965;12:17–30.

Hygnstrom SE, Craven SR. Electric fences and commercial repellents for reducing deer damage in cornfields. *Wildl Soc Bull* 1988;16:291–296.

Jones JM, Witham JH. Post-translocation survival and movements of metropolitan white-tailed deer. *Wildl Soc Bull* 1990;18:434–441.

Kammermeyer KE, Marchinton RL. Notes on dispersal of male white-tailed deer. *J Mamm* 1976;57:776–778.

Keiter RB, Boyce, WS. *The Greater Yellowstone Ecosystem.* New Haven, CT: Yale University Press, 1991.

Kellert SR. *The Public and Timber Wolf in Minnesota.* New Haven, CT: Yale University Press, 1985.

LaCava J, Hughes J. Determining minimum viable population levels. *Wildl Soc Bull* 1984;12:370–376.

Leopold AS, King SA, Cottam CM, Gabrielson IN, Kimball TL. Wildlife management in national parks. *N Am Wild Nat Resour Conf* 1963;28:29–42.

Macnab J. Carrying capacity and related slippery shibboleths. *Wildl Soc Bull* 1985;13:403–410.

Marchinton LR, Hirth DH. Behavior. In LK Halls (ed), *White-tailed Deer Ecology and Management.* Harrisburg, PA: Stackpole Books, 1984:129–168.

Masters RD, Mathews NE. Notes on reproduction of old ($\geq$ 9 years) free-ranging white-tailed deer, *Odocoileus virginanus,* in the Adirondacks, New York. *Can Field Nat* 1990;105:286–287.

Mathews NE. Social Structure, Genetic Structure and Anti-predator Behavior of White-tailed Deer in the Adirondacks (Ph.D. dissertation). Syracuse, NY: State University of New York at Syracuse College of Environmental Science and Forestry, 1989.

McCullough DR. *The George Reserve Deer Herd: Population Ecology of a K-selected Species.* Ann Arbor, MI: University of Michigan Press, 1979.

McLaren BE, Peterson RO. Wolves, moose and tree rings on Isle Royale. *Science* 1994;266:1555–1558.

Miller SG, Bratton SP, Hadidian J. Impacts of white-tailed deer on endangered and threatened vascular plants. *Nat Areas J* 1992;12:67–74.

Moen AN. *Wildlife Ecology.* San Francisco: Freeman, 1973.

National Park Service. *Management Policies.* Washington, DC: US Department of Interior, 1988.

Natural Park Service. Natural Resource Management Guideline. Publ. NPS- 77. Washington, DC: National Park Service, 1991.

O'Bryan MK, McCullough DR. Survival of black-tailed deer following translocation in California. *J Wildl Manage* 1985;49:115–119.

O'Connell AF, Sayre MW. White-tailed deer management study: Fire Island National Seashore. *USDA For Serv Gen Tech Rep* NPS/NAR, 1989.

Pimm SL. *The Balance of Nature?* Chicago: University of Chicago Press, 1991.

Porter WF. A baited electric fence for controlling deer depredation in orchards. *Wildl Soc Bull* 1983;11:325–328.

Porter WF. White-tailed deer in eastern ecosystems: Implications for management and research in national parks. *US Dept Inter Nat Res Rep* (NPS/NRSUNY/NRR-91/05), 1991.

Porter WF. Burgeoning Ungulate Populations in the National Parks: Is Intervention Warranted? In DR McCullough, RH Barrett (eds) *Wildlife 2001: Populations.* New York: Elsevier 1992:304–312.

Porter WF, Coffey MA, Hadidian J. In search of a litmus test: Wildlife management in U.S. National Parks. *Wildl Soc Bull* 1994;22:301–306.

Porter WF, Mathews NE, Underwood HB, Sage RW, Behrend DF. Social organization in deer: Implications for localized management. *Environ Manage* 1991;6:809–814.

Putman RJ. *Grazing in Temperate Ecosystems: Large Herbivores and the Ecology of the New Forest.* Portland, OR: Timber Press, 1986.

Schaberl JE. Assessment of Hunting Adjacent to Park Boundaries on the Survival and Population Dynamics of White-tailed Deer (M.S. thesis). Syracuse, NY: State University of New York at Syracuse College of Environmental Science and Forestry, 1995.

Schafer JA, Penland ST. Effectiveness of Swareflex reflectors in reducing deer-vehicle collisions. *J Wildl Manage* 1985;49:774–776.

Severinghaus CW, Moen AN. Prediction of weight and reproductive rates of a white-tailed deer population from records of antler beam diameter among yearling males. *NY Fish Game J* 1983;30:30–38.

Short HL. Habitat suitability index models: White-tailed deer in the Gulf of Mexico and South Atlantic coastal plains. *US Dep Inter Biol Rep* 82, 1986.

Sillings JL. White-tailed Deer Studies in a Suburban Community: Ground Counts, Impacts on Natural Vegetation, and Electric Fencing to Control Browsing (M.S. thesis). St. Paul, MN: University of Minnesota, 1987.

Storm GL, Yahner, RH, Cottom DF, Vecellio GM. Population status, movements, habitat use and impact of white-tailed deer at Gettysburg National Military Park and Eisenhower National Historic Site, Pennsylvania. *Tech Rep* NPS/MAR/NRTR-89/043, 1989.

Swihart RK, Conover MR. Reducing deer damage to yews and apple trees: Testing Big Game Repellent, Ro-Pel, and soap as repellents. *Wildl Soc Bull* 1990;18:156–162.

Taber RD. Development of the cervid antler as an index of late winter physical condition. *Proc Montana Acad Sci* 1958;18:27–28.

Theberge JB, Gauthier DA. Models of wolf-ungulate relationships: When is wolf control justified. *Wildl Soc Bull* 1985;13:449–458.

Thomas JW, Teer JG, Walker EA. Mobility and home range of white-tailed deer on the Edwards Plateau in Texas. *J Wildl Manage* 1964;28:463–472.

Tilghman NG. Impacts of white-tailed deer on forest regeneration in northwestern Pennsylvania. *J Wildl Manage* 1989;53:524–532.

Turner JW, Liu IKM, Kirkpatrick JF. Remote delivered immunocontraception of captive white-tailed deer. *J Wildl Manage* 1992;56:154–156.

U.S. Department of Interior. *Wolves for Yellowstone? Report to the United States Congress* (Vol 2) Yellowstone National Park, WY: 1990.

U.S. Department of Interior. *Draft Environmental Impact Statement: White-tailed Deer Management Plan for Gettysburg National Military Park and Eisenhower National Historic Site.* National Park Service, Gettysburg National Park, PA: 1994.

Underwood HB. Population Dynamics of White-tailed Deer in a Fluctuating Environment (Ph.D. dissertation). Syracuse, NY: State University of New York at Syracuse College of Environmental Science and Forestry, 1990.

Underwood HB, Austin KA, Porter WF, Burgess RL, Sage RW. White-tailed deer and vegetation at Saratoga National Historical Park. *Natl Park Serv Tech Bull,* 1995.

Underwood HB, Porter WF. Values and science: White-tailed deer management in the national parks. *N Am Wildl Nat Resour Conf* 1990;56:67–73.

Verme LJ, Ullrey DE. Physiology and Nutrition. In L K Halls (ed), *White-tailed Deer Ecology and Management.* Harrisburg, PA: Stackpole Books, 1984:91–118.

Vogel WO. Response of deer to density and distribution of housing in Montana. *Wildl Soc Bull* 1989;17:406–413.

Walters C. *Adaptive Management of Renewable Resources.* New York: Macmillan, 1986.

Warren RJ. Ecological justification for controlling deer populations in eastern national parks. *N Am Wildl Nat Resour Conf* 1991;56:56–66.

Wathen WG, New CJ. The white-tailed deer of Cades Cove: Population status, movements and survey of infectious diseases. Atlanta: U.S. Dep. Inter. Natl. Park Serv. Rep. SER-89/01, 1989.

Weise TF, Robinson WL, Rook RA, Mech LD. Eastern timber wolf. *Natl Audubon Soc Rep* 1975;5:1–28.

Witham JH, Jones JM. Deer-Human Interactions and Research in the Chicago Metropolitan Area. In LW Adams, DL Leedy (eds), *Proceedings of a National Symposium on Urban Wildlife.* Columbia: National Institute for Urban Wildlife, 1987:155–159.

Woolf A, Harder JD. Population dynamics of a captive white-tailed deer herd with emphasis on reproduction and mortality. *Wildl Monogr* 67, 1979.

Wright RG. *Deer Management Alternatives for Gettysburg National Military Park and an Associated Environmental Analysis.* Moscow, ID: Cooperative Park Study Unit, University of Idaho, 1990.

Wright RG. *Wildlife Research and Management in the National Parks.* Urbana, IL: University of Illinois Press, 1992.

13

Managing and Understanding Wild Ungulate Population Dynamics in Protected Areas

James G. MacCracken

Wild ungulates occur in most parks and protected areas in the United States and throughout the world. The diversity of ungulate species in these areas can be quite high. For example seven species, wapiti (*Cervus elaphus*), mule deer (*Odocoileus hemionus*), white-tailed deer (*Odocoileus virginianus*), moose (*Alces alces*), bison (*Bison bison*), pronghorn (*Antilocapra americana*), and mountain sheep (*Ovis canadensis*), occur in both Grand Teton and Yellowstone National Parks. Conversely, in other areas only one species may be present (e.g., white-tailed deer, which occur in many of the historic parks discussed in Chapter 12 as well as in Great Smoky Mountains and Shenandoah National Parks [Fox and Pelton, 1974; Bratton, 1979]).

A fundamental issue in the management of wild ungulates in protected areas is their potential for reaching high population densities and the effects these high densities have on individuals in the population and on associated species of plants and animals, soils, and ecosystem processes. At the heart of this issue is whether there exists a set of controlling mechanisms that serves to limit or control population density in the absence of human influence. Peek (1980) has called this concept *natural regulation*, and by definition it is a phenomenon that can occur only in areas where human influence is absent (Caughley, 1981a) or minimal (Van Ballenberghe & Ballard, 1994). In general, these areas restrict consumptive resource activities, including hunting, and fall into International Union for Conservation of Nature and Natural Resources (IUCN) categories I–IV (see Table 1.1).

The potential for a population to be naturally regulated is contingent on many factors, the most important being: (1) the size of the area occupied by the population, i.e., its ability to provide an adequate forage base and security from outside influences, and (2) the integrity of the ecosystem, including an intact complex of predators. Outside of Alaska, it appears that few U.S. parks meet the requirement of containing a complete ecological unit.

Concern over the status of ungulate populations in national parks, particularly in the West, goes back many decades. Wright (1992) documented changes in National Park Service (NPS) policy toward wild ungulates during the last 80 years, beginning with efforts to increase populations to provide more viewing opportunities for visitors, followed by culling to mitigate the alleged effects of high density populations, to current efforts to allow natural factors to regulate populations. This history is exemplified by wapiti management in Yellowstone. For the first 60 years of the park, predators were routinely killed to protect wapiti populations. This effort was so successful that the wolf (*Canis lupis*) was exterminated and populations of other predators, e.g., mountain lions (*Felis concolor*) and coyote (*Canis latrans*), were greatly reduced. The resultant increase in wapiti numbers was such that by 1934 an annual culling program was initiated to avoid what some scientists considered to be winter-range deterioration (Skinner, 1928; Cahalane, 1943). The culling program lasted until 1967 and resulted in the killing and translocation of over 16,000 wapiti from the park (Houston, 1982). Adverse public reaction to the culling program, combined with changes in NPS policy, brought an end to the culling program (Leopold et al., 1963; Cole, 1971; Houston, 1971).

Natural areas within the U.S. national park system today are managed with the goal of allowing natural processes that regulate the growth and decline of animal populations to take place unimpeded by direct human intervention (NPS, 1988). Underlying this policy is the presumption that the number of animals of a given species will fluctuate, with minor oscillations, about a long-term mean (similar to Fig. 12.2). The fluctuations in population numbers are normally attributed to changes in the availability of food and cover, climate severity, predation, and competition. Accordingly, the need for human intervention in the form of harvest or habitat manipulation should be minimal.

This policy, however, often is confronted by the ecological and socioeconomic realities of a park's environment. As described in several previous chapters, parks are subjected to a variety of stressors that impact species and ecological processes. Some processes that formerly contributed to the control of ungulates in parks, such as predation, are now largely absent. Also associated with this policy is the idea that naturally regulated ungulate populations will not irreversibly alter the natural vegetation structure and composition of their ranges (Houston, 1982). However, this factor is difficult to evaluate because of the lack of consensus as to what constitutes the "natural" condition of a range occupied by large numbers of ungulates (MacNab, 1985).

Robisch and Wright (1995) surveyed 29 U.S. national parks, containing 95 populations of 11 ungulate species, and examined population sizes, trends, management techniques, and other factors in an effort to better understand those factors believed to regulate park ungulate populations. They found that 56 percent were regulated by one or more anthropogenic factors and 14 percent appeared to be naturally regulated. Anthropogenic factors included hunting outside the park, poaching, accidents, and NPS translocations. In most cases, the populations lacking natural controls were those growing in number, those not adequately controlled by anthropogenic factors, and those whose impacts (e.g., bison and white-tailed deer) have been documented in previous chapters. In the case of wapiti in Yellowstone, it has been alleged that high densities have eliminated aspen (*Populus tremuloides*) and, to a lesser extent, willows (*Salix* spp.) from the park, causing the extirpation of beaver (*Castor canadensis*) and preventing their recolonization (Chadde & Kay, 1991). In turn, beaver can have substantial effects on stream hydrology, fisheries, and the structure and function of riparian areas (Naiman et al., 1986).

In summary, it is recognized that the potential for wild ungulates, particularly at high population densities, to alter ecosystem structure and processes through intense herbivory is a real concern in protected areas (Coughenour & Singer, 1991; Chadde & Kay, 1991; Wright, 1992). The most salient aspects of this concern are its spatial and temporal extent, whether it is a "natural phenomenon," and the extent to which this herbivory interferes with the goals and objectives of a particular reserve. Because most of these concerns are difficult to measure accurately or objectively, there should be little wonder why ungulate management in parks is fraught with controversy (Peek, 1986; Boyce, 1991). Given this background, the next section examines the ecological aspects of ungulate population regulation in protected areas.

Factors Limiting and Regulating Populations

There are a number of intrinsic and extrinsic factors that can potentially influence ungulate population numbers over time, such as competition for forage, predation, disease, social behavior, immigration and emigration, severe weather, accidents, and genetics. These factors, which are discussed in detail in later sections, are termed *limiting* when they impede the growth of a population (Sinclair & Norton-Griffiths, 1982; Messier, 1991; Van Ballenberghe & Ballard, 1994). In general, if the major limiting factor is removed, the population will increase dramatically (see Sinclair et al., 1985).

Limiting factors are termed *regulating* when they exhibit a positive density-dependent response, i.e., as density increases so does the effect of the limiting factor (Sinclair & Norton-Griffiths, 1982; Messier, 1991; Van Ballenberghe & Ballard, 1994). An inverse density-dependent response indicates an antiregulatory effect that, in combination with a regulating factor, may be responsible for population cycles (Messier, 1991) or recurrent population fluctuations (Van Ballenberghe, 1987; Van Ballenberghe & Ballard, 1994).

The distinction between limiting and regulating factors can be thought of as hierarchial because all mortality factors can be limiting, but they are regulatory only when density dependent. Caughley and Sinclair (1994) noted that "[R]egulation is a biotic process which counteracts abiotic disturbances affecting an animal population." Biotic processes can be both limiting and regulating, while abiotic factors can be only limiting. However, they both can interact in complex ways, and an abiotic disturbance could trigger changes that cause a biotic process to move from one of limitation to one of regulation.

For these reasons, it is easier to demonstrate limitation than regulation. For example, it is very difficult to measure directly intraspecific competition for forage among individuals when compared to indices such as winter mortality due to malnutrition. Even so, several studies have found "density-dependent effects" of food limitation (Skogland, 1986), some more specifically related to competition for forage (caribou/reindeer [*Rangifer tarandus*]: Messier et al., 1988; moose: Messier, 1991; white-tailed deer: Messier, 1991), while predation also has been found to regulate some moose populations (Messier & Crête, 1985; Van Ballenberghe, 1987; Van Ballenberghe & Ballard, 1994).

Factors Influencing Ungulate Population Densities

Competition for Forage

Most large predators have been eliminated or persist only at low densities in many protected areas with wild ungulates. In addition, the possible compensatory role provided by sport hunting is precluded in protected areas by definition. Even though in some protected areas, large predators are either recolonizing vacant habitat, being reintroduced, or experiencing population in-

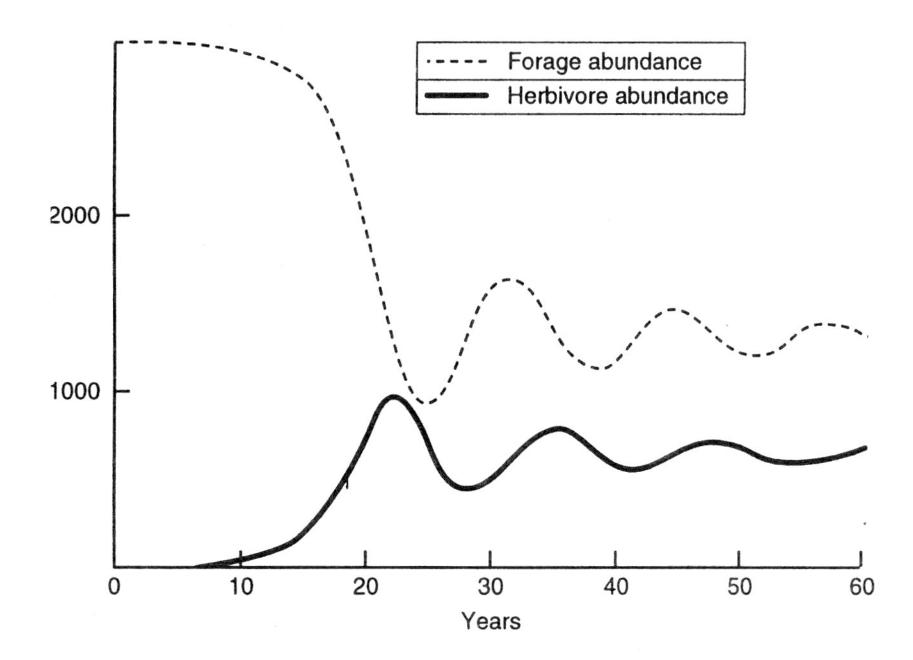

Figure 13.1 The relationship between forage and herbivore abundance during an irruptive sequence. (From G. Caughley, JH Lawton, Plant-Herbivore Systems. In RM May [ed], *Theoretical Ecology.* Sunderland, MA: Sinauer Associates, 1981:132–166.)

creases, the ecological and political barriers to predator restoration in other areas are immense (Curlee et al., 1994).

In protected areas without efficient predators (i.e., those capable of handling the ungulate), ungulate populations often are thought to be food regulated, comparable to Caughley's (1970, 1976) classic herbivore-forage equilibrium model (Fig. 13.1). Even so, Caughley believed that in such a system, the population changes he predicted would occur with or without predators. Variations of this model typically are invoked to justify the natural regulation of ungulates in protected areas (Cole, 1971; Houston, 1982).

Caughley and Lawton (1981) elaborated on the basic model by dividing grazing systems into those that were not interactive (i.e., the herbivores did not influence forage renewal) and those that were. Wild ungulates more likely exist in interactive systems that can be subdivided further into laissez-faire (i.e., those where herbivores do not interfere with each others' acquisition of food) and interference systems. Nonterritorial ungulates may fit the laissez-faire system, but generalizations are difficult to make, and direct evidence of interference among herbivores is rare (Caughley & Lawton, 1981).

The primary features of the plant-herbivore equilibrium model are in initial ungulate eruption followed by a decline, an inverse trend in forage resources, and an eventual smoothing of the oscillations in both forage and herbivore abundance over time (see Fig. 13.1). The initial herbivore eruption can be triggered by any number of events that eliminate or reduce the effectiveness of a limiting or regulating factor, such as a reduction in predators (Gasaway et al., 1983, Bergerud & Ballard, 1988 [see also Van Ballenberghe, 1985]), a rapid increase in forage resources following wildfire (Spencer & Hakala, 1964; Peek, 1974; Irwin, 1975; MacCracken & Viereck, 1990), or a combination of factors (Caughley, 1970).

The reintroduction or translocation of an ungulate into unoccupied, high-quality habitat can also result in a population eruption (Klein, 1968; Caughley, 1970; McCullough, 1979; MacCracken, 1992). Peek (1980) termed the initial increase, subsequent decline, and eventual leveling off of an ungulate population an *irruptive sequence.* He also noted that all of the documented ungulate eruptions were a result of human activities or occurred in altered ecosystems, and thus questioned whether the irruptive sequence was a feature of natural systems. However, the colonization of Isle Royale National Park by moose and their subsequent eruption would have to be considered a natural event. Furthermore, the establishment of moose on the island was most likely a prerequisite to wolf colonization.

Wild ungulate population eruptions in protected areas could occur under the following conditions: (1) the sudden exclusion of sport hunting in a new reserve with a resident population; (2) the natural (or otherwise) colonization of a protected area that contains high quality habitat and lacks an effective predator; (3) a disturbance such as wildfire that results in a rapid increase in forage and an ungulate adapted to exploit those conditions; and (4) the disappearance of an exotic disease.

During the irruptive sequence, one could expect herbivory-driven changes in plant community composition; changes in ungulate diet composition and quality, body condition, mortality, fecundity and recruitment rates, and population age and sex structure; and similar effects on other herbivores that use the same forage resources. Predator populations would also exhibit changes in population dynamics and social behavior.

Under the plant-herbivore equilibrium model, ungulate populations are regulated at forage-based ecological carrying capacity (K) by intraspecific competition for forage (Caughley & Sinclair, 1994). Interspecific competition is not considered to be important because of assumed niche separation among coevolved, coexisting herbivores.

Although Caughley (1967, 1970) proposed an equilibrium system determined by intraspecific competition, there are reasons to believe this conceptualization has limited use. Caughley and Lawton (1981) discussed examples in which interactions among the species altered plant demography in ways that both enhanced or reduced plant diversity. Furthermore, the responses of plants to herbivory (Bryant et al., 1989; McNaughton, 1985) and herbivores to plant defenses (Hofmann, 1986) suggest that no single equilibrium exists (Houston, 1982; Boyce, 1991).

Equilibrium concepts do not adequately explain the variation in herbivore-plant systems attributable to abiotic factors, time lags in herbivore-plant responses, stochastic events, and other density-independent factors (DeAngelis & Waterhouse, 1987). Nonequilibrium models may better describe ungulate populations when environments are highly seasonal and the animals are migratory (Coughenour & Singer, 1991). For example, severe winters and drought have short-term effects on both wapiti and their forage but the effect of a 300-year-fire-return interval on the long-term persistence of wapiti in Yellowstone is unknown. Likewise, a common winter range but segregated summer pastures and fall rutting areas may be essential to the persistence of the individuals herds (Coughenour & Singer, 1991).

The concept of overgrazing is a manifestation of human values, experiences, and land-use activities (Coughenour & Singer, 1991). How one views the ecological consequences of intense herbivory depends on one's goals and operating paradigms. The predicted response of wapiti managers in Yellowstone, using different paradigms of herbivore-plant relationships, to various phenomena potentially resulting from intensive herbivory is shown in (Table 13.1). Ironically, many park managers have embraced a model (natural regulation) that is relatively incompatible with the ostensible goal of nonintrusive management.

For herbivores, forage is the ultimate limiting and/or regulating factor (i.e., food-limitation; see Lack, 1954). As the effects of other factors are ameliorated naturally or by human activities, each population increase brings the herbivore closer to a forage-based population ceiling and regulation by competition for digestible forage (Messier, 1991). Changes in forage resources only shift the population ceiling up or down. Even so, other factors such as predation can regulate herbivore populations for long periods.

Predation

The primacy of food or predation as limiting and/or regulating factors of ungulates has long been a focus of debate (Keith, 1974; Caughley, 1976). It could be argued that the debate remains unresolved to the extent that simplistic approaches have been taken. The question is no longer which factor is more important than the other but, rather, under what set of circumstances is a factor regulating or

Table 13.1 Comparison of the acceptability of various ecological observations and responses to intense herbivory based on five theoretical management paradigms

	Paradigm				
Ecological observation-response	Range manager	Wildlife manager	Natural regulation	Caughley model	Persistence model
Plant effects					
Reduced mass	Accept	Accept	Accept	Accept	Accept
Reduced vigor	Accept	Accept	Accept	Accept	Accept
Reduced production	Limited[a]	Limited	Limited	Limited	Limited
Browse decrease–death	Conditional[b]	Reject	Reject	Conditional	Conditional
Zootic climax	NA[c]	Conditional	Accept	Accept	Accept
Zootic disclimax					
Ungulates maximized	Accept	Conditional	Reject	Accept	Reject
Ungulates not maximized	Reject	Reject	Reject	Accept	Reject
Declining range condition					
Irreversible	Reject	Reject	Reject	Reject	Reject
Transient, cultural	Reject	Reject	Reject	Conditional	Reject
Transient, natural	NA	Accept	NA	Accept	Accept
Plant species extinction	Conditional	Reject	Reject	Reject	Reject
Ungulate effects					
Irruptive sequence					
Predator limitation, but absent	NA	Reject	NA	Accept	Reject
No predator effects	NA	Accept	Accept	Accept	Accept
Immigration, reduced emigration					
Cultural	NA	Conditional	NA	Accept	Reject
Natural	NA	Conditional	NA	Accept	Accept
Interspecific exclusion	Accept	Reject	Reject	Conditional	Conditional
Reduced vigor, fecundity	Reject	Reject	Accept	Accept	Accept
Soil effects					
Increased runoff, reduced infiltration	Reject	Reject	Reject	Limited	Limited
Increased erosion	Reject	Reject	Reject	Reject	Conditional

NA = not applicable.
[a]Outcome accepted to a limited degree by some.
[b]Outcome accepted under certain conditions by most.
[c]Outcome considered impossible.
Source: Modified from MB Coughenour, FJ Singer, The Concept of Overgrazing and Its Application to Yellowstone's Northern Range. In RB Keiter, MS Boyce (eds), *The Yellowstone Ecosystem: Redefining America's Wilderness Heritage*. New Haven, CT: Yale University Press, 1991:209–230.

limiting and what conditions change those relationships. Predator-prey theory has a relatively rich history compared to herbivore-plant models, even though the latter is often viewed simply as a special case of the former (May, 1981).

Ungulate ecologists have attempted to incorporate predator-prey models (Haber, 1977) into ungulate management, but their utility has been questioned by some (Van Ballenberghe, 1980; 1987). Four different conceptual models may explain ungulate population dynamics under various levels of predation (Fig. 13.2). These models all assume that human influence on both predator and prey is minimal and that the ungulate of interest is the primary prey of the predator(s). These models also are specific to moose-wolf-bear interactions in boreal ecosystems (Haber, 1977; Messier & Crête, 1985; Van Ballenberghe, 1987; Gasaway et al., 1992), and their applicability to other ungulates is yet to be evaluated.

The first model (*predator free*) is applicable to areas without predators and follows Caughley's premise of food regulation. Ungulate populations fluctuate around a single equilibrium at a forage-based ecological carrying capacity (K_1), and intraspecific competition for forage regulates the ungulate population at high densities.

In the second model (*predator limited* or recurrent fluctuations), there is only one predator, or predator guild (e.g., black bears [*Ursus americanus*] and brown bears [*U. arctos*]), and ungulate populations still are regulated ultimately by competition for forage. Predation limits population growth, but over time the ungulate population would be expected to increase gradually to K_2 (see Fig. 13.2). The moose-wolf system of Isle Royale National Park can be characterized by this model. This system has been described as one of recurrent fluctuations, exhibiting mild periodicity (Keith, 1983; Van Ballenberghe, 1987; Van Ballenberghe & Ballard, 1994).

A variant of this model proposed for Isle Royale invokes cyclic-like fluctuations (stable-limit cycles), in which moose and wolf densities are inversely related (Peterson et al, 1984; Messier, 1991). However, the strict periodicity that is required for a "cycle" may not exist and, even if present, may be dynamic with time (Van Ballenberghe & Ballard, 1994).

The third model (two-state, multiple equilibrium, or "predator pit") exhibits two equilibria (K_4 and K_3) (see Fig. 13.2). The difference between K_4 and K_u has been described as a "predator pit." This model represents a system in which one predator can regulate ungulates at low density (K_4), with the possibility that ungulates can escape the regulatory effect of predation by an as yet undescribed mechanism, and reach high densities, at which point they would be regulated by food (K_3). However, there is no empirical evidence to support this model under "natural" conditions (Van Ballenberghe & Ballard, 1994). Ungulates probably do not possess the reproductive potential to escape predation (however, see Haber, 1977; and Walters et al., 1981). Sinclair (1981) similarly described a situation in which a culled wildebeest population in Africa moved from K_3 to a habitat-structure–predator-maintained K_4 when culling ceased.

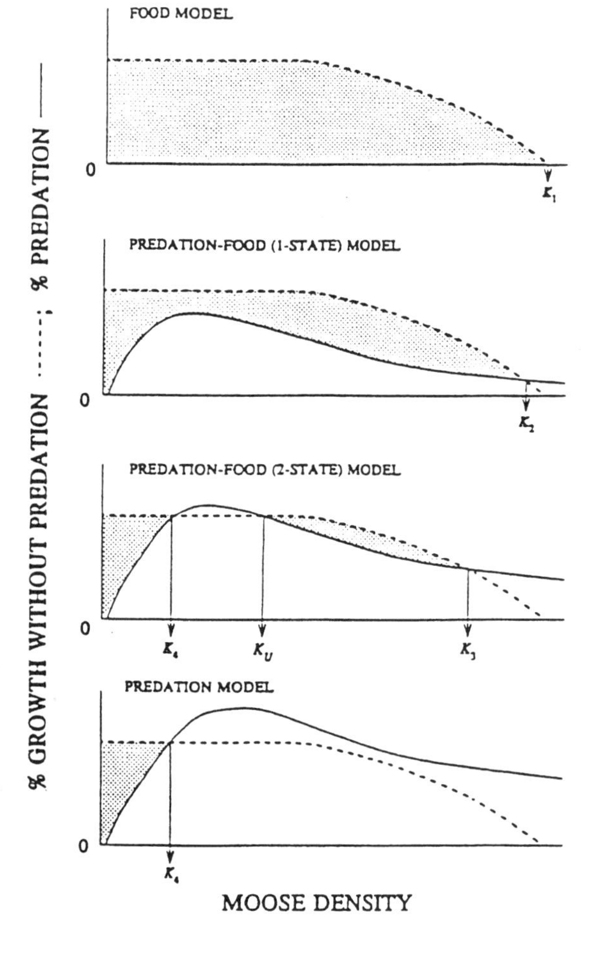

Figure 13.2 Four conceptual models of ungulate population regulation. The density relationship of wild predation (—) and the growth rate of the prey without predation (----) are illustrated. When the two lines cross, an equilibrium condition is possible. K_1 to K_4 are stable equilibria, whereas K_u is unstable. The shaded areas represent the net population growth rate after consideration of predation. In the predator free or food model, ungulates are regulated only by competition for forage and are food stressed at K_1. In the predation-food model ungulates are food stressed at K_2. Population growth rate would not increase monotonically if the density of ungulates is lowered, and predation is density dependent at low ungulate densities. In the 2-state model, ungulates are food stressed at K_3. Predation is density dependent at the lower range of ungulate density (below K_4). Predator removal would cause an immediate growth of the ungulate population up to K_3 after recolonization by predators. In the predator model, predation is density dependent below K_4. Ungulate density will revert to K_4 after termination of a predator control program. Habitat improvement would not cause an increase in ungulates from heavy predation. (From Messier F. Ungulate population models with predation: A case study with the North American Moose. *Ecology* 1994;75:478–488.)

Table 13.2 The predictions and prerequisite conditions for five conceptual models of ungulate-habitat-predator relationships.

	Models			
Conditions and predictions	Predator free	Predator limitation	Two-state	Low-density equilibrium
Predators	None	One	One	Multiple
Human influence	Minimal	Minimal	?	Minimal
Alternate prey	NA	Scarce	?	Scarce
Equilibria	One	One	Two	One
Density at equilibrium	High	High	Low-high	Low
Regulating factor	Food	Food	Predator (low) Food (high)	Predator
Predator pit	NA	No	At low density	No
Escape predation	NA	Yes	Yes	No
Population trends	Eruption, crash, stability	Recurrent fluctuations	Stability at two densities	Stability
Ungulate-predator densities	NA	Inverse	Noninverse	Noninverse

NA = not applicable.

The fourth model (*predator regulated* or low-density equilibrium) represents a system with multiple predators in which ungulates are regulated by predation and persist at low density ($< K_l/2$) for extended periods (see Fig. 13.2). Competition for forage is not important, and the number of predators must decline before the ungulate population can increase. Alternate prey may be present but are scarce. There are a number of case histories described by Van Ballenberghe (1987) and Van Ballenberghe and Ballard (1994) that appear to support this model (but see Boutin, 1992). The predicted behavior of predator and prey under conditions necessary for each of these four models is given in Table 13.2.

These models are relatively simple and their utility may therefore be limited. For example, contrary to all four models, many protected areas have abundant alternate prey. In practical terms, the presence of just one more prey species (moose) led to conflicting opinions about the effects of wolf predation on caribou (Van Ballenberghe, 1985; Bergerud & Ballard, 1988).

Disease and Parasites

Disease is invariably present in a population and often accounts for a continuous, but low-level source of mortality. In the hierarchy of limiting factors, disease outbreak is often viewed as a symptom of nutritional stress or a factor predisposing individuals to predation. In many cases, disease outbreaks among wildlife in protected areas occur because of imbalances in resistance that normally are held in equilibrium by factors inherent to the host, the disease agent, and the environment (Aquirre & Starkey, 1994).

Although uncommon, there are diseases that can have major effects on ungulate populations (Worley, 1979; Yuhill, 1987), even to the extent of limiting growth (Gavin et al., 1984; Sinclair et al., 1985). Most notably, populations of mountain sheep are locally limited by the lungworm-pneumonia complex, which may be the factor limiting overall abundance and distribution of this species in North America (Forrester, 1971). It is also theoretically possible for disease to be positively density-dependent (Wehausen et al., 1987), even though this has not yet been documented. For example, Anderson and May (1986) illustrated the theoretical importance of host density, parasite virulence, and host recovery and immune responses.

The introduction of exotic diseases to wild ungulates by livestock appears to be the most pervasive disease problem in and out of protected areas. Examples include the domestic sheep–lungworm–mountain sheep–pneumonia complex in western North America (Uhazy et al., 1973), the cattle-rinderpest-wildebeest association on the Serengeti of Africa (Talbot & Talbot 1963), and the cattle-brucellosis-bison/wapiti relationship in Yellowstone (Thorne et al., 1979; Meagher & Meyer, 1994). It is ironic that after eliminating a disease in livestock, livestock producers pressure wildlife managers to do the same, when the disease in wildlife likely originated with the introduction of livestock (Meagher & Meyer, 1994).

Transmission of disease among natural hosts may also be important. For example, white-tailed deer may transmit the meningeal worm (*Parelaphostrongylus tenuis*) to moose and cause subsequent moose population declines and range contraction (Karns, 1967; Gilbert, 1974). However, Nudds (1990, 1992) and Gilbert (1992) suggest that such conclusions are premature and substantiated more by repetition than by empirical evidence.

Managers of ungulates in protected areas could use parasites and disease to influence ungulate population growth either directly or indirectly (Caughley & Sinclair, 1994). However, success with this type of management approach has been much greater in limiting insect populations than vertebrates (Spratt, 1990). Furthermore, the deliberate introduction of an exotic organism is risky, both biologically and politically.

Disease can be a major management concern for populations of imperiled species. For example, in the early 1980s most of the endangered

Columbian white-tailed deer (*Odocoileus virginianus leucurus*) (n = 200) existed on a 2000-hectare refuge in southwest Washington (Gavin et al., 1984). Gavin et al. (1984) concluded that "foot rot" caused by necrobacillosis (*Fusobacterium necrophorum*) was a major mortality factor in this population because of environmental conditions ideal for transmission of the organism. This disease and female territoriality appeared to limit the population, as food was abundant and nutritious and predation was insignificant. However, in recent years, the population has more than doubled even though hoof rot continues to be a major mortality factor for adult deer (A. Clark, personal communication).

Managers may need to control or mitigate the effects of disease in some protected areas (Aguirre & Starkey, 1994). A key consideration seems to be whether the disease is native or exotic (NPS, 1988); often this determination is not easy to make (Aguirre & Starkey 1994). Specific strains of an organism may be native while others may not, yet they both manifest similar symptoms. In addition, disease organisms readily hybridize, beneficial mutations may rapidly become a part of the genotype, and a disease complex, such as lungworm–mountain sheep–pneumonia, may have both native and exotic components. These circumstances could be used both to justify and oppose intervention.

Disease management programs have taken many forms, ranging from habitat manipulation, to medicating individual animals, to killing an entire population in which some individuals have tested positive for the disease. Outside of host eradication, most of these programs have met with limited success. Effective disease management programs still remain to be proven and will depend on thorough research, good documentation, and use of an adaptive management approach when information is lacking and unlikely to be available in the near future (Peterson, 1991; Aguirre & Starkey 1994).

Accidents

Accidents often are viewed as an endemic but relatively infrequent source of mortality in most wild ungulate populations. Mortality from accidents can be almost instant (e.g., vehicle collisions, drownings) or gradual (e.g., a fracture of the jaw that interferes with foraging). However, it is unclear whether the rate of mortality caused by accidents is density dependent. For example, deer-vehicle collisions seem to be disproportionately greater along highways that traverse protected areas (Wright, 1992).

Accidental deaths of wild ungulates often can be dramatic. There are reports of wapiti and mule deer falling off cliffs and dying (Reed et al., 1979). Deep snow can concentrate wild ungulates along plowed highways and railroad tracks, resulting in unusually large numbers of deaths (Del Frate & Spraker, 1991; Child et al., 1991; Andersen et al., 1991). It is likely that deer-vehicle collisions will increase as human populations continue to expand but especially if ungulate populations are also increasing.

In most protected areas, accidents will remain an infrequent cause of mortality. However, there are exceptions. For example, Key deer (*Odocoileus virginianus clavium*) in Florida are confined to a small refuge that is fragmented by subdivisions and intersected by a major state highway (Hardin et al., 1984). Collisions with vehicles are a major source of mortality and an important limiting factor (Hardin et al., 1984; Humphrey & Bell, 1986).

Attempts have been made to reduce ungulate-vehicle collisions through public education, highway fencing, artificial lighting, underpass construction, off-road snow removal, establishment of winter feeding stations away from highways, reduced vehicle speeds, increased law enforcement, and installation of chatter strips and small speed bumps. However, efficacy of these approaches is often unknown as untreated "controls" are rarely established and monitoring is often lacking (see Pojar et al., 1975; Reed & Woodard, 1981; Reed et al., 1982; Singer & Doherty, 1985; Feldhamer et al., 1986; Foster & Humphrey, 1995).

Behavior

Wild ungulate behavior can limit population growth primarily through adult territoriality and the dispersal of subadults. These mechanisms are often related to sex, with males dispersing farther and more frequently than females (Jarman, 1974; Sinclair, 1992; Wolff, 1993, 1994) and females more often exhibiting territorial behavior (Hirth, 1977; Owen-Smith, 1977). These mechanisms may also be influenced by habitat stability. Populations in habitats with relatively frequent disturbances may need to track habitat changes through dispersal (Geist, 1971), while those in more stable habitats may be able to optimize fitness by in situ competition with conspecifics (Gavin et al., 1984).

Protected area managers may significantly influence these processes through habitat management involving, for example, adoption of a prescribed natural fire ("let-burn") policy, or even prescription human-set fires. Periodic fire can benefit species such as boreal moose by stabilizing the availability of early seral forest plant communities that are favored for browsing (Spencer &

Chatelain, 1953; Peek, 1974; MacCracken & Viereck, 1990). Gavin et al. (1984) concluded that several decades of haying and livestock grazing had stabilized Columbian white-tailed deer habitat. This stability facilitated the establishment and defense of territories by female deer. Finally, any land use or habitat modification, such as fencing, that blocks migration and dispersal can contribute to or trigger an ungulate population eruption.

Genetics

There are two situations in which genetics should be of concern to managers of ungulates in protected areas: (1) in reserves where populations are small and isolated, and (2) in the reintroduction of an extirpated species. In these situations the potential for genetic drift, inbreeding depression, founder effects, and outbreeding depression is real and should be addressed.

Even so, there is little to no empirical evidence that these genetic problems occur in free-ranging wild ungulates or that these problems require management intervention. For example, moose were introduced into the Copper River Delta in Alaska through the translocation of 22 calves and one yearling from 1949 to 1958 (MacCracken, 1992, unpublished data). The population grew rapidly because of high-quality habitat. The mechanics of the translocation, moose breeding behavior, and antler morphology suggest the existence of a founder effect. Males exhibited exceptionally large antler spreads at all age classes (Gasaway et al., 1987) and no spike or forked antlers in yearlings, but lacked antler brow palmation otherwise typical of the subspecies (MacCracken, 1992, unpublished data). Because the population was small and isolated, inbreeding was probably high, possibly evident in fecundity less than other Alaskan moose populations (MacCracken & Stephensen, unpublished data). Nonetheless, the population is currently limited, primarily by legal hunting, and is able to meet the sustained-yield management goals set by the Alaska Department of Fish and Game (MacCracken, 1992).

Even though inbreeding and outbreeding depression can theoretically limit population growth, it appears that drift and inbreeding problems are easily mitigated by the periodic introduction of a small amount of new genetic material (Simberloff, 1980). In small populations where inbreeding is potentially a serious problem and augmentation is not possible, intensive management of matings can influence genetic diversity (Schoenwald-Cox et al., 1983), even though this would be difficult with a free-ranging population. However, the rapid purging of recessive lethal alleles may be common among wild ungulates and may explain why many populations survived the genetic bottlenecks created by market hunting. Founder effects can furthermore be avoided by reintroduction of individuals from different populations, manipulation of breeding individuals through selective removals, as well as future augmentation.

Management Issues

Maximum Sustained Yield

Maximum sustained yield (MSY) theory (Caughley, 1976) has produced several concepts that may be beneficial to managers of protected areas. The idea of a "residual base population" and a "harvestable surplus" can be applied to any population if one assumes that the "harvest" due to natural mortality factors is similar to that which would occur under a sport hunting regime.

In general, MSY occurs somewhere between 0.5 and 0.8 K, depending largely on the species reproductive potential. The lower the reproductive potential of the species, the closer MSY will be to K (McCullough, 1987). Protected area managers need to understand the consequences of changing residual population size (McCullough, 1987). A large residual population may provide more public viewing opportunities, while small residual populations could reduce impacts on forage. However, both levels could lead to unforeseen consequences.

There appear to be gradients in population stability, resilience, additive-compensatory mortality, and the threat of extinction in relation to residual population levels and MSY. Population stability is low for small residual populations, reaches an optimum around MSY, and then declines at K. However, the other factors change consistently as the residual population increases; resilience increases, threat of extinction decreases, and mortality moves from additive to compensatory.

However, difficulties in reliably estimating K make application of these concepts problematic. Although the effort required to measure K is substantial and the variety of approaches can be confusing (McCullough, 1987), a number of studies apparently have derived relatively good estimates (Hobbs et al., 1982; Regelin et al., 1987; Crête, 1989; MacCracken, unpublished data). Accuracy is probably less important for protected areas without hunting or culling, and managers should be able to use most reasonable estimates.

Rarity

Two cases from North America illustrate some of the problems that can be associated with the conservation of rare ungulates. Both are subspecies of white-tailed deer (Columbian and Key), and both appear to be rare due to past unregulated hunting and habitat loss (Gavin 1978; Gavin et al., 1984; Hardin et al., 1984). Small refuges (2100 and 2400 ha) have been established for each, and both subspecies were listed as endangered under legislation preceding the Endangered Species Act (ESA) of 1973.

The current situation has improved substantially for the Columbian subspecies (Clark, 1995; personal communication). Two distinct populations exist, one in southwest Washington and another near Roseburg, Oregon (Gavin, 1984). The population on the Washington refuge and surrounding lands has doubled since the early 1980s, and its distribution has expanded through translocation, land acquisitions, and conservation easements with other landowners. The other population, in southwest Oregon, may consist of as many as 9000 animals on a 1200-square-kilometer area (Clark, 1995). This subspecies has met the population targets specified in its recovery plan (U.S. Fish and Wildlife Service, 1983), and the only criterion left unmet is the creation of additional secure habitat through further land acquisition or landowner agreements, in both the lower Columbia River Basin and southwest Oregon.

Coyote predation on fawns currently appears to be limiting the Columbian refuge population, although hoof rot is still a major cause of mortality for adults, and competition with wapiti for forage and cover is also a management concern.

Currently, there is less reason to be optimistic about the Key deer. The National Key Deer Refuge is composed of a number of closely spaced oceanic islands (i.e., keys) (Hardin et al., 1984). The island land ownership patterns, and intense suburbanization of private lands within the refuge, have fragmented deer habitat and created a number of mortality sources. Humphrey and Bell (1986) suggested that reconfiguration of the refuge, and further land acquisition may be necessary for conservation, but it also would be very difficult to accomplish.

Feeding of deer by residents of suburbs within the refuge has also created problems. This feeding may have allowed the population to exceed the forage-based carrying capacity as suggested by distinct browse lines on most of the core refuge. Public education may help alleviate this situation. In addition, such activity could be illegal under the broad definition of *take* in Section 9 of the ESA and the implementing regulations.

In 1974, the National Key Deer Refuge population was estimated to be about 300 (Hardin et al., 1984) and declining (Humphrey & Bell, 1986). However, the usefulness of these unreliable population estimates is questioned by refuge personnel, and population status and trend are currently indexed by the distribution of Key deer on the refuge. Recent range contractions and extinctions from some islands are cause for concern, and the subspecies is likely to remain listed as endangered for these reasons.

In summary, both subspecies appear to have become endangered for similar reasons. Recovery efforts have also taken similar routes, including protection from harvest and establishment of refuges. The difference in success between the two programs appears to be directly related to land-use patterns and human populations levels. Both populations of the Columbian white-tailed deer inhabit rural areas where opportunities to devote lands to recovery are much greater. In addition, low human populations also reduce the impacts of human-related mortality factors (e.g., vehicle collisions, poaching, domestic dogs, etc.). In contrast, the National Key Deer Refuge is limited physically and by human activity. These factors precipitate high mortality as well as limited opportunities to secure additional habitat.

Overabundance

The most significant problem that protected area managers face is high-density ungulate populations (Jewell & Holt, 1981). However, it is difficult to determine when an ungulate population is too abundant, and the problem is complicated by the lack of clear management objectives for protected areas (Noy-Meir, 1981; May & Beddington, 1981; Laws, 1981). (See the discussion on overabundance in Chap. 12.)

Caughley (1981b) identified four definitions of wildlife overabundance: (1) when wildlife are a nuisance to humans, (2) when wildlife negatively impact other species, (3) when wildlife negatively impact themselves, and (4) when the herbivore-plant relationship is outside the equilibrium range. The first three definitions of overabundance are rooted primarily in human perceptions and values, which led Caughley to focus on the ecologically based fourth definition. In addition, alien ungulates in protected areas may always be judged too abundant, simply by being present.

All of these definitions are related to the management objectives of a given refuge. For example, if a protected area is established to conserve an endangered plant species that is limited to habitats around mineral springs, and if those springs are also used by ungulates, there is undoubtedly an ungulate population level at which trampling would become a serious threat to those plant populations. In this situation, Caughley's second definition would specify overabundance.

Noy-Meir (1981) suggested that most observers would agree that species extinction as a result of high-density wildlife populations should not be allowed to occur in protected areas. However, this view conflicts with some predicted outcomes of the herbivore-plant equilibrium theory (Caughley, 1981b; Sinclair, 1981) and overlooks the potential for systems to exhibit two or more stable states (May & Beddington, 1981; Sinclair, 1981). These outcomes can include major changes in species composition, with the potential extinction of certain species.

Management of Overabundant Populations

Several options are available for managing and controlling ungulate populations in protected areas. Culling, translocation, and fertility control were discussed in Chapter 12. The effects of restoring predators and subsistence hunting are discussed in more detail in the following sections.

Restoration of Large Predators

Large predators (e.g., wolves, bears, mountain lions) have been eliminated or their populations reduced in many protected areas, at least in the United States (Wright, 1993; Curlee et al., 1994). These same areas also often have high-density wild ungulate populations that some perceive as a problem (Vecellio et al., 1994; Wagner et al., 1995). The restoration of predators to most of these areas in numbers sufficient to affect ungulate populations may be impossible, simply because the protected area is too small. Even in areas judged to be large enough, there are a number of social and cultural barriers complicating large predator restoration (Curlee et al., 1994; Clark & Minta 1994; MacCracken et al., 1994). Restoration of predator populations will be significantly hampered and may be impossible, if the fears and mistrust of local people are not allayed.

Nonetheless, predators are currently being restored to some large protected areas in the western United States. Wolves from Canada have colonized Glacier National Park, and they have begun to be reintroduced into Yellowstone National Park and the central Idaho wilderness complex as discussed in Chapter 16. The effects that wolves will have on ungulates in these areas are unknown. Even though Garton et al. (1990) and Vales and Peek (1990) estimated minimal impacts on elk populations in and surrounding Yellowstone, surprising responses are always possible from complex natural systems.

Successful restoration of one or more large predators has the potential to limit ungulate populations in a protected area. Whether population reductions will be enough to solve problems of ungulate overabundance depends on many factors. Predator effectiveness is a major factor. For example, black bears seem to limit moose populations on the Kenai Peninsula (Franzmann et al., 1980; Schwartz & Franzmann, 1991) but not in other areas of Alaska (Van Ballenberghe, 1987; MacCracken 1992). Black bears prey primarily on moose calves during a relatively short period following parturition, and their effectiveness follows from their ability to exploit this small window of opportunity.

Predator preferences and the abundance of alternative prey are also important (Andrewartha & Birch, 1984). Bergerud and Ballard (1988, 1989) argued that wolves preferred caribou over moose. This type of predator preference may be a function of ungulate body size and may be related to handling time, risk of physical injury, and food reward. Thus, wolves may prefer deer over elk in Yellowstone.

Habitat conditions can also influence predator-prey relationships through complex feedback loops and time lags in population responses. Schwartz and Franzmann (1991) found that rates of black bear predation on moose calves were similar between high- and low-quality moose habitat. However, the absolute number of calves available to bears was greater in the high-quality habitat, which resulted in significantly greater bear densities. One could speculate that as high-quality moose habitat diminishes with forest maturation, the bear population may remain high even though moose production declines. Thus, there may be several years when rates of calf predation increase before the bear population begins to decline.

Subsistence Hunting

The concept of subsistence hunting was codified into U.S. law with passage of the Alaska Native Claims Settlement Act and the Alaska National Interests Lands Conservation Act. These laws recognized the importance of subsistence hunting to the economy and culture of the indigenous Alaskan natives and attempted to provide a means to maintain that way of life. The concept has been highly controversial, difficult to implement, and deemed unconstitutional at the federal and state levels because it was racially based and excluded segments of the population from a common resource. State-based policy went through many polarizing iterations until the Alaska legislature finally abandoned attempts to find a solution and deferred wildlife management to the federal government on federal lands. Subsistence hunts thus are a current feature of many protected areas in Alaska. In general, subsistence hunting is restricted to local residents and follows established seasons and bag limits, although in some places seasons and limits may be extended. The widespread creation of national preserves in Alaska occurred largely in response to this difficult issue (Wright, 1984).

Despite difficulties of implementation, subsistence hunting has potential as a management tool for reducing overabundant ungulate populations. However, this usefulness is primarily defined by the purposes and goals of the protected area. Two examples illustrate the conditional nature of this potential:

1. Imagine that a relatively small refuge has been established in a unique area that contains several endemic endangered plants. The purpose of the reserve is to maintain the ecosystems on which the plants depend. The deer population has increased and is threatening the persistence and recovery of many of the plant species. The refuge managers decide to allow a subsistence hunt targeting specific age-sex cohorts of the deer population. Participation in the hunt is by permit, and the distance of the applicant's home to the refuge and income level are major selection criteria. There are a number of precedents that allow for discrimination based on residency and income in the allocation of wildlife resources. In most states, nonresident hunter permits often are limited and more expensive than the unlimited residence licenses. In addition, the salvage of road-killed wild ungulates is sometimes restricted to low-income families.
2. A number of protected areas also have been created to preserve historic sites. The recreation of the historic scene with people dressed in clothing of the period, using traditional tools, and living in customary housing is a feature of many of these areas. Subsistence hunting was a reality of those times and could be incorporated into the reliving of history in many areas such as national military parks. The seasonal use of the Yellowstone area by indigenous peoples has interesting implications. It could be argued that the effects of indigenous peoples on those ecosystems prior to European settlement were as natural as those of wolves and bison. Can ecosystem processes be restored and maintained without these inputs? Who might better restore the role of indigenous peoples to those ecosystems than their descendants?

Conclusions

Wild ungulates present protected area managers with a number of challenges as well as opportunities. Both can be related to the purpose and objectives of the reserve. It seems unlikely that many protected areas will have the singularity and clarity of purpose, which many have called for, that would greatly simplify the manager's job. Since the creation of Yellowstone National Park more than 100 years ago, the national parks of the United States have been charged with the dual and sometimes conflicting missions of preserving nature and providing for public enjoyment (NPS, 1988). These broad mandates are unlikely to change any time soon. In addition, many people see the large national parks as integral for the conservation of biodiversity (Boyce, 1991; Wright et al., 1994; Noss & Cooperrider, 1994; Wagner et al., 1995).

The ambiguous purpose(s) of many protected areas allows for a wide variety of policies to be adopted to deal with changing political realities and ecological understanding. In most cases, policies have and will continue either to mirror the consensus of public and professional opinions or, lacking consensus, will reflect the judgments of the policymakers. Controversy may be even greater in protected areas with well-defined purposes that do not accommodate changes promulgated by increased understanding of ecosystems or public opinion. It is much harder to redefine the purpose of a refuge than to modify a specific management policy.

Equilibrium theory has driven the conceptualization of ungulate population dynamics for the last two decades and has been applied in developing management policies for protected areas. These policies often were justified as tests of equilibrium theory, but it appears that little effort was expended in defining potential controls and replicates, and monitoring has been inconsistent. However, opportunities still abound in this area.

Equilibrium theory currently is being challenged (Botkin & Sobel, 1975; Botkin, 1990) along with its application to protected area management (Coughenour & Singer, 1991). The possibility of multiple stable states resulting from shifts among ecological domains in ungulate-plant dynamics (Sinclair, 1981), scale effects on stability (Coughenour & Singer, 1991), chaotic responses (Gleick, 1987), and continuums in dynamic behavior greatly complicate the issue and at present leave no clear direction for managers. However, the same opportunities to "experiment" exist as under equilibrium theory.

The vision of a nationwide, connected, biodiversity-reserve system with large core protected areas, such as national parks, surrounded by buffer areas of increasing human impact (Noss & Cooperrider, 1994) is likely far in the future. However, such a system is unlikely to reduce controversy or make the land manager's job any easier. Undoubtedly, wild ungulates will be a major feature of such a system, and the same types of real or perceived problems will continue to exist. Nontheless, one major benefit may be realized if such a system can truly function as envisioned—wild ungulates and the complete ecological units on which they depend will persist despite our best intentions.

Literature Cited

Aguirre AA, Starkey EE. Wildlife diseases in U.S. national parks: Historical and coevolutionary perspectives. *Cons Biol* 1994;8:654–661.

Andersen R, Wiseth B, Pederson PH, Jaren V. Moose-train collisions: Effects of environmental conditions. *Alces* 1991;27:79–84.

Anderson RM, May RM. The invasion, persistence, and spread of infectious diseases within animal and plant communities. *Phiol Trans Royal Soc London Ser B: Biol Sci* 1986;314:533–570.

Andrewartha HG, Birch LC. *The Ecological Web.* Chicago: University of Chicago Press, 1984.

Bergerud AT, Ballard WB. Wolf predation on caribou: The Nelchina herd case history, a different interpretation. *J Wildl Manage* 1988;52:344–357.

Bergerud AT, Ballard WB. Wolf predation on the Nelchina caribou herd: A reply. *J Wildl Manage* 1989;53:251–259.

Botkin DB. *Discordant harmonies: A new ecology for the twenty-first century.* New York: Oxford University Press, 1990.

Botkin DB, Sobel MJ. Stability in time-varying ecosystems. *Am Nat* 1975;109:625–645.

Boutin S. Predation and moose population dynamics: A critique. *J Wildl Manage* 1992;56:116–127.

Boyce MS. Natural Regulation or the Control of Nature? In RB Keiter, MS Boyce (eds), *The Greater Yellowstone Ecosystem: Redefining America's Wilderness Heritage.* New Haven, CT: Yale University Press, 1991:183–208.

Bratton SP. Impacts of white-tailed deer on the vegetation of Cades Cove, Great Smoky Mountain National Park. *Proc Ann Conf Southeast Assoc Fish Wildl Agen* 1979;33: 305–312.

Bryant JP, et al. Biogeographic evidence for the evolution of chemical defense by boreal birch and willow against mammalian browsing. *Am Nat* 1989;134:18–24.

Cahalane VH. Elk management and herd regulation—Yellowstone National Park. *Trans N Am Wildl Conf* 1943;8:95–101.

Caughley G. Eruption of ungulate populations, with emphasis on Himalayan thar in New Zealand. *Ecol* 1970;51:53–72.

Caughley G. Wildlife management and the dynamics of ungulate populations. *Appl Biol* 1976;1:183–246.

Caughley G. Comments on natural regulation of ungulates (what constitutes a real wilderness?). *Wildl Soc Bull* 1981a;9:232–233.

Caughley G. Overpopulation. In PA Jewell, S Holt (eds), *Problems in Management of Locally Abundant Wild Mammals.* New York: Academic Press, 1981b:7–19.

Caughley G, Lawton JH. Plant-herbivore Systems. In R M May (ed), *Theoretical Ecology.* Sunderland, MA: Sinauer Associates, 1981:132–166.

Caughley G, Sinclair ARE. *Wildlife Ecology and Management.* Boston: Blackwell Science, 1994.

Chadde SW, Kay CE. Tall-willow Communities on Yellowstone's Northern Range: A Test of the "Natural-regulation" Paradigm. In RB Keiter, MS Boyce (eds), *The Greater Yellowstone Ecosystem: Redefining America's Wilderness Heritage.* New Haven, CT: Yale University Press, 1991:231–262.

Child KN, Barry SP, Aitken DA. Moose mortality on highways and railways in British Columbia. *Alces* 1991;27:41–49.

Clark A. Columbian White-tailed Deer: Status Update. *Int Union Conserv Nature, Deer Specialist Group Newsletter,* 1995.

Clark TW, Minta SC. *Greater Yellowstone's Future.* Moose, WY: Homestead Publishing, 1994.

Cole GC. An ecological rationale for the natural and artificial regulation of native ungulates in parks. *Trans N Am Wildl Conf* 1971;36:417–425.

Coughenour MB, Singer FJ. The Concept of Overgrazing and Its Application to Yellowstone's Northern Range. In RB Keiter, MS Boyce (eds), *The Greater Yellowstone Ecosystem: Redefining America's Wilderness Heritage.* New Haven, CT: Yale University Press, 1991:209–230.

Crête M. Approximation of K carrying capacity for moose in eastern Quebec. *Can J Zool* 1989;67:373–380.

Curlee PA, Clark TW, Casey D, Reading RP. Large carnivore conservation: Back to the future. *Endang Spec Upd* 1994;11(1):1–4.

DeAngelis DL, Waterhouse JC. Equilibrium and non-equilibrium concepts in ecological models. *Ecol Monog* 1987;57:1–21.

Del Frate G, Spraker TH. Moose-vehicle interactions and an associated public awareness program on the Kenai Peninsula, Alaska, *Alces* 1991;27:1–7.

Feldhamer GA, Gates JE, Harman DM, Loranger AJ, Dixon KR. Effects of interstate highway fencing on white-tailed deer activity. *J Wildl Manage* 1986;50:497–503.

Forrester DJ. Bighorn Sheep Lungworm-Pneumonia Complex. In JW Davis, RC Anderson (eds), *Parasitic Diseases of Wild Mammals.* Ames, IA: Iowa State University Press, 1971:158–173.

Foster ML, Humphrey SR. Use of highway underpasses by Florida panthers and other wildlife. *Wildl Soc Bull* 1995;23:95–100.

Fox JR, Pelton MR. Observations of a white-tailed deer die-off in the Great Smoky Mountain National Park. *Proc Ann Conf Southeast Game Fish Comm* 1974;27:297–301.

Franzmann AW, Schwartz CC, Peterson RO. Moose calf mortality in summer on the Kenai Peninsula, Alaska. *J Wildl Manage* 1995;59:764–768.

Garton EO, Crabtree RL, Ackerman BB, Wright RG. The Potential Impact of a Reintroduced Wolf Population on the Northern Yellowstone Elk Herd. In *Wolves for Yellowstone? A report to the U.S. Congress* (Vol II Research and Analysis). Washington, D.C: US Government Printing Office, 1990:3-59–3-92.

Gasaway WC, Stephenson RO, Davis JL, Shepherd PEK, Burris OE. Interrelationships of wolves, prey, and man in interior Alaska. *Wildl Monogr* 84, 1983.

Gasaway WC, Preston DJ, Reed DJ, Roby DD. Comparative antler morphology and size of North American moose. *Swedish Wildl Res* (Suppl) 1987;1:311–325.

Gasaway WC, Boertje RD, Grangaard DV, Kellyhouse DG, Stephenson RO, Larson DG. The role of predation in limiting moose at low densities in Alaska and Yukon and implications for conservation. *Wildl Monogr* 120, 1992.

Gavin T. Status of Columbian white-tailed deer *Odocoileus virginianus leucurus:* Some quantitative uses of biogeographic data. In *Threatened Deer.* Morges, Switzerland: International Union of Conservation of Nature National Research 1978:185–202.

Gavin T. Pacific Northwest. In LK Halls (ed), *White-tailed Deer: Ecology and Management.* Harrisburg, PA: Stackpole Books, 1984:487–496.

Gavin T, Suring LH, Vohs PA, Meslow E C. Population characteristics, spatial organization, and natural mortality in the Columbian white-tailed deer. *Wildl Monogr* 91, 1984.

Geist V. *Mountain Sheep: A study in Behavior and Evolution.* Chicago: University of Chicago Press, 1971.

Gilbert FF. *Parelaphostrongylus tenuis* in Maine: II-prevalence in moose. *J Wildl Manage* 1974;38:42–46.

Gilbert FF. Retroductive logic and the effects of meningeal worms: A comment. *J Wildl Manage* 1992;56:614–616.

Gleick J. *Chaos: Making a New Science.* New York: Penguin Books, 1987.

Haber GC. Socio-ecological Dynamics of Wolves and Prey in a Subarctic Ecosystem. (Ph.D. thesis). Vancouver: University of British Columbia, 1977.

Hardin JW, Klimstra WD, Silvy NJ. Florida Keys. In LK Halls (ed), *White-tailed Deer: Ecology and Management.* Harrisburg, PA: Stackpole Books, 1984:381–390.

Hirth DH. Social behavior of white-tailed deer in relation to habitat. *Ecol Monogr* 1977:53:1–55.

Hobbs NT, Baker DL, Ellis JE, Swift DM, Green RA. Energy- and nitrogen-based estimates of elk winter-range carrying capacity. *J Wildl Manage* 1982;146:12–21.

Hofmann RR. Evolutionary steps of ecophysiological adaptation of ruminants: A comparative view of their digestive system. *Oecologia* 1986;78:443–457.

Houston DB. Ecosystems of national parks. *Science* 1971;172:648–651.

Houston DB. *The Northern Yellowstone Elk: Ecology and Management.* New York: Macmillan, 1982.

Humphrey SR, Bell B. The Key deer population is declining. *Wildl Soc Bull* 1986;14:261–265.

Irwin LL. Deer-moose relationships on a burn in northeastern Minnesota. *J Wildl Manage* 1975;39:653–662.

Jarman PJ. The social organization of antelope in relation to their ecology. *Behav* 1974;48:215–267.

Jewell PA, Holt S. *Problems in Management of Locally Abundant Wild Mammals.* New York: Academic Press, 1981.

Karns PD. *Pneumostrongylus tenuis* in deer in Minnesota and implications for moose. *J Wildl Manage* 1967;31:399–403.

Keith LB. Some features of population dynamics in mammals. *Trans Int Congr Game Biol* 1974;11:17–58.

Keith LB. Population Dynamics of Wolves. In L N Carbyn (ed), *Wolves: Their status, Biology, and Management in Canada and Alaska.* Ottawa: Canadian Wildlife Service Report (Ser 45) 1983:66–77.

Klein DR. The introduction, increase, and crash of reindeer on St. Matthew island. *J Wildl Manage* 1968;32:350–367.

Lack D. *The Natural Regulation of Animal Numbers.* London: Oxford University Press, 1954.

Laws RM. Large Mammal Feeding Strategies and Related Overabundance Problems. In PA Jewell, S Holt (eds), *Problems in Management of Locally Abundant Wild Mammals.* New York: Academic Press, 1981:217–232.

Leopold AS, Cain SA, Cottam CM, Gabrielson IN, Kimball TL. Wildlife management in the national parks. *Trans N Am Wildl Conf* 1963;28:28–45.

MacCracken JG. The Ecology of Moose on the Copper River Delta, Alaska (Ph.D. thesis) Moscow, ID: University of Idaho, 1992.

MacCracken JG, Goble D, O'Laughlin J. Grizzly bear recovery in Idaho. *Ida For Wildl Range Policy Anal Group* (Rep 12) Moscow, ID: University of Idaho, 1994.

MacCracken JG, Viereck LA. Browse regrowth and use by moose after fire in Interior Alaska. *Northwest Sci* 1990;64:11–18.

MacNab J. Carrying capacity and other slippery shibboleths. *Wildl Soc Bull* 1985;4:403–410.

May RM (ed). *Theoretical Ecology.* Sunderland, MA: Sinauer Associates, 1981.

May RM, Beddington JR. Notes on Some Topics in Theoretical Ecology, in Relation to the Management of Locally Abundant Populations of Mammals. In PA Jewell, S Holt (eds), *Problems in Management of Locally Abundant Wild Animals.* New York: Academic Press, 1981:205–216.

McCullough DR. *The George Reserve Deer Herd.* Ann Arbor, MI: University of Michigan Press, 1979.

McCullough DR. The Theory and Management of *Odocoileus* Populations. In CM Wemmer (ed), *Biology and Management of the Cervidae.* Washington, DC: Smithsonian Institution Press, 1987:535–549.

McNaughton SJ. Ecology of a grazing ecosystem: The Serengeti. *Ecol Monogr* 1985;55:259–294.

Meagher M, Meyer ME. On the origin of brucellosis in bison of Yellowstone National Park: A review. *Cons Biol* 1994;8:645–653.

Messier F. The significance of limiting and regulating factors on the demography of moose and white-tailed deer. *J Anim Ecol* 1991;60:377–393.

Messier F, Crête M. Moose-wolf dynamics and the natural regulation of moose populations. *Oecologia* 1985;65:503–512.

Messier F, Hout J, Le Henaff D, Luttich S. Demography of the George River Caribou Herd: Evidence of population regulation by forage exploitation and range expansion. *Arctic* 1988;41:279–287.

Naiman RJ, Melillo JM, Hobbie JE. Ecosystem alteration of boreal forest streams by beaver (*Castor canadensis*). *Ecol* 1986;67:1254–1269.

National Park Service. *Management Policies.* Washington DC: U.S. Department of Interior, National Park Service, 1988.

Noss RF, Cooperrider AY. *Saving Nature's Legacy: Protecting and Restoring Biodiversity.* Washington, DC: Island Press, 1994.

Noy-Meir I. Responses of Vegetation to the Abundance of Mammalian Herbivores. In PA Jewell, S Holt (eds), *Problems in Management of Locally Abundant Wild Mammals.* New York: Academic Press, 1981:233–246.

Nudds TD. Retroductive logic in retrospect: The ecological effects of meningeal worms. *J Wildl Manage* 1990;54:396–402.

Nudds TD. Retroductive logic and the effects of meningeal worms: A reply. *J Wildl Manage* 1992;56:617–619.

Owen-Smith N. On territoriality in ungulates and an evolutionary model. *Q Rev Biol* 1977;52:1–38.

Peek JM. Initial response of moose to a forest fire in northeastern Minnesota. *Am Midl Nat* 1974;91:435–438.

Peek JM. The natural regulation of ungulates (what constitutes a real wilderness?). *Wildl Soc Bull* 1980:217–227.
Peek JM. *A Review of Wildlife Management.* Englewood Cliffs, NJ: Prentice Hall, 1986.
Peterson MJ. Wildlife parasitism, science, and management policy. *J Wildl Manage* 1991;55:782–789.
Peterson RO, Page RE, Dodge KM. Wolves, moose, and the allometry of population cycles. *Science* 1984;224:1350–1352.
Pojar TM, Prosence RA, Reed DF, Woodard TN. Effectiveness of a lighted, animated deer-crossing sign. *J Wildl Manage* 1975;39:87–91.
Reed DF, Beck TDI, Woodard TN. Methods of reducing deer-vehicle accidents: Benefit-cost analysis. *Wildl Soc Bull* 1982;10:349–354.
Reed DF, Kincaid KR, Beck TDI. Migratory mule deer fall from highway cliffs. *J Wildl Manage* 1979;43:272.
Reed DF, Woodard TN. Effectiveness of highway lighting in reducing deer-vehicle accidents. *J Wildl Manage* 1981;45:721–726.
Regelin WL, Hubbert ME, Schwartz CC, Reed DJ. Field test of a moose carrying capacity model. *Alces* 1987;23:243–284.
Robisch E, Wright RG. A survey of ungulate management in selected US national parks. *Nat Areas J* 1995;15:117–123.
Schwartz CC, Franzmann AW. Interrelationship of black bears to moose and forest succession in the northern coniferous forest. *Wildl Monogr* 113, 1991.
Schonewald-Cox CM, Chambers SM, MacBryde B, Thomas WL. *Genetics and Conservation.* Menlo Park, CA: Benjamin/Cummings, 1983.
Simberloff D. The contribution of population and community biology to conservation science. *Ann Rev Ecol Syst* 1988;19:473–511.
Sinclair ARE. Environmental Carrying Capacity and the Evidence for Overabundance. In PA Jewell, S Holt (eds), *Problems in Management of Locally Abundant Wild Mammals.* New York: Academic Press, 1981:247–257.
Sinclair ARE. Do large mammals disperse like small mammals? In NC Stenseth, WZ Lidicker Jr (eds), *Animal Dispersal: Small Mammals as Models.* London: Chapman Hall, 1992:229–242.
Sinclair ARE, Dublin H, Borner M. Population regulation of Serengeti wildebeest: A test of the food hypothesis. *Oecologia* 1985;65:266–268.
Sinclair ARE, Norton-Griffiths M. Does competition or facilitation regulate migrant ungulate populations in the Serengeti? A test of hypotheses. *Oecologia* 1982;53:364–369.
Singer FJ, Doherty, JL. Managing mountain goats at a highway crossing. *Wildl Soc Bull* 13:469–477.
Skinner MP. The elk situation. *J Mamm* 1928;9:309–317.
Skogland T. Density-dependent food limitation and maximal production in wild reindeer herds. *J Wildl Manage* 1986;50:314–319.
Spencer DL, Chatelain EF. Progress in the management of the moose in southcentral Alaska. *Trans N Am Wildl Conf* 1953;18:539–552.
Spencer DL, Hakala JL. Moose and fire on the Kenai. *Tall Timbers Fire Ecol Conf* 1964;3:11–33.
Spratt DM. The role of helminths in the biological control of mammals. *Int J Parasitol* 1990;20:543–550.

Talbot LM, Talbot MH. The wildebeest in western Masai-land. *Wildl Monogr* 12, 1973.
Thorne ET, Morton JK, Ray WC. Brucellosis, Its Effect and Impact on Elk in Western Wyoming. In MS Boyce, LD Hayden-Wing (eds), *North American Elk: Ecology, Behavior, and Management.* Laramie, WY: University of Wyoming Press, 1979:212–220.
Uhazy LS, Holmes JC, Stelfox JG. Lungworms in Rocky Mountain bighorn sheep of western Canada. *Can J Zool* 1973;51:817–824.
U.S. Fish and Wildlife Service. *Columbian White-tailed Deer Recovery Plan.* Portland, OR: U.S. Department of Interior, Fish and Wildlife Service, 1983.
Vales DJ, Peek JM. Estimates of the potential interactions between hunter harvest and wolf predation on the Sand Creek, Idaho, and Gallatin Montana elk populations. In: Yellowstone National Park, U.S. Fish and Wildlife Service, University of Idaho, Interagency Grizzly Bear Study Team, eds. *Wolves for Yellowstone? Report to the U. S. Congress.* Vol II. Research and analysis. Yellowstone, WY: Yellowstone National Park, 1990:3-93–3-167.
Van Ballenberghe V. Utility of multiple equilibrium concepts applied to population dynamics of moose. *Proc N Am Moose Conf* 1980;16:571–586.
Van Ballenberghe V. Wolf predation on caribou: The Nelchina herd case history. *J Wildl Manage* 1985;49:711–720.
Van Ballenberghe V. Effects of predation on moose numbers: A review of recent North American studies. *Swedish Wildl Res* (Suppl) 1987;1:431–460.
Van Ballenberghe V, Ballard WB. Limitation and regulation of moose populations: The role of predation. *Can J Zool* 1994;72.
Vecellio GM, Yahner RH, Storm GL. Crop damage by deer at Gettysburg Park. *Wildl Soc Bull* 1994;22:89–93.
Wagner FH, et al. *Wildlife Policies in the U.S. National Parks.* Washington, DC: Island Press, 1995.
Walters CJ, Stocker M, Haber GC. Simulation and Optimization Models for a Wolf-Ungulate System. In CW Fowler, TD Smith (eds), *Dynamics of Large Mammal Populations.* New York: Wiley, 1981:317–337.
Wehausen JD, Bleich VC, Blong B, Russi TL. Recruitment dynamics in a Southern California mountain sheep population. *J Wildl Manage* 1987;51:86–98.
Wolff JO. What is the role of adults in mammalian juvenile dispersal? *Oikos* 1993;68:173–176.
Wolff JO. More on juvenile dispersal in mammals. *Oikos* 1994;71:349–352.
Worley DE. Parasites and Parasitic Diseases of Elk in the Northern Rocky Mountain Region: A Review. In M S Boyce, L D Hayden-Wing (eds), *North American Elk: Ecology, Behavior, and Management.* Laramie, WY: University of Wyoming Press, 1979:206–211.
Wright RG. Wildlife resources in creating the new Alaskan parks and preserves. *Environ Manage* 1984;8:121–124.
Wright RG. *Wildlife Research and Management in the National Parks.* Urbana, IL: University of Illinois Press, 1992.
Wright RG. Wildlife management in parks and suburbs: Alternatives to sport hunting. *Renew Res J* 1993;11:18–22.
Wright RG, MacCracken JG, Hall J. An ecological evaluation of proposed new conservation areas in Idaho: Evaluating proposed Idaho national parks. *Cons Biol* 1994;8:207–216.
Yuhill TM. Diseases as components of mammalian ecosystems: mayhem and subtlety. *Can J Zool* 1987;65:1061–1066.

14

The Use and Role of Fire in Natural Areas

Stephen C. Bunting

The establishment of parks and protected areas has been largely driven by society's desire to protect species and/or natural landscapes from major human-caused changes. As a consequence, for many years the philosophy of management was to suppress or eliminate any process or action that disrupted or destroyed the species and landscapes of value. These views were consistent with those of ecological science and society in general at the time. In fact, for much of this century, fire management was considered to be synonymous with fire suppression (Pyne, 1982), and, with the exception of a few dramatic incidents, fire suppression activities, particularly in the last 40 years, have been very successful.

It has taken some time, but ecologists now recognize the important role that fire has played in shaping the landscapes of North America. In the past, fires set by lightning or by Native Americans controlled plant and animal communities over most of North America. However, as a result of fire exclusion, many of these communities have undergone major changes in composition and structure, both in natural and human-modified areas (Fig. 14.1). In many cases, the cumulative changes resulting from fire suppression are so profound they can no longer be ignored. In extreme cases, the successful exclusion of fire has resulted in the loss of the ecosystem managers were charged to protect. Learning from observation and scientific studies, protected area managers have begun to question the philosophy of fire exclusion in the protection of natural areas. Today, many managers now recognize that fire is an ecological process vital to the maintenance of many species communities and to the health of the ecosystem they are obligated to protect. On the other hand,

Figure 14.1 Ponderosa pine natural area in New Mexico where fire exclusion has resulted in change in community structure and fuel loading.

efforts to expand the use of fire and broaden its acceptance have been thwarted by years of successful public relations campaigns (e.g., Smokey Bear) that touted the evils of fire. Compounding the problem is the fact that the inclusion of fire in many contemporary protected ecosystems is difficult because of the size of the area and the management objectives of adjacent lands. This chapter addresses the questions of why and, if so, how to include fire in protected areas.

Fire as an Ecological Process

Fire has been an agent of change, shaping the composition and structure of nearly every terrestrial vegetation type on earth (Wright, 1974; Chandler et al., 1983). The probability and effects of these fires vary widely depending on community characteristics such as fuel loading, topography, and potential ignition sources. In extremely arid, mesic, and/or cold environments the probability of fire occurrence is usually minimal. However, when fires occur in these environments, they tend to be intense and often extensive. In most of the regions characterized by intermediate environmental conditions, fires are sufficiently frequent for the system and species found within it to evolve mechanisms that allow them not only to tolerate fire but often to take advantage of the changes that fire created. Thus, fire frequently becomes a dominant feature in the successional development of these communities. In some instances, this adaptation has been sufficient to result in a species or vegetation type dependent on fire for its very existence.

Community ecologists have only gradually come to recognize the importance of disturbance in general and fire in particular in the maintenance of naturally functioning ecosystems. Indeed, many early ecologists probably agreed with Frederic Clements (1935) who wrote that "[u]nder primitive conditions, the great climaxes of the globe must have remained essentially intact, since fires from natural causes must have been both infrequent and localized." However, some ecologists such as Aldo Leopold, Henry Cowles, and Robert Humpheries did recognize the influence of fire and other disturbances, and through their pioneering work and that of many others who followed in their footsteps, the crucial role of disturbance in vegetation dynamics is now more fully understood.

Most ecologists have come to accept the dynamic nature of communities with the equivalent understanding that few populations or communities are at equilibrium (Connell & Sousa, 1983; Sousa, 1984). In fact, given the abundance of information on the frequency and importance of disturbances to ecological function, it has become difficult to develop a concise definition of disturbance (Sousa, 1984; White & Pickett, 1985). Huston (1994) has defined *disturbance* as "any process or condition external to the natural physiology of living organisms that results in the sudden mortality of biomass in a community on a time scale significantly shorter by several magnitudes of order than that of the accumulation of biomass." This definition provides a framework with which to evaluate the role of disturbance in an ecosystem. Furthermore, because disturbance and vegetation composition and structure are interdependent, often it is not useful to consider a single disturbance event but, rather, a sequence of disturbance events through time. This sequence of disturbances is often referred to as a *disturbance regime.* A disturbance regime commonly contains the following descriptors: (1) areal extent; (2) magnitude, which includes both intensity and severity; (3) frequency; (4) *predictability,* the variance in the mean time between disturbance events; and (5) *turnover rate,* the mean time to disturb the entire area (Sousa, 1984).

The "fire regime" is a useful concept when considering fire effects on broad spatial or temporal scales. Climate, topography, and vegetation type influence fire regimes (Kessell, 1976; Heinselman, 1978; Romme & Knight, 1981; Swetnam & Betancourt, 1990; Clark, 1988, 1990; Agee 1993). A fire regime includes many characteristics of the typical fire, applied at landscape scales, over long periods (i.e., several burning cycles). Factors included within a fire

Table 14.1 Classification of fire regimes for various western vegetation types

Pristine fire regime	Degree of change due to fire exclusion	Facility of restoring original fire regime	Example vegetation types
Nonlethal, very frequent	Very high	Good	Ponderosa pine, tallgrass prairie, mountain grassland
Nonlethal, frequent	Moderate	Relatively difficult	Palouse grassland, whitebark pine, lodgepole pine
Mixed, frequent	Moderate	Good to relatively difficult	Wetlands, mixed conifer woodland
Mixed, infrequent	Moderate	Difficult	Grand fir, lodgepole pine
Mixed, very infrequent	Low	Difficult	Subalpine fir, western juniper
Stand-replacing, very frequent	High	Moderate	Mountain big sagebrush
Stand-replacing, frequent	High	Moderate to relatively difficult	Wyoming big sagebrush, western juniper, aspen
Stand-replacing, infrequent	Moderate	Difficult	Lodgepole pine, grand fir, aspen, curlleaf mountain-mahogany
Stand-replacing, very infrequent	Low	Difficult	Whitebark pine, subalpine fire, mountain hemlock

Intervals: very frequent, <25 year; frequent, 26–75 yr; infrequent, 76–150 yr; very infrequent, 151–300 yr; extremely infrequent, >300 yr.
Source: Modified from P Morgan et al., Fire regimes in the Interior Columbia River Basin: Past and present. Final project report to USDA Forest Service, Interior Columbia River Basin Assessment, 1995.

regime include those previously listed, as well as the typical effects on vegetation, and fire pattern (Heinselman, 1981; Kilgore, 1981, 1985; Rykiel, 1985; White & Pickett, 1985; Agee, 1993). Disturbance frequency tends to be inversely related to disturbance intensity (Sousa, 1984). Consequently, high-intensity fires tend to occur in areas with low frequency, and low-intensity fires occur in vegetation with high frequency (Heinselman, 1981; Swetnam, 1993; Huston, 1994). A fire regime allows the consideration of broad-scale fire effects on the landscape rather than a limited focus on the influences of a single fire event.

Several classification systems of fire regimes have been developed, primarily for forested ecosystems (Morrison & Swanson, 1990; Agee, 1993; Brown et al., 1994). A system applicable to forest, shrub, and grassland vegetation was developed by Morgan et al. (1995), based on the mean fire-free interval (MFI) and fire severity. MFIs were selected to include effects that were ecologically significant for all terrestrial vegetation types. Intervals ranged from very frequent (<25 yr) to extremely infrequent (>300 yr). Interpretation of fire severity was based on comparison of the preburn and postburn vegetation three years following the fire. If structure and composition of the burned vegetation was similar to the preburn conditions within three years following the burn, the fire was considered to be *nonlethal. Stand-replacement fires* were those that changed the physiognomy and overstory structure three years post-burn. Fires that resulted in a pattern of both stand-replacing and nonlethal patches were classified as *mixed.* Variable fire regimes, those in which the fire effects vary greatly through time (Brown et al., 1994), were not included in this classification. The combination of five MFI intervals and three categories of fire severity resulted in 12 fire regime classes, which are explained with example vegetation types and the probability of restoring natural fire in each regime in Table 14.1.

Community Function Relationships

Creation and Maintenance of Habitat Conditions

Some of the earliest observed and most direct relationships of fire with the ecosystem are in the creation and maintenance of favorable habitat conditions for a specific species. Many of these relationships have been reported, and many more are continually being discovered. The possible types of relationships with fire are numerous when all species are considered. For example, the seeds of redstem (*Ceanothus sanguineus*) and shinyleaf (*Ceanothus velutinus*) ceanothus germinate more readily if they receive a heat treatment. The seed coat is water impermeable with the exception of a small orifice that is blocked by a wax material. The heat from fire melts the wax so the seed can imbibe water, break dormancy, and germinate (Gratkowski, 1962; Dyrness, 1973; Geier-Hayes, 1989; Keeley, 1991). Other species may possess two types of seeds, some that germinate readily and others that require a heat treatment for germination (Keeley, 1991). The germination of other species, such as skunkbush (*Rhus trilobata*), manzanita (*Arctostaphylos* spp.), phacelia (*Phacelia* spp.), and chamise (*Adenostema fasciculatum*), responds to charcoal or smoke produced by the fire (Keeley, 1991).

The reproduction of other species may be affected more by fire-induced or fire-enhanced flowering. In the case of wiregrass (*Aristida stricta* Michx.), fire appears to be the only natural factor that initiates flowering (Clewell, 1989). This species, which once was the primary understory species of pine savannahs in the southeastern United States, may well become rare in one or two decades due to a number of factors, fire exclusion being one. Wiregrass may be a key species for a plant community that in turn provides habitat for a wide array of rare vascular plant species (Hardin & White, 1989).

Many species that increases in abundance following burning are simply responding to the decrease in competition for water, light, and nutrient resources that results from the fire's reduction of existing vegetation and opening of the community. Many early successional forbs may exist only in the earlier stages of succession in juniper woodlands of the western United States. Mountain big sagebrush (*Artemisia tridentata* subsp. *vaseyana*) occurs only in the more open stands of ponderosa pine (*Pinus ponderosa*) and Douglas-fir (*Pseudotsuga menziesii*) (Gruell et al., 1986). For these species, the relationship may not be specific to fire, as any disturbance that maintains an open conifer canopy will enhance reproduction. However, fire is one of the primary factors initiating secondary succession in pristine juniper woodlands, conifer forests, sagebrush steppe, and many other vegetation types.

Increased antelope bitterbrush (*Purshia tridentata*) reproduction with fire is related both to the reduction in overstory and the reduction in seed predators (Dealy, 1970; Sherman & Chilcote, 1972; Bunting et al., 1985). This large-seeded plant has a high seed loss due to rodent predation; rodents tend to develop seed caches. Fire alters the habitat for the rodents, which then makes the seed caches available for germination. For many populations of bitterbrush, abandoned seed caches are the primary source of reproduction (Sherman & Chilcote, 1972).

Reproduction of whitebark pine (*Pinus albicaulis*) has strongly coevolved with Clark's nutcracker (*Nucifraga columbiana*) and fire (Tomback, 1978, 1982; Hutchins & Lanner, 1982). Nutcrackers gather seeds from mature pine stands and create seed caches in open areas nearby. They frequently select recently burned areas on which to make their caches, and abandoned caches are the primary source of regeneration for the pines. Seeds have been reported to be transported up to 22 kilometers as the birds seek open areas on which to cache. Regeneration without assistance of the nutcrackers is uncommon. The large seeds suffer high predation from other animals. Furthermore, the seedling generally does not readily mature in communities with a dense tree overstory. Periodic fire is a critical factor because, on most sites, whitebark pine is seral to subalpine fire or other conifers. Fire may aid in the maintenance of whitebark pine in two ways: (1) it may burn as a nonlethal surface fire, killing the younger subalpine fir cohorts and thinning older fir that are more sensitive to fire than the pine; and (2) it may burn as a stand-replacement fire, killing the entire overstory, thus making the site attractive to the birds for caching pine seeds (Fig. 14.2).

Three factors have contributed to the decline in abundance of whitebark pine in this century. Fire suppression has resulted in successional changes over greater portions of the landscape, allowing subalpine fir to establish and replace

Figure 14.2 Previously burned whitebark pine stand with regeneration establishing in burned area.

seral whitebark pine. Mountain pine beetle (*Dendroctonus ponderosae*) epidemics, originating in lodgepole pine stands, have spread into adjacent whitebark pine stands. The human-caused introduction of a parasite, white pine blister rust (*Cronartium ribicola*), may cause significant mortality in some stands but more frequently kills the upper portion of the tree, which destroys it cone-bearing potential (Arno, 1986; Keane & Arno, 1993). The combination of these influences has resulted in the rapid decline of whitebark pine in some areas but may also threaten other species that are dependent on the pine nuts. For example, pine nuts may be an important source of food for grizzly bears (*Ursus arctos*) in areas where berries and other high-energy foods are limited (Kendall, 1980; Kendall & Arno, 1990).

Few species are totally dependent on fire, but many are readily capable of taking advantage of increased forage supplies or habitat provided by fire. For example, the three-toed woodpeckers (*Picoides arcticus* and *P. tridactylis*) often are uncommon residents of closed and boreal forests in North America (Bock & Bock, 1974). They are opportunistic and may concentrate in forests following fire or insect outbreaks (Blackford, 1955; West & Spiers, 1959; Baldwin, 1968; Bock & Bock, 1974). Following large fires, their populations have been recorded to have grown substantially, to a point where they have expanded outside their normal range and into less desirable and untypical habitats (Axtell, 1957; Yunick, 1985).

Creation and Maintenance of Beneficial Ecological Characteristics

The relationships of many species with fire have been well documented. Entire biotic communities have similar relationships and also may be dependent on the periodic occurrence of fire. The most obvious of these relationships occurs when frequent fire prevents further successional development of a site. The effects of frequent fires counterbalance natural successional change influencing the site's composition and maintain a quasistable species composition. Ecologists have termed these vegetations *fire climaxes* or *pyric* (pyral) *climaxes* (Daubenmire, 1968b; Kimmins, 1987). Examples of these communities might include the tall grass prairie adjacent to the eastern hardwood forest (Daubenmire, 1968a; Vogl, 1974; Bragg & Hulbert, 1976) or the open ponderosa pine stands of the northern Rocky Mountains (Weaver, 1967; Arno, 1980). Fire-climax sites often have a MFI of less than ten years; changes in plant composition following fire are minimal. Only the more fire-resistant species dominate the site. For these communities, an increase in the MFI can have major impacts on community composition and the effects of fire when they occur. MFIs often have lengthened fivefold, and many more fire-sensitive species may become established on the site. Early in this century, Leopold (1924) noted this change in Arizona ponderosa pine. Fuel-loading also changes, which in turn alters the characteristics of the fire. When fire has been excluded from these ecosystems, fires tend to burn more extensively and severely than when they are allowed to occur more frequently. Resulting changes in fire severity may thus change the responses of species present on the burn site.

Many other biotic communities with long MFIs also may be dependent on periodic fire. These tend to be seral communities where fire tends to act as the primary disturbance factor, returning the community to early seral stages of succession. The occurrence of fire often results in the replacement of the existing vegetation by a sequence of successional communities. For these communities, wide variation in species composition often occurs through the successional sere. It may require decades or even centuries for the prefire community to redevelop, but fire is as an essential part of the community, as it is for those with short MFIs and minor variations through time. Without the periodic occurrence of fire, succession will result in changes in composition and eventually the development of another vegetation type. Resulting changes in soil characteristics and propagule availability may prevent going backward in successional time or at least might require extremely long time periods to do so. An example of this is seral curlleaf mountain-mahogany (*Cercoparpus ledifolius*), which is sensitive to fire (Wright & Bailey, 1982) but often is replaced through succession by Douglas-fir, western juniper, and other conifers in the absence of disturbance (Gruell et al., 1985) (Fig. 14.3). Because curlleaf moun-

Table 14.2 Examples of seral vegetation types in which infrequent fire is an essential factor in the long-term maintenance of the community on the landscape

Vegetation type	Pristine MFI	References
Wetlands	—	Vogl, 1969; Cypert, 1973; Linde, 1985
Mesquite savannah	20–30	Wright & Bailey, 1982
Mountain big sagebrush steppe	20–70	Houston, 1973; Burkhardt & Tisdale, 1976; Gruell et al., 1986
Quaking aspen	—	Loope & Gruell, 1973
Douglas fir	5–67	Arno, 1976; Davis et al., 1980
Lodgepole pine	25–300	Brown, 1975; Arno, 1976; Tande, 1979; Romme, 1982
Whitebark pine	20–75	Arno, 1986; Morgan & Bunting, 1990
Black spruce, aspen	30–100	Aber & Melillo, 1991

MFI = mean fire-free interval.

tain-mahogany seldom resprouts, a century or more may be required for the stand to redevelop following fire. Nonetheless, it is dependent on periodic disturbance for continued survival when it occurs as a seral stage within coniferous sere. Other examples include many common forested, woodland, rangeland, and wetland-aquatic ecosystems (Table 14.2).

Figure 14.3 Curlleaf mountain-mahogany stand in Lewis and Clark Caverns State Park, Montana, being replaced successionally by Douglas-fir.

Maintenance of Diversity at the Community and Landscape Level

Conservation of biological diversity is currently a major national and global issue. Because diversity can be characterized in many different ways, as described in Chapter 2 (Magurran, 1988), this chapter considers diversity at the species level—characterized by species richness—probably the most easily interpreted and readily available index in the literature. It does not, however, convey all that diversity, in its broadest sense, may contribute to an ecosystem. Diversity may be measured at a variety of scales, from very fine-scale patch diversity, to broad-scale regional diversity. Disturbance, including fire, will affect diversity differently depending on the scale in which it is measured. Consequently, it is important to discuss this issue at different organizational levels. Species richness will be considered at both the community and landscape (multiple community) levels.

At the community level, disturbances such as fire tend to increase diversity when they occur at low to intermediate levels of frequency and intensity (Loucks, 1970; Huston, 1994). This tendency is more consistent for ecosystems that have high productivity and high growth rates (Huston, 1994). Therefore, this increase in diversity is more likely to occur in temperate grassland ecosystems that in desert ecosystems. Low to intermediate levels of fire disturbances will, through time, provide microsites suitable for early, mid, and late seral species. As the fire frequency decreases and/or as the intensity increases and the community's recovery rate declines, the community tends to become dominated by species that are resistant to fire or those that can recover between fire events. These species are herbs or sprouting shrubs, which are habitat generalists. The late seral species are reduced, and species diversity declines. Stability of community composition may occur when fires are frequent and all species present are resistant to fire and are affected little by a fire event. Examples of these communities include many savannahs or temperate grassland vegetations in which fire occur every 5 to 10 years. Composition stability also may occur when fires are very infrequent and the community becomes dominated by late seral species. Composition changes little in these areas, except following a disturbance such as fire.

When considering landscape-scale diversity, similar trends exist, but there is more consistency between environments of varying productivity than for community-scale diversity. Intermediate levels of fire occurrence tend to increase diversity because the fires create communities in varying successional stages and, therefore, suitable habitat for the broadest variety of species (Figure 14.4). Areas with low fire occurrence tend to be characterized by extensive fires of high intensity. Swetnam (1993) found the highest regional synchrony during

Figure 14.4 Multiple burn landscape near Shoshone, Idaho.

periods of low fire frequency in giant sequoia (*Sequioadendron giganteum*) forests in California parks. Synchrony may be achieved either through a few extensive fires or many smaller fires in the same year. In either case, large areas burned within a region during a single year , and these fires produced a course-grained pattern on the landscape. Periods with higher frequency fire occurrence produced a fine-grained pattern because less total area was burned. Landscapes dominated by extremely high fire occurrence tend to become more homogeneous, and late successional species and communities decline in the landscape (Figure 14.5). Habitat for late successional species is lost through frequent disturbance. Habitat for early successional species is lost via successional processes when disturbance rates are infrequent.

It has been suggested that a dynamic stability of communities occurs within the landscape when, on the average, the area burned each year is offset by the successional recovery (Bormann & Likens, 1979). The variability within the system is dependent on the variability in a number of factors such as fire severity, annual area affected, and pattern of fire. Turner et al. (1993) have suggested that landscape stability is a function of two ratios: (1) the ratio of disturbance interval to recovery interval (T), and (2) ratio of disturbance extent to landscape extent (S) (Figure 14.6). The landscape may become relatively stable either when T is high and S is low or when T is low and S is high. At moderate T/S ratios, the landscape behaves as an equilibrium system but exhibits high variation in community composition through time.

A

B

Figure 14.5 Landscape illustrating the results of extensive repeated burns in sagebrush steppe vegetation on Snake River Plains, Idaho. Two photographs, one showing sagebrush steppe (A), another showing annual grass-dominated rangeland (B).

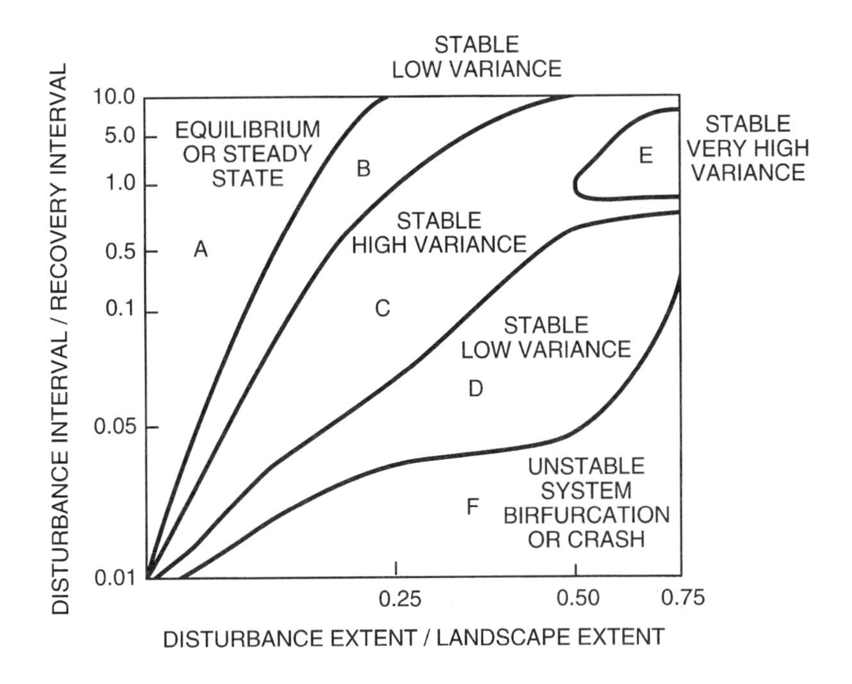

Figure 14.6 Diagram of the temporal and spatial parameters which illustrates regions that display qualitatively difference landscape dynamics. (From M G Turner et al., 1993.)

Management of Exotic Species

The vegetation of North America, and particularly the arid and semiarid West, has been modified by species that have been introduced intentionally or accidentally by humans from other portions of the world. Many of these exotic species have altered the functional relationships within these communities. In some instances, the entire vegetation has been displaced by the exotic species. The changes caused by cheatgrass, medusahead, and other exotics in the Great Basin sagebrush steppe, the canyon grasslands of the Pacific Northwest, and the grasslands of California's Central Valley are examples. The annual grasses have affected both interspecific competition and disturbance regimes (Peters & Bunting, 1994; Pellant, 1990). Protected area managers often are interested in reducing the influences of these exotic species in order to maintain the native composition, and to achieve this end fire is sometimes considered. Generally, the results of using fire to reduce the exotic component have not been encouraging. While short-term reductions are possible for some species, these results are ephemeral. Most of the exotic species are aggressive short-lived opportunists that are able to fill voids in the community caused by fire or other disturbances more rapidly than the natives. This weedy characteristic is often why these species are of concern in the first place.

Several studies (e.g., Pechanec & Hull, 1945; Patton et al., 1988) have shown the difficulty of establishing degraded native bunchgrass communities in areas dominated by exotic annual grasses like cheatgrass. Fire also has been used to attempt to alter environmental conditions and gradually displace exotic perennials such as Kentucky bluegrass (*Poa pratensis*), with only moderate success (Daubenmire, 1968a; Owensby & Smith, 1979; Engle & Bultsma, 1984; Steuter, 1987). Smooth brome, another prairie exotic, also has been reduced, but the timing of the fire has been shown to be a critical factor (Kirsch & Kruse, 1978; Blankespoor, 1987). It is clear from these studies that in most cases a long-term fire management effort is required to gradually shift the vegetation composition toward a greater abundance of native species, even with a native vegetation that is well adapted to repeated fires.

Recycling of Organic Matter and Nutrients

Many ecologists believe that nutrient availability within an ecosystem regulates many processes and determines many community characteristics (DeAngelis, 1980; Burrows, 1990). Nutrients within an ecosystem become available to the plant through many processes including herbivory, biochemical oxidation, organic decomposition, and oxidation by fire. Rates of nutrient decomposition and oxidation often are lower than rates of nutrient uptake by the community as secondary succession occurs and nutrients tend to become incorporated into biomass (Waring & Schlesinger, 1985; Putnam, 1994). This results in the accumulation of the system's nutrient pool into living and dead organic materials and a reduction in the amount of available forms. In forest, woodland, and shrub vegetation types, nutrients also are incorporated simultaneously into above-ground biomass, lowering the amounts in the soil and below-ground organic material. For example, in sagebrush steppe vegetation, more than 98 percent of the system's total nitrogen is contained below ground. As juniper woodlands develop through succession, organic nutrients are gradually moved above ground. Juniper woodland below-ground concentrations may be only 82 percent (Tiedemann, 1987). Thus, fire is particularly important in the incorporation of woody material and the nutrients they contain into forest soils in cold or dry environments (Harvey et al., 1979). In these ecosystems, rates of biomass production exceed rates of decomposition, and organic matter tends to accumulate until a fire occurs.

Initially, there usually are nutrient losses from the system as a result of fire. Most of the nitrogen, phosphorus, and sulfur that is contained in the burned organic material is lost through volatilization (Wright & Bailey, 1982; Chandler et al., 1983; DeBano, 1991; Hungerford et al., 1991). The degree of loss is directly related to the fire temperature and the degree of combustion of the surface organic material (Chandler et al., 1983; Hungerford et al., 1991). Nitrogen often is the most limiting nutrient in the ecosystem, and its availability regulates many processes including primary and secondary succession (Olsen, 1958; Lawrence et al., 1967; Tilman, 1984). Early successional species are more capable of taking advantage of high available soil nitrogen levels. The transition from early- to mid- and late-successional species may be determined in part by soil nitrogen, because later successional species have a lower nitrogen requirement (McLendon & Redente, 1994). Thus, low nitrogen availability on burned sites, either from nitrogen loss or from carbon input, which also regulates nitrogen availability, may decrease the length of time early-successional species dominate the site.

Depending on the heat of the fire and subsequent soil heating, substantial amounts of soil nitrogen also may be volatilized (White et al., 1973; Hungerford et al., 1991). At temperatures over 300°C, 50 to 100 percent of the soil nitrogen may be volatilized. However, the postfire environment may increase the availability of nitrogen through other mechanisms. Nitrogen fixation by blue-green algae may be stimulated by greater light intensity or higher soil temperature (Vogl, 1974; Hulbert, 1988). Hulbert (1988) also suggested that nitrogen mineralization may be stimulated in the postfire environment. Plant-available nitrogen may also increase on the site as microorganisms in the soil transform portions of the unavailable soil-nitrogen pool into more available forms. In forested vegetation, fires will reduce the amount of surface organic material, but movement into the soil often increases the amount of organic material in the soil. Thus, nutrients contained within these organic materials may increase in the soil and thereby increase availability to the plant despite an overall reduction on the site (Chandler et al., 1983). Youngberg and Wollum (1976) found that the symbiotic nitrogen-fixing species shinyleaf ceanothus provided for significant amounts of nitrogen replacement into coniferous forests of the Pacific Northwest following fire. Other species of ceanothus, alder (*Alnus* spp.) and lupine, probably provide the same function in other vegetation types.

Losses of other nutrients such as phosphorus, calcium, and potassium occur through the immediate transport of ash material during combustion (DeBano, 1991). Additional losses may occur as wind and water move ash material from the burned areas (Tiedemann, 1981; Woodmansee & Wallach, 1981). However, losses from the burned area result in nutrient deposition of adjacent vegetation

(Clayton, 1976). Soil concentrations of these nutrients generally remain unchanged or are increased due to additions from above-ground ash material (Stark, 1977; Hungerford et al., 1991). Losses of the cation nutrients occur only at extremely high soil temperatures (greater than 750°C), which do not normally occur during most fires (Wright & Bailey, 1982; Hungerford et al., 1991).

Fires also may initiate the development of postfire vegetation important for the long-term maintenance of site productivity. For example, fire in the taiga forests of interior Alaska results in increased soil temperature and rates of nutrient cycling and a more productive site (Van Cleve et al., 1983; Aber & Melillo, 1991). Reduction of the forest canopy and understory by fire results in reduced shading of the forest floor, increased depth to permafrost, and increased effective soil depth. Increased soil volume increases the mineralization of organic matter and the availability of nitrogen. The temporary increase of deciduous species such as aspen (*Populus tremuloides*) and paper birch (*Betula papyrifera*) produces less shade in the spring and fall, and leaf litter that is higher in nitrogen and lower in lignin than the coniferous tree litter. Thus, increased mineralization of nitrogen is correlated to increased net primary productivity. As black spruce (*Picea mariana*) increases with secondary succession, soil temperature, permafrost depth, and decay rates decline. Decreased permafrost depth also increases water saturation of the soil. This results in lower mineralization rates and decreases in net primary production (Aber & Melillo, 1991). It is likely that inputs of phosphorus, calcium, and other nutrients into the soil are also affected, and nutrients tend to be sequestered in the aboveground living and dead organic material. Viereck et al. (1979) found four to seven times more available phosphorus in floor layers of burned black spruce forest than prior to burning. Fires occurring on a 30- to 100-year interval start the cycle again (Aber & Melillo, 1991).

Reintroduction of Fire

It is easier to describe the ecological role of fire in a natural area than it is to reintroduce fire to restore ecosystem integrity and function. Many factors need to be considered when making a decision about fire management strategies. Perhaps two of the most critical are (1) the size of the natural area relative to the expected fire area, and (2) the fire regime. Relating the potential size of the fire to the size of the protected area helps provide a determination of the probability of the fire crossing onto adjacent land ownerships. In addition, fires that burn a major portion of a given protected area may be undesirable, even though the size and intensity of the fire are within the pristine range. Many parks, for example, are too small with respect to the potential fire size, making it difficult to permit any fires to burn without suppression.

Most regions of North America have been influenced by direct or indirect fire suppression for at least 100 years. Effective direct suppression of many types of fire has only been achieved since the middle of this century. However, other human-caused influences that began much earlier, such as intensive agriculture, livestock grazing, and urbanization, have resulted in vegetation changes such as multistory vegetation, horizontal and vertical continuity of fuels, and changes in species composition. These changes may make the reintroduction of fire difficult. Two broad categories of changes are common: (1) altered vegetation composition and structure through fire suppression, and (2) introduction of exotic species. The former is related to the effects historic fire suppression has had on vegetation and fuel structure. This is particularly prevalent in dry forest, savannah, and temperate grassland vegetation where the fire regime was typically dominated by frequent nonlethal fires. Lengthening the MFI through management, either intentionally or unintentionally, has resulted in a vegetation-fuel complex that now burns under a very different fire regime. The current fire regime is characterized by fires that are less frequent, more intense, and remove greater amounts of the overstory. In some natural areas, managers may need to use mechanical fuel and vegetation treatments, combined with prescribed fire, to reverse the changes that have occurred on the site (Brown, 1991). Prescribed fire may be used to lower fuel loads and adapt vegetation to withstand fires that burn under more pristine conditions. This has been the strategy employed at Sequoia and Kings Canyon National Parks (Kilgore, 1973). In other areas, however, the changes may well be irreversible.

The introduction of exotic species has altered the manner in which many natural systems function. Because most natural areas attempt to manage for natural vegetation, managers often try to minimize the influence of exotic species. Many of the exotic species are aggressive colonizers that can take advantage of the more open community following a fire. They may become established on the site more quickly than natives and, once established, can alter the natural composition and dominate the site for long periods of time.

When considering the role of fire, it is important to consider a protected area's primary purpose and the effect that fire would have on that element of the system. Many small parks were established to protect a particular vegetation type or species. While fire probably had been a part of the natural system, it may not be consistent with the reason the protected area was established. For example, small research natural areas have been established to protect relict areas of late seral Wyoming big sagebrush steppe. Fire was certainly a factor in the pristine environment of that vegetation (Wright & Bailey, 1982; Peters & Bunting,

1994). However, because relatively few undisturbed stands of that vegetation type currently exist and the recovery time is relatively long, it is not judicious to artificially reintroduce fire into these areas. Currently, there are extensive areas of recently burned sagebrush steppe.

The opposite may be true for natural areas established to protect western juniper (*Juniperus occidentalis*) savannahs. In this vegetation, fire is an integral factor in maintaining the species type. In the absence of fire, the stand will continue to close as more young junipers become established and, eventually, the very nature of the natural area will change to another community. Again, this would not be consistent with the intent of the natural area's designation. Often it is unclear whether a protected area was designated to preserve the state (i.e., plant community) or a natural process. When preservation of natural processes is the objective, often the vegetation will change. However, change may be socially or biologically unacceptable if the area supports rare species that once moved from refugia to refugia as areas burned, with few potential refugia now remaining.

A factor generally not considered in discussions of the reintroduction of fire in protected areas is the risk involved if fire is not reintroduced into the ecosystem. Risks might include (1) loss through successional replacement by another species or vegetation, of the very element the area was established to protect; (2) increased fire potential on the natural area and on adjacent lands; and (3) fires that burn under "unnatural" conditions, creating unprecedented effects for the site. Most natural systems are extremely dynamic, and it is virtually impossible to suspend factors causing change. Rather, it is essential to develop management strategies that take advantage of the dynamic nature of the landscape.

Once the determination is made to reintroduce fire into a natural area, the manager is faced with the decision of how to effectively accomplish this goal. Generally, there are two methods: prescribed burning and limited suppression of wildfires when they occur. Often this approach is referred to as a prescribed natural fire program. Prescribed burning, the deliberate ignition of areas under predefined conditions to achieve a management objective, provides managers with the most predictable circumstances. It allows them to select weather conditions and periods of the year when the fire behavior is suitable and personnel are available. The risk of fire escape and problems of smoke management, while still present, can be minimized. Prescribed burning also allows small units to be burned, which can be important for reintroducing fire into small-sized natural areas. The primary criticism of prescribed burning is that often weather conditions and geographic locations are selected that are relatively safe but not necessarily "natural." Usually, prescribed fires are not ignited under the more extreme burning conditions. (Conversely, wildfires burn during the pristine period under a variety of conditions, including those producing extensive, intense fires that subsequently have major impacts on the existing community.) Prescribed fire also can be employed until the fuels are reduced to a level that represents natural conditions more closely. Once this is achieved, naturally ignited fires can more safely be permitted to burn.

Fires in Protected Areas

There has always been significant concern about permitting fires, whatever their origin, to burn unchecked in protected areas. This concern was elevated by the large fires that burned in much of Yellowstone National Park in 1988. In the summer of 1988, these fires burned approximately 45 percent (400,000 ha) of the park and a total of 570,000 hectares of the greater Yellowstone ecosystem over a two-month period. The fires were portrayed by the national media and perceived by the general public to be a great tragedy. Park managers were soundly criticized for the park's fire policies and for not acting more quickly to suppress the fires in their early stages (Schullery, 1989a; Knight, 1991). One consequence of the Yellowstone fires was that NPS fire suppression policies, which had been gradually relaxed in recognition of fire's important role as a natural ecological process, were tightened considerably. Currently, fires are permitted in only a few large wilderness areas and in isolated portions of national parks. Even in these circumstances, the conditions under which wildfires are allowed to burn without suppression are usually strictly limited. If the conditions under which fires are permitted to burn without suppression are narrowly defined, it is believed that fire may be effectively limited in the park (Schullery, 1989b).

Much of our knowledge about the role of fire in maintaining community and landscape mosaics has come from the study of wildfires and prescribed fires in natural areas. One of the issues of contention about the Yellowstone fires, and in essence fires in many large protected areas, was whether they were natural or a consequence of past fire-control actions (Knight, 1991).

Yellowstone, as did most national parks, had a policy of complete fire suppression from 1886 to 1972. Between 1886 and the 1940s, fire suppression efforts were probably fairly effective along roads and major trails, but they likely had little effect in remote areas. After World War II, new fire-fighting methods and technologies became widely available, and suppression became much more effective. After 1972, revised NPS fire-management policies for western parks permitted some lightning-caused fires to burn without interference in backcountry areas (Despain & Sellers, 1977). However, of 235 lightning-caused fires that occurred between 1972 and 1987, only 5 burned an area greater than 40 hectares (Romme & Despain, 1989).

Indeed, Romme and Despain (1989) concluded that the 30 years of intensive fire exclusion only delayed the onset of a major fire event, which was probably inevitable given the nature of the fuels complex that had developed since the last extensive fires in the 1700s. In fact, the fires of 1988 were comparable, in total area burned, to fires in the late seventeenth and early eighteenth centuries. Therefore, they were not an abnormal event, but, rather, a consequence of a forest mosaic that develops over a long time period (250 years) (Romme, 1982), combined with the extreme drought and wind conditions that occurred in the summer of 1988.

Literature Cited

Aber JD, Melillo TM. *Terrestrial Ecosystems.* Philadelphia: Saunders College Publishing, 1991.

Agee JK. Fire ecology of Pacific Northwest Forests. Washington, DC: Island Press, 1993.

Arno SF. The historical role of fire on the Bitterroot National Forest. *USDA For Sev Gen Tech Rep* 187, 1976.

Arno SF. Forest fire history in the Northern Rockies. *J For* 1980;78:460–465.

Arno SF. Whitebark pine cone crops—A diminishing source of wildlife food. *West J Appl For* 1986;1:92–94.

Axtell HH. The three-toed woodpecker invasion. *Audubon Outlook* 1957;6:13–14.

Baldwin PH. Predator-prey relationships of birds and spruce beetles. *Proc Entomol Soc Am N Cent Br* 1968;23:90–99.

Blackford JL. Woodpecker concentration in a burned forest. *Condor* 1955;57:28–30.

Blankespoor GW. The effects of prescribed burning on a tall grass prairie remnant in eastern South Dakota. *Prairie Nat* 1987;19:177–188.

Bock CE, Bock JH. On the geographical ecology and evolution of the three-toed woodpeckers, *Picoides tridactylis* and *P. arcticus. Am Midl Nat* 1974;92:397–405.

Bormann FH, Likens GE. *Pattern and Process in a Forested Ecosystem.* New York: Springer-Verlag, 1979.

Bragg TB, Hulbert LC. Woody plant invasion of unburned Kansas bluestem prairie. *J Range Manage* 1976;29:19–24.

Brown JK. Fire Cycles and Community Dynamics in Lodgepole Pine Forests. In DM Baumgartner (ed), *Proceedings—Management of Lodgepole Pine Ecosystems Symposium.* Pullman, WA: Washington State University, Cooperative Extensive Service, 1975:430–456.

Brown JK. Should Management Ignitions Be Used in Yellowstone National Park? In RB Keiter, MS Boyce (eds), *The Greater Yellowstone Ecosystem: Redefining America's Wilderness Heritage.* New Haven, CT: Yale University Press, 1991:137–148.

Brown JK, Arno SF, Barrett SW, Menakis JP. Comparing the prescribed natural fire program with presettlement fires in the Selway-Bitterroot Wilderness. *Int J Wildl Fire* 1994;4:157–168.

Bunting SC, Neuenschwander LF, Gruell GE. Fire Ecology of Antelope Bitterbrush in the Northern Rocky Mountains. In JE Lotan, JK Brown (eds), Fire's Effects on Wildlife Habitat—Symposium Proceedings. *USDA For Serv Gen Tech Rep* 186, 1985:48–57.

Burkhardt JW, Tisdale EW. Causes of juniper invasion in southwestern Idaho. *Ecol* 1976;57:472–484.

Burrows CJ. *Processes of Vegetation Change.* London: Unwin Hyman, 1990.

Chandler C, Cheney P, Thomas P, Trabaud L, Williams D. Fire in Forestry. In *Forest Fire Behavior and Effects* (Vol I). New York: J Wiley, 1983.

Clark JS. Effect of climate change on fire regimes in northwestern Minnesota. *Nature* 1988;334:233–235.

Clark JS. Patterns, causes, and theory of fire occurrence during the last 750 years in northwestern Minnesota. *Ecol Monog* 1990;60:135–169.

Clayton JL. Nutrient gains to adjacent ecosystems during a forest fire: An evaluation. *For Sci* 1976;22:162–166.

Clements F. Experimental ecology in the public service. *Ecol* 1935;16:342–363.

Clewell AF. Natural history of wiregrass (*Aristida stricta* Michx.) *Nat Areas J* 1989;9:223–233.

Connell JH, Sousa WP. On the evidence needed to judge ecological stability or persistence. *Am Nat* 1983;121:789–824.

Cypert E. Plant succession on burned areas in Okefenokee Swamp following the fires of 1954 and 1955. *Proc Tall Timbers Fire Ecol Conf* 1973;12:199–217.

Daubenmire R. Ecology of fire in grasslands. *Adv Ecol Res* 1968a;5:209–266.

Daubenmire R. *Plant Communities: A Textbook of Plant Synecology.* New York: Harper & Row, 1968b.

Davis KM, Clayton BD, Fischer WC. Fire ecology of Lolo National Forest habitat types. *USDA For Serv Gen Tech Rep* 79, 1980.

Dealy JE. Survival and growth of bitterbrush on the Silver Lake deer winter range in central Oregon. *USDA For Serv Res* (Note PNW-133). 1970.

DeAngelis DL. Energy flow, nutrient cycling, and ecosystem resilience. *Ecol* 1980;61:764–771.

DeBano LF. The Effect of Fire on Soil Properties. In AE Harvey, LF Neuenschwnader (eds), Proceedings—Management and Productivity of Western-Montane Forest Soils. *USDA For Serv Gen Tech Rep* (INT-280). 1991:151–156.

Despain DG, Sellers R E. Natural fire in Yellowstone National Park. *Western Wildlands* 1977;4:20–24.

Dyrness CT. Early stages of plant succession following logging and burning in the western Cascades of Oregon. *Ecol* 1973;54:57–69.

Engle DM, Bultsma EM. Burning of northern mixed prairie during drought. *J Range Manage* 1984;37:398–401.

Geier-Hayes K. Vegetation response to helicopter logging and broadcast burning in Douglas-fir habitat types at Silver Creek, Central Idaho. *USDA For Serv Res* (INT-405). 1989.

Gratkowski HJ. Heat as a Factor in the Germination of Seeds of *Ceanothus velutinus* var. *laevigatus* T. and G. (Ph.D. dissertation). Corvallis, OR: Oregon State University, 1962.

Gruell GE, Bunting S, Neuenschwander L. Influence of Fire on Curlleaf Mountain-mahogany in the Intermountain West. In JE Lotan, JK Brown (eds), Fire's Effects on Wildlife Habitat—Symposium Proceedings. *USDA For Serv Gen Tech Rep* 186, 1985:58–72.

Gruell GE, Brown JK, Bushey CL. Prescribed fire opportunities in grasslands invaded by Douglas-fir. *USDA For Serv Gen Tech Rep* (INT-198). 1986.

Hardin ED, White DL. Rare vascular plant taxa associated with wiregrass (*Aristida stricta*) in the southeastern United States. *Nat Areas J* 1989;9:234–245.

Harvey AE, Jurgensen MF, Larsen MJ. Role of forest fuels in the biology and management of soil. *USDA For Serv Res* (Note INT-65). 1979.

Heinselman ML. Fire Intensity and Frequency as Factors in the Distribution and Structure of Northern Ecosystems. In HA Mooney, JM Bonnikson, NL Christiansen, JE Lotan, WA Reiners (eds), Fire Regimes and Ecosystems Properties. *USDA For Serv Gen Tech Rep* (WO-26). 1978:7–57.

Houston DB. Wildfires in northern Yellowstone National Park. *Ecol* 1973; 54: 1111–1117.

Hulbert LC. Causes of fire effects in tallgrass prairie. *Ecol* 1988;69:46–58.

Hungerford RD, Harrington MG, Frandsen WH, Ryan KC, Niehoff GJ. Influence of Fire on Factors That Affect Site Productivity. In AE Harvey, LF Neuenschwander (compilers), Proceedings—Management and Productivity of Western-Montane Forest Soils. *USDA For Serv Gen Tech Rep* (INT-280). 1991:32–50.

Huston MA. *Biological diversity: The Coexistence of Species on Changing Landscapes.* Cambridge: Cambridge University Press, 1994.

Hutchins HE, Lanner RM. The central role of Clark's nutcracker in the dispersal and establishment of whitebark pine. *Oecologia* 1982;55:192–201.

Keane RE, Arno SF. Rapid decline of whitepark pine in western Montana: Evidence from 20-year remeasurements. *West J Appl For* 1993;8:44–47.

Keeley JE. Seed germination and life history syndromes in the California chaparral. *Botan Rev* 1991;57:81–116.

Kendall KC. Use of pine nuts by grizzly and black bears in the Yellowstone area. *Int Conf Bear Res Manage* 1980;5:166–173.

Kendall KC, Arno SF. Whitebark Pine—An Important but Endangered Wildlife Resource. In WC Schmidt, KJ McDonald (compilers), Proceedings—Symposium on Whitebark Pine Ecosystems: Ecology and Management of a High Elevation Resource. *USDA For Serv Gen Tech Rep* (INT-270). 1990:264–274.

Kessell SR. Wildland inventories and fire modeling by gradient modeling analysis in Glacier National Park. *Proc Tall Timbers Fire Ecol Conf* 1976;14:115–162.

Kilgore BM. Impact of prescribed burning on a sequoia-mixed conifer forest. *Proc Tall Timbers Fire Ecol Conf* 1973:12:345–375.

Kilgore BM. Fire in Ecosystem Distribution and Structure: Western Forests and Scablands. In HA Mooney, JM Bonnikson, NL Christiansen, JE Lotan, WA Reiners (eds), Fire Regimes and Ecosystems Properties. *USDA For Serv Gen Tech Rep* (WO-26). 1981:58–89.

Kilgore BM. The Role of Fire in Wilderness: A State-of-Knowledge Review. In R C Lucas (compiler), Proceedings—National Wilderness Research Conference: Issues, State-of-Knowledge, Future Directions. *USDA For Sev Gen Tech Rep* (INT-220). 1985:70–103.

Kimmins JP. *Forest Ecology.* New York: Macmillan, 1987.

Kirsch LM, Kruse AD. Prairie fires and wildlife. *Proc Tall Timbers Fire Ecol Conf* 1978;12:289–303.

Knight DH. The Yellowstone Fire Controversy. In RB Keiter, MS Boyce (eds), *The Greater Yellowstone Ecosystem: Redefining America's Wilderness Heritage.* New Haven, CT: Yale University Press, 1991:87–103.

Lawrence DB, Schoenike RE, Quispel A, Bond G. The role of *Dryas drummondi* in vegetation development following ice recession at Glacier Bay, Alaska, with special reference to its nitrogen fixation by root nodules. *J Ecol* 1967;55:793–813.

Leopold A. Grass, brush, timber, and fire in southern Arizona. *J For* 1924;22:1–10.

Linde AF. Vegetation in Water Impoundments: Alternatives and Supplements to Water Level Control. In MD Knighton (compiler), Proceedings—Water Impoundments for Wildlife: A Habitat Management Workshop. *USDA For Serv Gen Tech Rep* (NC-100). 1985:51–60.

Loope LL, Gruell GE. The ecological role of fire in the Jackson Hole Area, northwestern Wyoming. *Quaternary Res* 1973;3:425–443.

Loucks OL. Evolution of diversity, efficiency, and community stability. *Am Zool* 1970;10:17–25.

Magurran AE. Ecological Diversity and Its Measurement. Princeton, NJ: Princeton University Press, 1988.

McLendon T, Redente EF. Role of Nitrogen Availability in the Transition from Annual-dominated to Perennial-dominated Seral Communities. In SB Monsen, SG Kitchen (compilers), Proceedings—Ecology and Management of Annual Rangelands. *USDA For Serv Gen Tech Rep* (INT-GTR-313). 1994:352–362.

Morgan P, Bunting SC. Fire Effects in Whitebark Pine Forests. In WC Schmidt, KJ McDonald (compilers), Proceedings—Symposium on Whitebark Pine Ecosystems: Ecology and Management of a High Elevation Resource. *USDA For Serv Gen Tech Rep* (INT-270). 1990:166–170.

Morgan P, Bunting SC, Black AE, Merrill T, Barrett S. *Fire Regimes in the Interior Columbia River Basin: Past and Present.* Final project report to USDA Forest Service, Interior Columbia River Basin Assessment, 1995.

Morrison PH, Swanson FJ. Fire history and pattern in a Cascade Range landscape. *USDA For Serv Gen Tech Rep* (PNW-GTR-254). 1990.

Olsen JS. Rates of succession and soil changes on southern Lake Michigan sand dunes. *Botan Gaz* 1958;119:125–170.

Owensby CE, Smith EF. Fertilizing and burning in Flint Hills bluestem. *J Range Manage* 1979;32:254–258.

Patton BD, Hironaka M, Bunting SC. Effect of burning on seed production of bluebunch wheatgrass, Idaho fescue and Columbia needlegrass. *J Range Manage* 1988;41:232–234.

Pechanec JF, Hull AC Jr. Spring forage lost through cheatgrass fires. *National Wool Grower* 1945;35(4)13.

Pellant M. The Cheatgrass-Wildfire Cycle—Are There Any Solutions? In ED McArthur, EM Romney, SD Smith, PT Tueller (compilers), Proceedings—Symposium on Cheatgrass Invasion, Shrub Die-off, and Other Aspects of Shrub Biology and Management. *USDA For Serv Gen Tech Rep* (INT-276). 1990:11–18.

Peters EF, Bunting SC. Fire Conditions Pre- and Postoccurrence of Annual Grasses on the Snake River Plain. In S B Monsen, S G Kitchen (compilers), Proceedings—Ecology and Management of Annual Rangelands. *USDA For Serv Gen Tech Rep* (INT-313). 1994:31–36.

Putnam RJ. *Community Ecology.* New York: Chapman & Hall, 1994.

Pyne SJ. *Fire in America: A Cultural History of Wildland and Rural Fire.* Princeton, NJ: Princeton University Press, 1982.

Romme WH. Fire and landscape diversity in subalpine forests of Yellowstone National Park. Ecol *Monog* 1982;52:199–221.

Romme WH, Despain DG. Historical perspective on the Yellowstone fires of 1988. *BioScience* 1989;39:695–699.

Romme WH, Knight DH. Fire frequency and subalpine forest succession along a topographic gradient in Wyoming. *Ecol* 1981;60:403–417.

Rykiel EJ, Jr. Towards a definition of disturbance. *Australian J Ecol* 1985;10:361–365.

Schullery P. The fires and fire policy. *BioScience* 1989a;39:686–694.

Schullery P. The Yellowstone fires: A preliminary report. *Northwest Sci* 1989b;63:44–54.

Sherman RJ, Chilcote WW. Spatial and chronological patterns of *Purshia tridentata* as influenced by *Pinus ponderosa. Ecol* 1972;53:294–298.

Sousa WP. The role of disturbance in natural communities. *Ann Rev Ecol System* 1984;15:353–391.

Stark NM. Fire and nutrient cycling in a Douglas-fir/larch forest. *Ecol* 1977;58:16–30.

Steuter AA. C_3/C_4 production shift on seasonal burns—northern mixed prairie. *J Range Manage* 1987;40:27–31.

Swetnam TW. Fire history and climate change in giant sequoia groves. *Science* 1993;262:885–889.

Swetnam TW, Betancourt JL. Fire-southern oscillation relations in the southwestern United States. *Science* 1990;249:1017–1020.

Tande GF. Fire history and vegetation pattern of coniferous forest in Jasper National Park. *Can J Bot* 1979;57:1912–1931.

Tiedemann AR. Regional impacts of fires. *USDA For Serv Gen Tech Rep* (WO-26). 1981:532–556.

Tiedemann AR. Nutrient Accumulations in Pinyon-juniper Ecosystems—Managing for Future Site Productivity. In RL Everett (compiler), Proceedings—Pinyon-juniper Conference. *USDA For Serv Gen Tech Rep* (INT-215). 1987:352–359.

Tilman GD. Plant dominance along an experimental nutrient gradient. *Ecol* 1984;65:1445–1453.

Tomback DF. Foraging strategies of Clark's nutcracker. *Living Bird* 1978;16:123–160.

Tomback DF. Dispersal of whitebark pine seed by Clark's nutcracker. *J An Ecol* 1982;51:451–467.

Turner MG, Romme WH, Gardner RH, O'Neill RV, Kratz TK. A revised concept of landscape equilibrium: Disturbance and stability on scaled landscapes. *Landscape Ecol* 1993;8:213–227.

Van Cleve K. et al. Taiga ecosystems in interior Alaska. *BioScience* 1983;33:39–44.

Viereck LA, et al. Preliminary results of experimental fires in the black spruce type of interior Alaska. *USDA For Serv Res* (Note PNW-332). 1979.

Vogl R. One hundred and thirty years of plan succession in a Wisconsin lowland. *Ecol* 1969;50:248–255.

Vogl R. Effects of Fire on Grasslands. In T T Kozlowski, C E Ahlgren (eds), *Fire and Ecosystems.* San Francisco: Academic Press, 1974:139–194.

Waring RH, Schlesinger W H. *Forest Ecosystems: Concepts and Management.* New York: Academic Press, 1985.

Weaver H. Fire and its relationship to ponderosa pine. *Proc Tall Timbers Fire Ecol Conf* 1967;7:127–149.

West JD, Spiers JM. The 1956–57 invasion of three-toed woodpeckers. *Wilson Bull* 1959;71:348–352.

White EM, Thompson WW, Gartner F R. Heat effects on nutrient release from soils under ponderosa pine. *J Range Manage* 1973;26:22–24.

White PS, Pickett STA. Natural Disturbance and Patch Dynamics: An Introduction. In STA Pickett, PS White (eds), *The Ecology of Natural Disturbance and Patch Dynamics.* New York: Academic Press, 1985:3–13.

Woodmansee RG, Wallach LS. Effects of fire regimes on biogeochemical cycles. *USDA For Serv Gen Tech Rep* (WO-26). 1981:379–400.

Wright HA. Landscape development, forest fires, and wilderness management. *Science* 1974;186:487–495.

Wright H A, Bailey A W. *Fire Ecology—United States and Southern Canada.* New York: Wiley, 1982.

Youngberg C T, Wollum A G. Nitrogen accretion in developing *Ceanothus velutinus* stands. *Soil Science Soc Am Proc* 1976;40:109–112.

Yunick R P. A review of recent irruptions of the black-headed woodpecker and three-toed woodpecker in eastern North American. *J Field Ornithol* 1985;56:138–152.

15

Restoring Aquatic Environments: A Case Study of the Elwha River

Catherine Hawkins Hoffman
Brian D. Winter

The lamp lighting this page manifests an extraordinary feat of ingenuity and invention. In many parts of the United States, the lamp also reflects historic anticipation of a clean and seemingly free source of electricity from rivers. However, with altered hydrologic regimes and water quality, inundation of wildlife habitat, and diminishing salmon populations, society is coming to recognize that hydropower is neither environmentally "clean," nor free (Hawkins Hoffman, 1992). As hundreds of hydropower facilities nationwide "come of age" (by virtue of design, license expiration, or other factors), choices must be made in managing these facilities in consideration of the full cost of energy.

In the early 1900s, two hydroelectric facilities were constructed on the Elwha River. The Elwha is a principal drainage of the Olympic Peninsula in Washington State and, historically, one of the primary salmon-bearing streams of the area (Figure 15.1). Local residents hailed construction of the Elwha and Glines Canyon Dams as a sign of great progress for the region, despite obvious devastation to salmon populations, as neither dam provided fish-passage facilities.

Now, less than a century later, the dams constructed in the name of progress are considered, also in the name of progress, for removal—a sign of changing societal values and at least, to some individuals, more enlightened management of natural resources. In addition to the few remaining salmon, the Elwha's

Figure 15.1 Fishing the Elwha River in the 1920s (From Department of Interior, Department of Commerce, Lower Elwha S'Klallam Tribe, The Elwha Report: Restoration of the Elwha River Ecosystem and Native Anadromous Fisheries. Washington, DC: U.S. Government Printing Office, 1994: 32. Courtesy of Washington State Historical Society, Tacoma, WA.)

current now carries diverse interests, such as those of the National Park Service (NPS), National Marine Fisheries Service, U.S. Fish and Wildlife Service, Bureau of Reclamation, Washington Department of Fish and Wildlife, Lower Elwha Klallam Tribe, Pacific Fisheries Management Council, and various environmental groups, all united in seeking dam removal as the only means to restore the river's fisheries and ecosystem.

This chapter describes the remarkable event in progress (i.e., potential dam removal on the Elwha River) and outlines considerations for other river restoration efforts nationwide. We discuss the history of the Elwha and Glines Canyon Dams, the process that engendered consideration of dam removal, potential impacts and benefits of dam removal, and implications for other areas.

Location Map for the Elwha and Glines Canyon Dams

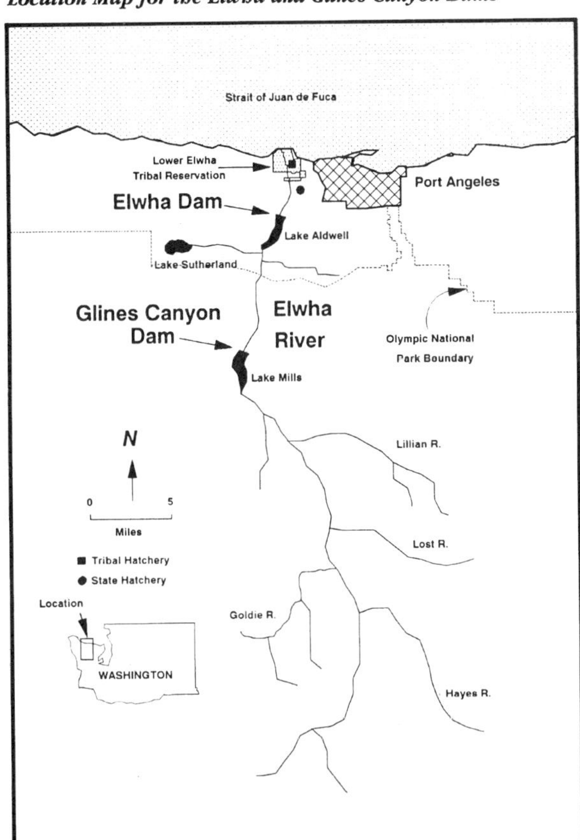

Figure 15.2 Location map of Elwha River basin. (From Department of Interior, Department of Commerce, Lower Elwha Klallam Tribe, The Elwha Report: Restoration of the Elwha River Ecosystem and Native Anadromous Fisheries. Washington, DC: U.S. Government Printing Office, 1994: 4.)

The Setting

The Elwha River is the fourth largest river on the Olympic Peninsula. Over 83 percent of the Elwha watershed lies protected within Olympic National Park, its habitat little changed from conditions hundreds of years ago. Seventy-two kilometers long, draining a basin of 831-square kilometers at an average rate of 42.7 cubic meters per second, the river flows northward from the slopes of Mount Olympus to the Strait of Juan de Fuca (Department of Interior [DOI] et al., 1994) (Fig. 15.2). Much of the river corridor downstream of Olympic National Park supports timber, agriculture, or low-density residential development (White et al., 1992). The Elwha River supplies municipal, industrial, and residential users of Port Angeles, Washington, with water, most of which is withdrawn downstream of the Elwha Dam.

The Elwha Dam (and associated reservoir, Lake Aldwell) block the river to anadromous fish about 8 kilometers from the river's mouth (Table 15.1). This lower reach is an impoverished remnant of the 121 kilometers of mainstem and tributary spawning habitat available prior to construction of the Elwha Dam. Located within the boundaries of Olympic National Park about 21 kilometers

Table 15.1 Data summary for Elwha and Glines Canyon Dams

Feature	Glines Canyon	Elwha
Dam		
Year constructed	1926–1927	1910–1913
Location	21.6 km (RM 13.4)	7.9 km (RM 4.9)
Length, base	14.0 m (46 ft)	30.5 m (100 ft)
Length, crest	82.3 m (270 ft)	137.2 m (450 ft)
Height	64.0 m (210 ft)	32.9 (108 ft)
Impoundment		
Reservoir name	Lake Mills	Lake Aldwell
Reservoir length	4.0 km (2.5 mi)	4.5 km (2.8 mi)
Surface area	167.9 ha (415 acres)	108.0 ha (267 acres)
Gross storage	49.3 mcm	10.0 mcm
Sediment	10.6 mcm	3.0 mcm
Power Plant		
Total capacity	13.1 megawatts	14.8 megawatts
Units	2	5
Gross average annual energy	102 gigawatt hours	70 gigawatt hours

RM = river mile; mcm = million cubic meters.

from the river mouth, Glines Canyon Dam (and associated reservoir, Lake Mills) is the larger of the two projects. Between the two dams and below Elwha Dam, the river bed is scoured and armored; cobbles and boulders are the predominant substrate. Much of the river's gravel—a resource desperately needed by spawning salmon—lies trapped within Lake Mills (Federal Energy Regulatory Commission, 1993).

History of the Issue

Environmental sensitivity is a concept generally not attributed to turn-of-the-century development, let alone to regulations of the time. However, salmon sustained native cultures for thousands of years in the Pacific Northwest and later supported a booming cannery business. Unquestionably, salmon were historically a highly valued natural resource. One of the earliest environmental laws codified in the state of Washington was intended to protect Pacific salmon. Lessons from the demise of the Atlantic salmon in eastern streams, largely due to hydropower facilities, prompted an 1890 Washington law requiring fish-passage facilities to accompany any obstruction on a salmon-bearing stream (Brown, 1982).

By the early 1900s, hydroelectric projects—most owned by private corporations—were developed on at least a dozen of Washington's rivers (Zubalik, 1988). These corporations regarded fishways as an unfair requirement that significantly reduced profits. In 1913, the Olympic Power and Development Company completed the Elwha dam (DOI, 1994). Not surprisingly, the president of the company resisted the fish-passage requirement. Financial backers of the project were influential, and the state fish commissioner, Leslie Darwin, proposed a creative solution to meet concerns of both fish and industry. Darwin's reasoning successfully convinced the governor and the legislature that hatcheries were suitable substitutes for passage facilities, and the fish-passage law was so modified (Brown, 1982).

Supported by powerful financial interests, "industry-friendly" regulations, and a governor enamored with hydropower, the hydroelectric industry found an open door. Environmental concerns quickly became second to the desire for "growth and development as the way to a better world" (Brown, 1982).

Prior to construction of Elwha Dam, the Elwha River was one of the primary salmon-bearing streams of the region. Renowned for chinook salmon (*Oncorhynchus tshawytscha*), some attaining 37 kilograms, the Elwha was one of few rivers outside Alaska supporting all five species of Pacific salmon (chinook; coho, *O. kisutch*; pink, *O. gorbuscha*; chum, *O. keta*; and sockeye, *O. nerka*), as well as steelhead trout (*O. mykiss*) and searun cutthroat trout (*O. clarki*) and char (*Salvelinus* spp.). Although passage was not required at the Elwha project, as described in the *Olympic Leader* (1911), long-time river watchers were dismayed at the impending losses:

> Game Warden Pike and Sheriff Gallagher have been having their troubles this week over the shutting off of the salmon from their spawning grounds by the big dam at the Aldwell Canyon, the same being contrary to the statutes made and provided. True, there is a flume up which the fish may go, but the trouble is, they can't. Hundreds of them have gathered just below the dam during the last few days until they are packed in together like a school of herring, or sardines in a box. Every few moments a big fellow makes a jump clear of the water that shoots out of the flume as tho from a hydraulic nozzle and strikes square in the flume above, only to be thrown back to the pool below.

In lieu of fish passage, the Olympic Power Company funded a hatchery, which was completed in 1915 at the base of Elwha Dam (Brown, 1982). Initially, the hatchery seemed successful as fish which had reared upriver before construction of the dam returned to spawn and were captured for eggs. However, fewer and fewer wild fish returned to the river and, in 1922, the state Department of Fisheries abandoned the hatchery (Maib, undated). The once-abundant salmon runs of the Elwha were decimated, cut off from over 90 percent of their historic habitat (DOI et al., 1994).

In 1927, the Olympic Power Company (reorganized as the Northwestern Power and Light Company) completed Glines Canyon Dam on the Elwha River, 13 kilometers upstream of Elwha Dam (Crown Zellerbach, 1973) (Figure 15.3). Because Elwha Dam had blocked fish passage, fishways were not considered for Glines Canyon Dam. Through corporate name changes and consolidations, both the Elwha and Glines Canyon projects eventually were owned by the Crown Zellerbach Corporation. The current owner is James River Paper, Incorporated. Daishowa America Company Limited, a pulp and paper mill in Port Angeles, Washington, operates both dams; all power produced by the dams is utilized by Daishowa.

In 1968 and 1973, Crown Zellerbach (now James River) submitted license applications to the Federal Energy Regulatory Commission (FERC) for the Elwha and Glines Canyon projects, respectively. The Glines Canyon application was for relicensing, as the original license expired in 1976. The Elwha project

Figure 15.3 Glines Canyon Dam during construction, December 28, 1926. (From Department of Interior, Lower Elwha Klallan Tribe, The Elwha Report: Restoration of the Elwha River Ecosystem and Native Anadromous Fisheries. Washington, DC: U.S. Government Printing Office, 1994: 34.)

was constructed prior to passage of the Federal Power and Water Act and has never been licensed.

Recent Events

License applications for the Elwha and Glines projects entered a much different regulatory environment than that which sanctioned the dams' construction initially, and the fish argument which has persisted since their construction was renewed. Environmental sensitivities are now heightened through statutes such as the National Environmental Policy Act, the Endangered Species Act, and the Clean Water Act. Also, the Federal Power Act requires FERC consultation with federal and state resource agencies for impacts to fish and wildlife before licensing hydropower facilities. Over two dozen intervenors joined in the Elwha and Glines licensing proceedings, including fish and wildlife agencies, Lower Elwha Klallam Tribe, and James River and Daishowa America, among many others. Inevitable litigation lay ahead regarding jurisdiction, as the Glines project is within Olympic National Park. Additionally, responses by the FERC to agency submissions indicated considerable disagreement regarding licensing conditions. The Elwha Report (DOI et al., 1994) describes the conflict:

> During the 1980's, the FERC licensing process became extremely contentious and drawn out, due primarily to national policy implications of licensing a project within a national park, the inability to design fish and wildlife mitigation measures capable of meeting Federal, State, and Indian Tribe resource goals, and legal challenges by conservation groups . . . Continued attempts to resolve FERC licensing issues were certain to result in protracted litigation, and considerable delay and expense for all parties, including the Federal Government. Failure to reach consensus would lead to the courts deciding vital issues without the opportunity for rational compromise. Verdicts would be narrowly defined by the issues taken before the courts, resulting in a piecemeal approach to the problem, when a comprehensive solution is needed.

To resolve that issue, the U.S. Congress passed the Elwha River Ecosystem and Fisheries Restoration Act in 1992 (Elwha Act). The Elwha Act stipulates a goal of "full restoration of the Elwha River ecosystem and native anadromous fisheries," and authorizes the Secretary of the Interior to acquire and remove the projects if he or she determines this is necessary to fully restore the ecosystem. Additionally, the Secretary was directed to prepare a report to Congress analyzing dam removal and other alternatives for river restoration. In June 1994, the Secretary submitted the Elwha Report to Congress. The report describes the Secretary's analysis and his determination that removal of the Elwha and Glines Canyon dams is necessary to meet the objectives of ecosystem restoration (DOI et al., 1994). Environmental impact analyses under the National Environmental Policy Act (NEPA) provided more detailed analyses of alternatives and designs for dam removal.

A

Figure 15.4A Elwha Dam.

Dam Removal

Structure Removal

Options for removing the dams vary greatly in river diversion methods that will allow demolition in dry conditions. The Elwha Dam diverts surface water to a powerhouse located downstream between two spillways (Figure 15.4A). The gravity-type, concrete section of the Elwha Dam now stands over the historic course of the Elwha River. The dam was constructed on unconsolidated material (gravels and cobbles) and was inadequately anchored to bedrock. In 1912, as the reservoir began to fill, the foundation of the dam blew out. Construction began a second time, utilizing rock blasted from the canyon walls, "pine mattresses" (fir, cedar, and hemlock boughs lashed together), and earth fill, all capped with a gunnite seal to close the breach (Reineking, 1914). One option for removal of this dam includes additional reinforcement actions to avoid another blowout during deconstruction. Under this option, diversion of the river would initially be over the north abutment, then through the sluiceways of the gravity section, and lastly over the remaining gravity section (NPS, 1996).

Another river diversion option would entail excavation of a channel through bedrock underlying the north abutment, and construction of a temporary coffer dam to direct the river to the channel. Following dam removal, this channel would be filled and graded to match the surrounding landscape. Free flow of the river would follow removal of the gravity section, appurtenances, hydraulic fill, and coffer dams (Figure 15.4B).

Demolition of Glines Canyon Dam can be accomplished without diverting the river around the dam site. The deconstruction plan involves slowly dropping the reservoir to the level of the turbine intake, located about 25 meters below the crest of the dam. The structure remaining above the waterline would be removed. Thereafter, gated notches would be constructed at progressively lower levels in the dam to pass the river, as layer after layer of the concrete structure is removed. This option is not possible at Elwha Dam because of the fill material placed behind it (DOI et al., 1994).

Deconstruction of both dams could be by conventional drill-and-blast techniques or by using a boom-mounted hydraulic hammer. Both methods would result in broken concrete rubble. If the concrete is desired for use in a stabilization project elsewhere or in an artificial marine reef, diamond sawcutting would be used to produce smooth-faced blocks (NPS, 1996).

B

Figure 15.4B Computer simulation of Elwha Dam removed. (Courtesy of Visual Presentations Group, Technical Service Center, U.S. Bureau of Reclamation.)

Sediment Management

During the past 80 years, an estimated 13.5 million cubic meters of sediment, consisting of coarse bedload materials (i.e., sand, gravel) and fine sediment (i.e., clay, silt), has accumulated in Lakes Mills and Aldwell (NPS, 1996). Of the two reservoirs, most of the accumulated sediment is located within Lake Mills. The bedload deposited at the heads of both reservoirs form deltas; the Lake Mills delta is up to 21 meters thick (DOI et al., 1994). Fine sediments are fairly evenly distributed throughout the reservoirs; however, much of the clay fraction naturally transported by the river remains in suspension and passes through the two reservoirs (NPS, 1995).

Three methods of managing the sediment have been identified: (1) mechanical removal of material to a terrestrial or marine site; (2) redistribution and stabilization of sediment within the confines of the existing reservoirs; or (3) river erosion of the sediment, with or without active manipulation (DOI et al., 1994; Randle & Lyons, 1995). A combination of methods is also possible. For example, fine sediments could be dredged from the reservoirs and slurried downstream (via pipeline) to minimize water-quality impacts, while the deltas are allowed to erode naturally. Further discussion of these methods follows.

Impacts

Potential Adverse Impacts and Anticipated Mitigation

Removal of the two dams and reservoirs would result in both short- and long-term environmental impacts. Short-term impacts are primarily associated with sediment management.

Mechanically removing all of the sediment would minimally impact downstream water users and resources but would affect the disposal site and prevent the bedload from reaching the river below the dams where it is needed for fish spawning. Stabilizing the material within the reservoirs would protect downstream water quality but would constrain the river between armored (e.g., riprap) banks, and would likely result in future bank failures and erosion of the material, essentially delaying impacts to a later date (NPS, 1995). Natural river erosion of the material would provide quickest physical recovery of the river but would result in high suspended sediment levels, which would significantly affect water quality and instream biota, including remaining Elwha River anadromous fish stocks. To limit the impact period, river erosion could be accelerated by mechanically moving the material into the river during high-flow events, or the dams could be removed at a rate to "meter" sediment releases and provide periods of clearer water. Dredging fine sediments with release immediately below the dam(s) would not eliminate water-quality impacts but would allow the fines to be released at selected times and concentrations to reduce these impacts. Alternatively, dredging fine sediments and moving them through a pipeline to the Strait of Juan de Fuca would minimize impacts to water quality, while river erosion of the deltas would allow the bedload to replenish the lower river (NPS, 1996).

Regardless, the Elwha Act requires the Department of the Interior to protect municipal and industrial water users prior to initiating dam removal. Protection may include new well systems, water treatment facilities, facility protection against floods, or other measures (DOI et al., 1994). Impacts to remaining anadromous fish stocks could be partially mitigated by removing fish to clean water sources (e.g., hatcheries, tributaries, the upper river) prior to and during dam removal.

The two dams are operated in run-of-the-river mode and are not managed for flood control benefits. Nonetheless, increased flooding may result in the long term if the dams are removed. Restoration of the natural sediment transport regime would eventually result in gravel aggradation of 1 to 2 meters in some sections of the river, similar to historic streambed elevations. Areas subject to flooding now may become vulnerable to flooding on a more frequent basis. Some of these sites may have cultural and historical significance, such as the Elwha Ranger Station Historic District within Olympic National Park. The U.S. Army Corps of Engineers has recommended the upgrade of the few existing levees and spur dikes in the Elwha River floodplain to protect facilities. Nonstructural measures (e.g., flood insurance) are also possible.

Noise associated with dam demolition may affect federally protected species occurring near the dams, such as the northern spotted owl (*Strix occidentalis caurina*) and marbled murrelet (*Brachyramphus marmoratus*). Appropriate timing of deconstruction activities and standard noise abatement procedures would seek to mitigate this impact. Other effects would be associated with disposal of the rubble and loss of registered historic structures (i.e., the powerplants). Beneficial uses of the rubble, such as construction of an artificial reef in the Strait of Juan de Fuca to increase bottomfish habitat, are possible. Mitigation for loss of the historic powerplants would likely include onsite interpretation and complete documentation of the projects according to the standards of the Historic American Engineering Record (NPS, 1995).

Loss of the reservoirs would eliminate resident fish habitat. Together with restoration of anadromous fish, resident fish populations, most notably rainbow trout (*Oncorhynchus mykiss*), would decline to historic levels (DOI et al., 1994).

However, full restoration of the ecosystem and anadromous fish would largely mitigate this loss. Draining the reservoirs would also affect animals that prefer lake habitat such as scaup and some other waterfowl. More than 60 wintering trumpeter swans (*Cygnus buccinator*) have been observed on Lake Aldwell. Although these birds also frequent rivers, estuaries, bays and farm ponds, a study will identify the most appropriate mitigation for the loss of Lake Aldwell trumpeter swan habitat. Overall, wildlife species would benefit from restoration of anadromous fish and the food these fish represent, and from recovery of 289 hectares of terrestrial habitat inundated by the reservoirs (NPS, 1995).

Removal of the dams would limit slack water recreational activities in the Elwha basin to Lake Sutherland, but other lakes (e.g., Lake Crescent) are also available in the area. Recreational access to the Elwha Valley would be restricted for 18 to 24 months during dam demolition, but a shuttle service to various trailheads would mitigate this impact. Additionally, some could consider the exposed reservoir bottoms to be an eyesore (during the 3 to 5 years before vegetation is largely reestablished). Others could find the process of ecosystem restoration to be fascinating. Interpretive displays and programs would explain the process to the public.

The two dams produce 18.7 average megawatts of electricity, or about 38 percent of the power used by Daishowa America, Inc. The Elwha Act stipulates that replacement power would be provided to the mill by the Bonneville Power Administration (BPA) through Port Angeles City Light at a cost equal to that of all other City Light industrial customers. Because the dams produce a relatively small amount of power, no single source would be developed to make up the loss. A mix of generation sources already being pursued in the region, including end-use conservation, would counteract elimination of the dams (DOI et al., 1994).

The Elwha Report (DOI, 1994) estimated dam removal and associated costs (i.e., sediment management, water quality protection, fish restoration, dam acquisition, etc.) at $148 to $203 million. Investigations since production of The Elwha Report indicate dam acquisition and removal would cost $111 to $128 million (NPS, 1996).

Potential Benefits

The dams have blocked fish migration since 1914. Due to the age and configuration of the dams and associated reservoirs, even state-of-the-art fish passage would not restore the native anadromous fisheries (FERC, 1993; DOI et al., 1994). Removal of the dams would allow unobstructed upstream and downstream fish migration, and full restoration of at least nine of the ten anadromous fish stocks affected; only sockeye salmon restoration is in doubt (NPS,

1995; Wunderlich et al., 1994). It is estimated that restoration would result in the production of over 380,000 adult salmon and steelhead within 20 to 25 years (FERC, 1993).

The reservoirs, particularly Lake Mills, absorb solar radiation, retaining it as heat. Release of heated water results in a 2° to 4° C elevation in water temperature in the lower river during late summer and early fall low-flows. Elevated water temperatures exacerbate fish diseases such as *Dermocystidium salmonis* and *Ichthyophthirius* spp. Also, confining fish to the lower river creates conditions promoting disease transmission by increasing fish densities. More than 60 percent of chinook salmon returning in 1992 died prior to spawning due to disease (DOI et al., 1994). Removal of the dams would reduce river temperatures and allow the chinook salmon to naturally distribute within cooler waters of the upper river, thus reducing prespawning mortalities.

Studies conducted during the FERC licensing process indicate that overall, wildlife will benefit from removal of the dams, including all federally listed species in the project area such as the bald eagle (*Haliaeetus leucocephalus*), spotted owl, and marbled murrelet. Restoration of anadromous fish runs would benefit at least 22 species of animals that feed on salmon carcasses (Cederholm et al., 1989). These carcasses would contribute over 300,000 kilograms of biomass to the ecosystem. Recovery of areas inundated by the reservoirs would restore important terrestrial habitat for elk and other animals (DOI et al., 1994).

The sediment trapped in the reservoirs is a resource prevented from reaching the river below the dams, which results in a coarsening (armoring) and lowering of the riverbed and restriction of natural geomorphological processes (DOI et al., 1994). Spawning gravel for anadromous and resident salmon and trout is thus limited and rearing habitat (side channels and off-channel pools) reduced. The dams also block the natural downstream transport of large woody debris, an important component of fish habitat. Nutrient transport is similarly reduced (FERC, 1993). Additionally, the reservoirs inundate over 8.5 kilometers of important low gradient, river habitat. Removal of the dams would allow full recovery of fish and river habitat.

Lack of sediment recruitment contributes to a diminished estuary, to eroded beaches, and to a nearshore habitat altered from sandy to rocky substrate. Natural replenishment of material to Ediz Hook, the spit that forms Port Angeles Harbor, is also precluded. Restoration of the natural sediment regime would reverse these effects.

The entire Elwha River Valley is a cultural resource, including the location of the Creation Site of the Lower Elwha Klallam Tribe. Many other significant cultural sites and historic villages are inundated by the two reservoirs. Removal of the dams would allow access to these sites once again.

Full Elwha River restoration would result in a positive benefit to cost ratio (Meyer et al., 1995). Restoration of the native anadromous fisheries would greatly benefit commercial and recreational fishing and the local economy through direct sales of fish and the purchase of fishing gear and associated amenities. Within ten years following dam removal, recreationists and tourists could generate increased spending of $28.5 million per year in Clallam County, supporting 446 additional jobs locally, while the annual nonmarket value (i.e., the value of satisfaction consumers enjoy above the amount they pay) of a restored ecosystem would be worth a staggering $3.5 billion (Meyer et al., 1995).

Regulatory, Legal, and Political Process

Although the Elwha Act authorizes the Secretary of the Interior to remove the two dams, Congress must appropriate the funds. Additionally, over 30 permits or authorizations are required to remove the Elwha and Glines Canyon projects, including Sections 401 and 404 of the Federal Clean Water Act (for water

Table 15.2 Examples of permits and authorizations needed for dam removal

Permit/process	Statutory authority	Implemented by
National Environmental Policy Act (environmental review)	42 USC 4321	Federal (state may have companion process)
Federal Clean Air Act (new source construction approval)	RCW 43.21A; RCW 70.94	State
Federal Clean Water Act (discharge of dredge and fill material)	Section 404 (33 USC 1344)	U.S. Army Corps of Engineers
Federal Rivers and Harbors Act (obtained jointly with Section 404 permit)	33 USC 403	U.S. Army Corps of Engineers
National Historic Preservation Act (review of impacts to cultural/ historic properties)	Section 106	State
Endangered Species Act (review of impacts to listed species)	Section 7 (USC 1531 et seq.)	Consultation with U.S. Fish and Wildlife Service or National Marine Fisheries Service (if anadromous fish)
Federal Clean Water Act (water quality certification)	Section 401 (RCW 90.48.260)	State
U.S. Coastal Zone Management Act (certification)	16 USC 1451 et seq.	State (may be relegated to county)

quality certification and for discharge of dredge and fill material), Section 106 compliance with the National Historic Preservation Act (for modification/removal of historic structures), and Section 7 consultation (for impacts to listed species pursuant to the Endangered Species Act) (Table 15.2).

The Elwha Act is a negotiated settlement providing a wide array of interests with the opportunity to settle a contentious and complicated issue. Provisions within the Act are supported by all parties involved in the FERC proceeding. Nonetheless, local opposition to removal of the dams is (predictably) vocal. Short-term costs of removal are difficult for some opponents and politicians to justify in the current deficit-reduction climate; the long-term benefit-to-cost ratio does not seem to be considered. Yet, short-term costs must be weighed against continued fisheries losses, violations of Indian treaty rights, erosion of estuarine and marine habitat, and litigation.

If implementation of the Elwha Act is not funded, the negotiated settlement will dissolve and all parties will return to court. For example, the U.S. Department of Justice, representing the Departments of Commerce and Interior, the Lower Elwha Klallam Tribe, and various environmental groups (Sierra Club, Seattle Audubon Society, Olympic Park Associates and Friends of the Earth) filed suit in 1992 challenging FERC's authority to license the Glines Canyon project within a national park. The United States Court of Appeals for the Ninth Circuit dismissed this case based on the Elwha Act, thereby removing FERC's authority to issue a long-term license for the project. This suit can be refiled if the federal government does not acquire the projects and the FERC licensing process is reinstated. Other suits would likely address licensing conditions, treaty right abrogations, inconsistency with the Endangered Species Act, and/or the "taking" of private hydroelectric projects.

Conclusions

Many dams have been removed across the United States, some by the simple expedient of dynamiting with no active sediment management programs (Winter, 1990). More dams will have to be removed as they reach their design life, are no longer economical to operate, or to satisfy environmental needs. Unfortunately, there is no national policy regarding retirement of existing projects, and dam owners are not required to set aside funds for the costs of project retirement. A national policy must be implemented to provide for long-term maintenance of facilities that will be retained beyond their operational capabilities, or to pay costs of removal and habitat restoration (Tyus & Winter, 1992).

Full restoration of the Elwha River ecosystem provides an exceptional opportunity to fully document the removal of two high-head dams, management of accumulated sediments, and restoration of the physical and biological processes and fish communities of a large river. Restoration techniques learned in the Elwha may be applied to other ecosystem restoration efforts. However, Elwha River restoration should not be viewed as precedent-setting for the removal of other projects. As noted, many other dams have already been removed. Also, it is unlikely that issues surrounding other dams are similar to those of the Elwha. Some of these unique conditions are:

1. The Elwha River is one of few rivers in the contiguous United States that supported all species of Pacific salmon.
2. One of the facilities (Glines Canyon) is located within a national park and is inconsistent with park policy.
3. Because most of the Elwha watershed is located within a national park, the majority of the river basin is in pristine condition, providing excellent habitat for restoration of fisheries.
4. Treaty rights of several Native American tribes have been affected by the reduction in fisheries and elimination of access to cultural and usual and accustomed areas, and one tribe's reservation is located at the mouth of the river.
5. An alternative source of power (Bonneville Power Administration) is readily available to replace the relatively small amount produced by the dams.
6. Project acquisition and dam removal satisfy the interests of each of the parties to the FERC proceeding.
7. The interests of other parties, including the City of Port Angeles, Dry Creek Water Association, and the Rayonier pulp mill, are protected.

The Elwha River offers one of the best chances for full ecosystem restoration in the Pacific Northwest (Fig. 15.5). Much of the Elwha basin lies within a national park; no other candidate for restoration of anadromous fish are similarly protected. National park protection of the vast majority of the basin means that the river's productive potential will not be significantly affected by disturbances that continue to limit other northwest rivers (e.g., timber harvest, water withdrawals, dams, urbanization). Nature will maintain the system in health; thus, the cost of removing the Elwha and Glines Canyon Dams and restoring the ecosystem is a one-time expense. Other northwest rivers will require continual mitigation.

Within the Elwha, all levels of the biological diversity hierarchy (e.g., genetic, species, ecosystem, landscape; see Winter & Hughes, 1995) can be restored and contribute to the biological health of the Olympic Peninsula.

Figure 15.5 Elwha River at Geyser Valley, May 27, 1907 (From Department of Interior, Department of Commerce, Lower Elwha Klallam Tribe, The Elwha Report: Restoration of the Elwha River Ecosystem and Native Anadromous Fisheries. Washington, DC: U.S. Government Printing Office, 1994: 142. Courtesy of Washington State Historical Society, Tacoma, WA.)

Additionally, restoration offers an outstanding opportunity to establish a "living laboratory" and document restoration methods and the recovery of a large ecosystem.

The biggest impediments to ecosystem restoration are not environmental; environmental issues can be satisfied. For most who contemplate the issue, the remaining question is cost and concern about who pays. Currently, there is no requirement for dam owner/operators to set aside costs for retiring and removing private hydroelectric facilities. In the absence of such a requirement, the Elwha Act stipulates that should the Secretary of Interior determine dam removal is necessary for full ecosystem recovery, he or she may acquire the projects, remove them, and restore the river at federal expense. This seems an almost impossible venture at a time of deficit reduction and dwindling agency budgets. However, supporters view the estimated costs of dam removal ($111–128 million) as being minuscule relative to long-term benefits.

Moreover, substantial costs also are associated with not removing the projects. If the dams remain, the answer to the question "who pays?" is the same. If removed, the public, who will pay the costs, will reap the benefits of a restored ecosystem (in fisheries, wildlife, cultural resources, and a restored watershed within a national park). Alternatively, if the dams remain, for lack of appropriated funds within the Congress, the people of the nation will bear the costs of a lost salmon fishery, extended litigation, and attendant loss of jobs.

As clearly illustrated in other cultures, the health of the people is inextricably tied to the health of the land (Hawkins Hoffman, 1989). An impoverished ecosystem will cost us dearly; restoring the ecosystem is an investment with high returns. The answer to "who pays?" is "we do." The answer to "when?" is "either now, or later at higher cost." There is no option without cost. There is no better place to accomplish river restoration; there is no better time than now.

Literature Cited

Brown B. *Mountain in the Clouds: A Search for the Wild Salmon.* New York: Simon & Schuster, 1982.

Cederholm CJ, Houston DB, Cole DL, Scarlett WJ. Fate of coho salmon (*Oncorhynchus kisutch*) carcasses in spawning streams. *Can J Fish Aquat Sciences* 1989;46:1347–1355.

Crown Zellerbach Corporation. Application for License of a Major Project. Submitted to Federal Power Commission, Washington, DC: 1973.

Department of Interior, Department of Commerce, and Lower Elwha S'Klallam Tribe. *The Elwha Report: Restoration of the Elwha River Ecosystem and Native Anadromous Fisheries.* Washington, DC: U.S. Government Printing Office, 1994-590-269.

Federal Energy Regulatory Commission. Draft Staff Report, Glines Canyon (FERC No. 588) and Elwha (FERC No. 2683) Hydroelectric Projects. Washington, DC: Federal Energy Regulatory Commission, 1993.
Hawkins Hoffman C. First Asian School on Conservation Biology—A trip report. *George Wright Forum* 1989;6(2):7–12.
Hawkins Hoffman C. The Elwha issue: A Fish problem that just won't die. *George Wright Forum* 1992;9(2):45–52.
Maib CW. A Historical Note on the Elwha River, Its Power Development and Its Industrial Diversion. Washington State Department of Fisheries, Olympia, WA: Stream Improvement Division. No date.
Meyer PA, et al. Elwha River Restoration Project: Economic Analysis. Port Angeles, WA: Lower Elwha S'Klallam Tribe, 1995.
National Park Service. *Final Environmental Impact Statement, Elwha River Ecosystem Restoration, Olympic National Park, Washington.* Port Angeles, WA: Olympic National Park, 1995.
National Park Service. *Draft Environmental Impact Statement, Elwha River Ecosystem Restoration Implementation, Olympic National Park, Washington.* Port Angeles, WA: Olympic National Park, 1996.
"Power Company Will Fish with A Derrick." Port Angeles, WA: *Olympic Leader* September 1, 1911.
Randle TJ, Lyons JK. Elwha River Restoration and Sediment Management (Paper submitted to U.S. Committee on Large Dams). Denver: Bureau of Reclamation, 1995.
Reineking VH. Reconstruction of the Elwha River Dam. *Engineer Record* 1914;69(13):372–375.
Tyus HM, Winter BD. AFS draft position statement: Hydropower development. *Fisheries* 1992;17(1):30–32.
White W, Stalheim D, James R. Clallam County Profile. Port Angeles, WA: Clallam County Department of Community Development, 1992.
Winter BD. A brief review of dam removal efforts in Washington, Oregon, Idaho and California. *U.S. Dept Comm NOAA Tech Memo* (NMFS F/NWR-28). 1990.
Winter BD, Hughes RM. AFS draft position statement: Biodiversity. *Fisheries* 1995;20(4):20–25.
Wunderlich RC, Winter BD, Meyer JH. Restoration of the Elwha River ecosystem. *Fisheries* (Bethesda) 1994;19(8):11–19.
Zubalik S. The Rising Cost of Falling Water: The Case for Hydropower Planning in Washington State (M.S. thesis). Seattle: University of Washington, 1988.

16

Wolf Restoration in the Northern Rocky Mountains

James M. Peek
John C. Carnes

In protected areas, the restoration of any major ecosystem process that has been eliminated by anthropocentric activity clearly is important in reestablishing a functional ecosystem. In most protected areas in the continental United States one major missing ecosystem dynamic is the large mammal predation attributable to wolves (*Canis lupus*). Historically, wolves were originally present throughout North America, except in California and the southeastern United States (Hall & Kelson, 1959). Within its range, the gray wolf occupied the top rung in a major food chain involving the larger mammals. When we consider that this food chain occurred in virtually all of the forests and rangelands that provided food and cover for the larger mammals, then we begin to recognize the wolf's importance.

The wolf preys on the larger herbivores including white-tailed deer (*Odocoileus virginianus*), mule deer (*O. hemionus*), elk (*Cervus elaphus*), moose (*Alces alces*), woodland caribou (*Rangifer tarandus*), mountain sheep (*Ovis canadensis*), mountain goat (*Oreamnos americanus*), pronghorn (*Antilocapra americana*), muskox (*Ovibos moschatus*), and bison (*Bison bison*). Only the gray wolf and the mountain lion (*Felis concolor*), however, prey predominantly on ungulates, because all other predators typically take smaller prey or feed extensively on plants.

A comparison between the cougar and the gray wolf further illustrates the uniqueness of the wolf as a generalist predator of larger mammals. The cougar primarily inhabits highly dissected and rugged terrain, where its stealth and solitary behavior is most suited for predation on large prey. Its short rushes at prey, often after extensive periods of waiting, are facilitated by terrain that provides adequate opportunities for surprise. While the wolf also may benefit from such terrain, it is more of a pursuit predator and is capable of running its prey for long distances over all kinds of terrain. The original range of the gray wolf, which included desert, mountain, prairie, sagebrush, forest, and tundra, attests to its adaptability.

Thus, it is clear that the gray wolf has been a significant influence on its prey (antlered and horned game) throughout evolutionary time in which wolf and prey have coexisted. This is not meant to minimize the collective influence of the entire predator complex on these larger mammals, as each species has undoubtedly had an influence over evolutionary time. However, the wolf, being an obligate predator with a wide range, probably has exerted the most significant influence in the context of evolutionary time.

The demise of the wolf in the West came quickly. It was poisoned, shot, and trapped for bounty and sport. The carnage was awesome, relentless, and effective, and the magnitude stunning. For example, between 1883 and 1918, 80,730 wolves were bountied for $342,764 in Montana alone (Lopez, 1978). As a result, by the 1960s wolves were extirpated in the United States except in northeastern Minnesota, with some "wanderers" coming into the Glacier National Park area in Montana (Mech, 1970).

Elsewhere (i.e., outside the contiguous United States)—although persecuted—wolves remain relatively abundant. An estimated 40,000 to 50,000 wolves, or about 90 percent of the total North American population, occupy about 80 percent of their former range in Canada, having been extirpated from the Maritime Provinces around 1870 (Carbyn et al., 1987). In Alaska, wolves occupy most of their original range, and in the late 1980s their population numbered between 5200 and 6500 (Alaska Department of Fish and Game, 1988).

Wolves have been recolonizing northwestern Montana since the early 1980s (Ream et al., 1991). In 1989, three breeding packs were present in the Flathead River drainage, suggesting that the recolonization occurring in this region might result in a resident population. A small population persists on Isle Royale, and wolves are recolonizing the Lake States and North Dakota from adjacent Minnesota. It is estimated that between 1500 and 1750 wolves presently occupy Minnesota (Fuller et al., 1992).

Endangered Species Status

If a species is approaching extinction or is extinct on all or a portion of its original range, it may be classified as endangered under the Endangered Species Act.

While there are substantial populations farther north, the wolf occupies less than 3 percent of its original range in the contiguous states. Thus, as with the woodland caribou and grizzly bear (*Ursus arctos*), the wolf is classified endangered in the contiguous United States, except in Minnesota where it is classified as threatened (U.S. Fish and Wildlife Service, 1987).

The potential for long-distance dispersal (Fritts, 1983; Ballard et al., 1983) has implications to wolf systematics. While 24 subspecies are officially recognized by Hall and Kelson (1959), Nowak (1983) suggested that as few as 5 subspecies are likely valid. A wolf captured at Glacier National Park was killed near Dawson Creek, British Columbia, suggesting at least some population interchange from the southern limits of occupied range to the northern reaches of the Rocky Mountains (Ream et al., 1991).

Ecotypic variations related to terrain and prey are probably more important for wolf restoration efforts than are subspecific definitions. Wolves that prey on the same species existing in the recovery areas and occupying similar terrain may be better adapted to recovery-area conditions and, consequently, have the highest probability of establishing a viable population. Griffith et al. (1989) reported that successful restorations were characterized by at least 20 conspecifics captured from similar habitats and released to locations well within the original range. In any case, the Endangered Species Act refers to species and not subspecies.

Even as wolves are returning to northwestern Montana and surrounding areas, breeding packs have not been found outside of the northwestern Montana region. Success of dispersing individuals in finding a mate and establishing a den depends on age, with dispersing adults being most successful and pups least (Gese & Mech, 1991). Mortality rates of individual wolves that disperse into new surroundings are likely to be high, as evidenced by the fate of wolves dispersing from Minnesota packs (Mech, 1987). The risks likely increase with increasing dispersal distances (Gese & Mech, 1991). Therefore, for example, while wolves have been known to occur at least sporadically in the North Fork of the Clearwater River of Idaho for a decade or more, breeding has not been documented. Though some people have claimed pack activity, no definitive evidence has been produced, and the inability of searches to locate pack activity suggests that these wolves are dispersers who most likely have not been able to locate a suitable partner. The pack activity in northwestern Montana is in contiguous territories, implying that dispersers remaining close by their natal range are forming adjacent packs.

The lack of breeding-pack formation in Yellowstone National Park and the central Idaho wilderness was one of the considerations behind the current restoration efforts in these areas. It was also believed that an artificial restoration program, with radio-collared animals that could be readily followed and their locations accurately determined, would provide more control over where the wolves established. A new capture-recapture collar device allows the recapture of individuals that move into areas where they are not wanted. It also allows managers to identify wolves in risky areas and provide them with more protection (Mech & Gese, 1992).

The Role of Parks and Protected Areas

An ecological value can be attributed to the restoration of a major ecosystem process, large mammal predation. However, locations suitable for species like the wolf are restricted to areas where conflicting human uses are minimal. These are primarily the larger national parks and wilderness areas, which have significant populations of ungulates. Indeed, the legislation and directives for maintaining and restoring natural ecosystem processes within these parks and wilderness areas would seem to mandate wolf recovery. However, many other public lands, and some of the more extensive private holdings as well, are also suitable for wolves, as experience in Canada and Minnesota illustrates (Fritts, 1993). Only where livestock grazing is extensive and human presence is all-pervasive would the wolf not be welcome. Today, opportunities to restore the gray wolf to the northern Rocky Mountains depend on (1) effective management of populations to minimize conflicts and (2) areas where humans are willing to tolerate them.

Wolf Ecology

Behavior and Dispersal

Dispersal of wolves after release into a recovery area could create problems and is thus a topic of considerable relevance. Wolf dispersal is attributed to resource competition, mate competition, and inbreeding avoidance (Gese & Mech, 1991). Lone wolves can make up 2 to 40 percent of a wolf population and constitute a pool of potential dispersers or recent immigrants (Mech, 1977b; Peterson et al., 1984; Ballard et al., 1987; Fuller & Snow, 1988; Peterson & Page, 1988; Fuller, 1989; Thurber & Peterson, 1993). Dispersers are typically yearlings, but may be as old as 3 years (Mech, 1987; Ballard et al., 1987; Potvin, 1988; Fuller, 1989). Conversely, it is uncommon for pups to disperse (Fritts & Mech, 1981; Peterson et al., 1984; Messier, 1985b; Fuller, 1989).

Social stress apparently is the primary cause of dispersal. Van Ballenberghe (1983), Peterson et al. (1984), Messier (1985b), Fuller (1989), and Gese and

Mech (1991) found decreased pack cohesiveness and higher rates of dispersal during breeding and denning season, when social stress appears to be highest. Similarly, Rabb et al. (1967) and Zimen (1976) saw the most aggression in captive packs during these seasons. Food scarcity also seems to lead to greater dispersal (Peterson & Page, 1988), apparently working in combination with social factors to precipitate pack splitting.

Dispersal seems to vary with wolf population density and level of exploitation. In low-density, heavily exploited populations, there may be little difficulty finding unused territory, and in this case wolves tend to disperse early rather than stay in the natal pack (Packard & Mech, 1980; Fritts & Mech, 1981; Packard et al., 1983). Conversely, high population density may cause food stress within a pack and also increase dispersal rates. Peterson and Page (1988) found that lone wolves were most common on Isle Royale when both wolf density and food stress were high.

Dispersers have potentially varied effects on populations. Lone wolves are commonly accepted into existing packs and can stabilize social structure, prevent reproductive failure, and minimize inbreeding in populations with high mortality (Rothman & Mech, 1979; Fritts & Mech, 1981; Van Ballenberghe, 1983; Fritts et al., 1985; Ballard et al., 1987; Fuller 1989). These contributions can be limited by increased vulnerability to human harvest and intraspecific aggression (Mech, 1972, 1977b; Fritts & Mech, 1981; Peterson et al., 1984; Messier, 1985a), although during their studies Fuller (1989) and Hayes et al. (1991) found no difference in survival between dispersers and territorial residents. Packard and Mech (1980) speculated that high-density populations may be partly regulated by increased mortality of dispersers and that low-density populations may also increase due to establishment of new breeding units. Peterson and Page (1988) found support for this speculation on Isle Royale, where abundant food and restricted opportunity for dispersal created a high-density population. An unusual level of intraspecific aggression led to high dispersal and mortality rates and a corresponding population decline.

Disassociation from a pack typically precedes dispersal and is thus an important underlying phenomena. Messier (1985b) found that yearlings were the most loosely associated pack members and, along with adults, showed lower pack affinity in areas with low prey density. Females of both age classes were more solitary than males. As winter progressed (December–March), the frequency of pack dissociation increased for all age classes and may have been due to higher social stress during breeding season (Messier, 1985b). Messier (1985b) speculated that females most often dissociated from their packs because of lower social status and increased nutritional stress. He concluded that dispersal appears to be a dynamic and gradual process, with a wolf breaking ties with its pack over a period of a few months to a few years. Extraterritorial movements can be interpreted as predispersal forays to assess territory vacancy. However, Gese and Mech (1991) found that extraterritorial movements may be common and unrelated to dispersal.

As part of an ongoing study begun in 1958 on Isle Royale, Thurber and Peterson (1993) found that lone wolves were not restricted to marginal areas between packs and ranged throughout the island to establish territories when wolf densities were low. Small packs and single wolves were common, and food was abundant during winter. Under these circumstances, wolves may have been attempting to maximize food intake and immediate reproductive opportunities (Thurber & Peterson, 1993).

Pack Sizes

Pack sizes are highly variable among populations through time. Packs vary from two to as many as 15 animals, with size declining typically from early to late winter (Carbyn, 1975; Carbyn, 1980; Mech, 1986). Packs can decline by 40 percent over a given winter and increase by as much as 86 percent from spring to winter (Ream et al., 1991). Harrington et al. (1983) reported that April pack size was related to December litter size, conditional on prey availability. Pack size also seems to reflect the size of prey (Peterson, 1977). For example, wolves feeding on moose maintain larger packs than those feeding on elk, which maintain larger parks than those feeding primarily on deer.

Pack size is also dependent on ungulate biomass, wolf population densities, and level of exploitation. There is a positive correlation between pack size and food availability (Van Ballenberghe et al., 1975; Haber, 1977; Mech, 1977a, 1977b; Messier, 1985a; Peterson & Page, 1988). Furthermore, in a low-density, heavily exploited population, subordinate wolves tend to disperse at a young age (Packard & Mech 1980; Fritts & Mech, 1981; Packard et al., 1983). Thus, a harvested population should probably consist of a higher density of smaller packs compared to an unharvested population (Rausch, 1967; Peterson et al., 1984; Fuller, 1989).

Wolf Predation

Wolves kill, on average, 16.6 (range of 12–28) deer or elk and 8.5 moose per wolf (Keith, 1983) (Table 16.1). Kill rates for caribou (Holleman & Stephenson, 1981) are comparable to rates reported for deer and elk. These estimates assume that the winter kill rates were maintained throughout the year, as the likely reductions in kill rates in summer would be offset by the increased number of calves and fawns in the kill. Nevertheless, these kill rates may be considered high. Wolves also may exhibit a functional response to prey numbers and increase their kill rates when prey populations increase in size or vulnerability (Holling, 1959).

Table 16.1 Wolf kill rate summary[a]

Mean winter kill rate (days/kill)				
Location	Ungulate prey	Per pack	Per wolf	Ref
Isle Royale	Moose	3.1	47	1
Isle Royale	Moose	3.3	36	2
NE Alberta	Moose	4.7	45	3
Tanana AK	Moose	3.4	53	4
Nelchina, AK (summer)	Moose, caribou	7.3–15.7	22.5–117.5	5
Nelchina, AK (winter)	Moose, caribou	4.9–10.8	44–62	5
Kenai, AK	Moose, caribou	3.1–21.4	58–42	6
Brooks Range AK	Caribou	1.9–4.6	13.4–32.3	4
Yukon	Sheep, moose, caribou	7.7–9.0	31–54	13
Southwest Quebec	Moose	19–90	91–250	7
Southeast Alaska	Black-tailed deer	—	24	4
Ontario	White-tailed deer	2.2	18	8
Manitoba	Elk	3.6–6.9	14–21	9
Western Minnesota	White-tailed deer	7	32	10
Northeast Minnesota	White-tailed deer	7.8	25	11
Brazeau, Alberta	Elk, deer, moose, sheep	2.5	25	12

[a]After Vales and Peek (1993).
References: (1) Mech (1966); (2) Peterson (1977); (3) Fuller and Keith (1980); (4) Holleman & Stephenson (1981); (5) Ballard et al. (1987); (6) Peterson et al. (1984); (7) Messier and Crete (1985); (8) Kolenosky (1972); (9) Carbyn (1983); (10) Fritts and Mech (1981); (11) Mech (1977); (12) Gunson (1986); (13) Sumanik (1987).

Estimates of biomass consumed by individual wolves can be calculated from kill rates, assuming an average weight for each prey type and a constant percent of each carcass consumed. With this approach, a pack in northeastern Minnesota was estimated to have consumed 1.5 to 5.8 kilograms per wolf per day (Mech, 1977b). This amounts to 5.7 to 22.1 percent of the body mass for a 26-kilogram wolf. Mech (1977b, personal communication 1989) estimated that 1.7 kilograms per day was required for maintenance and that, although wolves will eat as much as three times this amount (5.1 kg), 2.5 kilograms per wolf per day (or about 8 percent of body weight) is a good average across the wolf range. Thus, an adult wolf weighing 45 kilograms would eat approximately 1314 kilograms of food per year. Annual intake would range from 730 to 1700 kilograms per wolf, compared to an annual minimum requirement of 621 kilograms (1.7 × 365 days), an average annual intake of 913 kilograms per wolf (2.5 kg × 365 days), and a likely maximum intake of 1862 kilograms per wolf (5.1 × 365 days). Carbyn et al. (1987) calculated an annual intake of 2175 kilograms (4785 lbs) per wolf, of which 1820 kilograms (4004 lbs) was from elk and deer. However, the total annual biomass consumption per wolf may be low because other prey could be more important, or may be higher because light prey-body-weights and heavy wolf-body-weights were used to estimate consumption.

Although selection of prey species has not been studied in areas where moose are abundant, wolves seem to select deer over elk where these two ungulates exist. Carbyn (1975) reported that wolves in Jasper National Park took elk at a ratio of 0.50–2.75 : 1 (percent occurrence in scat compared to relative abundance) and mule deer at a ratio of 2.4–13.2 : 1 in winter and summer respectively. Annually, 30 percent of the wolf diet was elk and 43 percent deer. Mule deer were thus the most frequent food item even though they were outnumbered by elk by a ratio of 8 to 1. This trend also was evident in Riding Mountain National Park (i.e., wolves favored white-tailed deer over elk).

The sex and age of wolf kills varies among prey species, presumably for reasons related to differences in prey behavior. Mule deer often rely on erratic bounding and rapid bursts of speed to escape wolves, while elk do not (Carbyn, 1975). In contrast to deer, wolves usually encounter elk in herds. Because mass flight may impede the escape of otherwise healthy individuals, wolves have a greater opportunity to kill vigorous elk. Conversely, wolves encountering many prey at once may have a greater chance to detect the more vulnerable individuals. Carbyn (1975) thus expected wolves to take fewer prime mule deer than prime elk. In any case, wolves have little difficulty killing either elk or mule deer species once they are run down, especially in contrast to the difficulty they have killing moose.

Carbyn (1975, 1983) thought there were indications that bull elk were more vulnerable to predation than cows. Kolenosky (1972) reported a male-female sex ratio of 250 : 100 for white-tailed deer killed by wolves in Ontario. Bucks constituted 57 percent of the adults killed by wolves in Algonquin Park (Pimlott et al., 1969) and 71 percent in northeastern Minnesota (Mech & Frenzel, 1971). Carbyn (1975) found six adult males and two adult female mule deer killed by wolves in Jasper. Boyd et al. (1994) reported that in Glacier National Park, Montana, male white-tailed deer and elk were killed at a higher rate than would be indicated by their observed proportion in the population by a colonizing wolf population.

In Riding Mountain National Park, Carbyn (1983) reported that calves constituted 34 percent of the wolf-killed elk, and animals over 7 years old constituted 40 percent. Elk calves were killed most frequently in early- to midwinter while, by late winter, adult cows were more frequently taken (Carbyn, 1983). Carbyn (1983) reported that 39 of 57 adult elk killed by wolves were cows, a female-male sex ratio of 46 : 100. Cows often are found in large groups, which may facilitate detection of predators and minimize the probability of predation. Bulls occur in smaller groups, are often injured in fights, and enter the winter in poor condition, which make them more vulnerable to predation than cows.

The relative proportions of calves, adults, and older elk taken by wolves in Jasper and Riding Mountain National Parks were quite similar (Table 16.2). In Riding Mountain National Park, calves were taken 1.79 times more frequently than their estimated occurrence in the population, while adults were taken 0.63 times as much, and old animals were taken in proportion to their occurrence. Carbyn (1983) concluded that more elk calves and fewer adult elk (1–11.5 years old) were killed by wolves than by hunters.

Relationships with Other Predators

Wolves will occupy designated recovery areas with coyotes, mountain lions, black bears, and grizzly bears. However, coyotes, unlike the other predators, are preyed on by wolves (Thurber et al., 1992). In south-central Alaska, coyote home ranges overlapped wolf home ranges, but coyotes appeared to frequent areas closer to open roads more often than wolves; in other areas, coyotes appeared to be less abundant in wolf habitat (Berg & Chesness, 1978; Carbyn, 1982). Interactions between wolves and the two bear species should not result in any significant demographic effect on either wolf or bear populations, although some instances of mortality and displacement from kills are known (Servheen & Knight, 1993).

Table 16.2 Age structures of elk and deer in wolf-killed samples compared with proportions in the population

		Young		Adult		Old		
Area	Species	Wolf-killed	Population	Wolf-killed	Population	Wolf-killed	Population	References*
Northeast Minnesota	White-tailed deer	17	26	68	73	15	1	1
Northwest Minnesota	White-tailed deer	34	33	35	62	31	6	2
Eastern Ontario	White-tailed deer	30	35	65	63	5	2	3
Western Ontario	White-tailed deer	17	20	61	52	22	28	4
Jasper	Mule deer	62	—	31	—	27	—	5
Jasper	Elk	41	—	32	—	27	—	5
Riding Mountain	Elk	37	19	33	41	33	40	6,7

*1. Mech & Frenzel (1971). Population from hunter harvest; 2. Fritts & Mech (1981). Population from hunter harvest; 3. Kolenosky (1972). Population from hunter harvest; 4. Pimlott et al. (1969). Population from road kills; 5. Carbyn (1975). No population estimates available; 6. Carbyn (1980). Population from hunter harvest; 7. Carbyn et al. (1987). Wolf kills, 1975–1986.

Adjustments by coyote, black bear, grizzly bear, and cougar populations to the presence of wolves in newly occupied ranges will inevitably occur. Perhaps the most significant adjustment may be behavioral rather than demographic, although it is possible that coyote populations may decline in areas where wolves are most common. Females with young may shift their habitat use to more secure habitats, and mountain lions may be further confined to the more rugged portions of their habitats. However, such responses are likely to be subtle, will occur over extended periods of time, and will shift constantly in relation to many factors, not merely the presence of wolves.

All investigations of wolf-prey relationships in North America have been done in more northerly areas, where the relevant relationships may differ from those occurring farther south in U.S. recovery areas. Prey populations may exhibit more resiliency to wolf predation where growing seasons are longer and climate is generally less severe. Furthermore, predation rates are variable and may be modified by weather, wolf management, pack sizes, wolf population turnover, prey nutrition, prey dynamics, and numbers of prey species.

As yet, there is no evidence of merit from the literature that wolf predation is compensatory with other forms of mortality, a conclusion also reached by Gauthier and Theberge (1987). Carbyn (1975) compared elk cow-to-calf ratios from Jasper and Elk Island National Parks and demonstrated a higher ratio where wolves were present (Jasper) compared to where they were not (Elk Island). However, he did not consider prey density or general habitat conditions, both factors likely to interact with predation. Nevertheless, if predation indirectly improves overall nutritional status of a prey population, increased production and survival might be expected. A reduced population may exhibit higher fecundity, or be less susceptible to disease, other predators such as bears and coyotes, or malnutrition. The magnitude, timing, and nature of the functional and numerical responses to predation may be the most significant unknown factors influencing our understanding of this phenomena. It is probable that lightly hunted prey populations, such as northern Yellowstone elk, would more likely demonstrate compensatory responses to wolf predation than would hunted populations maintained at lower densities by harvest.

Demography and Recruitment

While survival of transplanted stock will be the initial major determinant of reintroduction success, breeding and recruitment soon become important. Percentage of pups in a wolf population during fall or winter is a good index of annual recruitment (Fritts & Mech, 1981). Percentage of pups reflects both reproduction and pup survival through summer relative to adults, with pup survival alone probably more important to population growth and reproduction (Keith, 1983). Numerous studies have suggested that pup percentage is higher in an exploited population than in an unexploited one. In unexploited populations, pups are reported to comprise 13 to 31 percent of the population, and in exploited populations, 35 to 73 percent (Pimlott et al., 1969; Mech, 1970). However, Pimlott et al. (1969) reported no change in pup-adult ratios after the cessation of wolf trapping, and Gasaway et al. (1983) found that the percent of pups in a kill did not change when winter harvest was dramatically increased. These unexpected results may have been caused by a lack of breeding-age animals of one or both sexes, or by low ungulate biomass (Gasaway et al., 1983).

There is not much data available on ratios of pups to adults in unharvested populations simply because unharvested populations are rare. Investigations of four unexploited populations revealed only two with lower pup-adult ratios than in exploited populations (Fuller & Novakowski 1955; Kelsall 1968). This may have been caused by biases in data collection. Public harvest (i.e., hunting, trapping) may select for young, inexperienced animals (Van Ballenberghe & Mech, 1975) and may explain the 42 to 74 percent of pups reported by Ballard et al. (1987) from harvest records in the study population.

Given these uncertainties, numerous theories have been advanced to explain the purported larger percentage of pups in exploited populations: (1) increased productivity and pup survival; (2) increased proportion of females that breed; (3) higher proportion of pups due to constant litter size and decreased number of adults; and (4) increased ungulate-biomass per wolf available in exploited populations. The suggestion that larger litters in exploited populations were at least partly responsible for the increased proportion of pups should therefore be viewed with caution (Kelsall, 1968; Van Ballenberghe, & Mech, 1975). For example, Pimlott et al. (1969) suggested that low reproduction and high mortality of pups and yearlings were the most important factors maintaining a low pup-adult ratio in unharvested populations. In a review of five studies, Mech (1970) similarly found higher pup survival rates in exploited than in unexploited populations. Conversely, Van Ballenberghe et al. (1975) compared their data to Rausch (1967) and Pimlott et al. (1969) and found pup survival rates similar to those of the first winter under varying regimes of exploitation.

The greater percentage of pups in harvested populations may be due to social disruptions that allow an increased proportion of females to breed. In exploited populations, 58 to 83 percent of adult females either breed or were found to be pregnant (Rausch, 1967; Van Ballenberghe et al., 1975; Gasaway et al., 1983; Parker & Luttich, 1986). The proportion of breeding wolves appears to vary with population age, structure, and density, and with pack sizes (Fuller, 1989), or, proximally, with variation in intrasexual aggression and mate preferences (Rabb et al., 1967; Harrington et al., 1983). In a single, unexploited population, 59 percent of observed adult females had bred (Pimlott et al., 1969). Peterson et al. (1984) found that harvest reduced pack and territory size correspondingly, and increased the number of packs. However, because pup numbers and litter frequency varied independently of pack size, Peterson et al. (1984) suggested that the increased proportion of pups and breeding females in exploited populations may simply result from constant litter size and more numerous packs. Thus, the proportion and total number of pups increase as additional packs develop and the number of adults per pack declines.

Although one litter per pack is generally the rule, numerous studies have documented more than one litter being produced by a pack in one year (Murie, 1944; Rabb et al., 1967; Rausch, 1967; Haber, 1977; Paquet et al., 1982; Van Ballenberghe, 1983, Peterson et al., 1984; Ballard et al., 1987). Although Packard et al. (1983) found that only one litter was born in 94 percent of 101 cases, regardless of pack size, Harrington et al. (1983) found more than one litter per pack per year in 22 to 41 percent of packs with more than one adult female. They considered these figures minimal as early mortality, combining of litters, and observational difficulties reduce the chances of locating all litters. It should be remembered that some data on multiple litters comes from examination of carcasses, with the possibility that many fetuses and young pups fail to survive (Packard & Mech, 1980). Although Peterson et al. (1984) documented a case in which a subordinate female, trying to raise a litter alone, died from malnutrition along with her pups, Murie (1944), Haber (1977) and Van Ballenberghe (1983) documented cases in the wild in which pups from multiple litters survived.

Disruption of pack social order may allow for more varied breeding by increasing mating combinations in a pack (Rabb et al., 1967). Packard et al. (1983) and Paquet et al. (1982) suggested that multiple litters occur in the absence of dominant pair-bonded breeders. Haber (1977) contended that harvest, especially aerial hunting, tends to randomly fragment packs and thereby permits subordinates to escape sexual repression by dominant animals and increases the proportion of multiple litters. This may be how heavily exploited populations maintain their numbers, as well as higher pup-adult ratios, and thus constitutes a form of compensatory natality (Ballard et al., 1987).

Exploitation may affect recruitment by lowering wolf densities, increasing per capita food supplies, and thereby increasing both birth rate and pup survival. Ungulate-biomass per wolf is positively correlated with both the percentage of pups and the mean number of pups per pack (Fuller, 1989), and may, furthermore, allow higher pup survival in exploited populations where ungulates are abundant. High prey density seems to lead to greater reproductive activity and success, presumably as a result of better physical condition (Messier, 1985a; Boertge & Stephenson, 1992).

Reproductively inactive adult females had less subcutaneous fat than reproductively active females, and average litter size (based on fetuses) was significantly smaller with low ungulate-biomass per wolf than with high ungulate-biomass per wolf.

However, evidence for increased reproductive activity is compromised by uncontrolled differences in study methods. Boertge and Stephenson (1992) reported reproductively active animals from a population with low per capita ungulate biomass having "enlargement and thickening of the uteri from increased vascularization . . . (i.e., in proestrus, in estrus or pregnant)," in comparison to Rausch (1967), who reported only animals that were pregnant in a population with high per capita ungulate biomass. If Boertge and Stephenson (1992) had only reported pregnancies, their estimate of reproductively active females would undoubtedly have been lower for moderate and low ungulate-biomass per wolf. This would have provided added support for their conclusions and could have led to a significant difference between estimates for moderate and high ungulate-biomass per wolf areas, in addition to high and low areas. Boertge and Stephenson (1992) assumed that the differences they saw were due to nutrition. Rausch (1967) collected data after a period of intensive federal predator control, and the high percentage of pregnant females may have been partly due to the effects of exploitation on pack social structure. "Enlargement and thickening of uteri" is a subjective indicator at best, and it is unknown whether suppression of estrus was the actual mechanism behind lower reproduction.

As a final note, the method of calculating litter size should be kept in mind when comparing different studies. Perinatal mortality is high in domestic dogs. In wolves, this could lead to overestimation of litter sizes based on numbers of fetuses or placental scars in carcasses. Two of the three studies of unharvested populations estimated litter-size observations of pups after they had emerged from dens, whereas most of the studies of exploited populations relied on examination of carcasses. The relationship of fetuses and placental scars to pups observed outside a den is not known, although Lindstrom (1981) documented 19 percent intrauterine mortality in red foxes. Persistence of scars from earlier pregnancies can also cause overestimation of litter size. In any case, Lindstrom (1981) and Kelsall (1968) noted that the lowest mean litter sizes came from pup counts at dens for red fox and exploited wolves, respectively.

Survival

Human activity, including trapping, hunting, and automobile collisions, is the primary cause of adult mortality in most studies. Mech (1977b), Peterson et al. (1984), and Fuller (1989) found that human-caused mortality, legal or illegal, was in turn greatly influenced by accessibility. Roads facilitate and allow humans greater opportunity to deliberately or incidentally kill wolves. Thus, Mech et al. (1988) and Fuller (1989) found a dissociation between wolf and road densities greater than 0.58 and 0.79 per square kilometer, respectively. Fuller (1989) also noted that wolves were killed more frequently by humans in 2-kilometer areas with great road densities, than in areas where wolves died from intraspecific strife and disease. In general, this lethal human presence translates into an inverse relationship between human and wolf density in the Great Lakes region (Weise et al., 1975). Moreover, several studies have shown that state and federal protection did not significantly reduce human-caused mortality (Van Ballenberghe et al., 1975; Robinson & Smith, 1977; Mech et al., 1988; Fuller, 1989). In contrast, Fritts and Mech (1981) found that before the Endangered Species Act (ESA), 80 percent of mortality was human caused; after passage of the ESA, mortality dropped to 33 to 50 percent.

Survival of colonizing wolves would be expected to be lower than for wolves that have established territories. Little is known about early pup survival in colonizing wolves because of difficulties obtaining data. Therefore, Mech (1970) and Van Ballenberghe et al. (1975) calculated pup survival rates based on (1) autumn age structure, (2) adult sex ratio, (3) average litter size, and (4) percent of adult females bearing litters.

Early survival could more accurately be determined by counting pups at the den shortly after whelping and comparing these numbers with the number of pups observed with the pack in autumn. Occasionally, researchers are also in a position to observe litters early on. Mech (1975) examined four litters shortly after whelping, and Murie (1944) entered a den, apparently with no adverse effects on the survival of the pups. However, there are few other such reports in the literature, and if an investigator caused abandonment of a litter he or she would probably be unwilling to report it. Time-lapse video cameras are another option for obtaining data; they could be mounted near a known den site well-before denning, although this method would have a good chance of failure due to technical difficulties and the likelihood the pack will use a different den.

Food availability is probably the most important factor affecting pup survival. Mech (1977a, 1977b) documented decreased production and pup survival during a decline in the primary prey population. In addition, Van Ballenberghe and Mech (1975) found a positive correlation between annual variation in pup weights and survival and food supply. They found that pups weighing less than 65 percent of standard weight had a lower survival than pups heavier than 80 percent of standard weight.

Given the importance of human-caused mortality, food supply and social status can also affect adult wolf survival. Mech (1977a, 1977b) found that less dominant adults were more vulnerable to malnutrition than were more dominant animals. Under these conditions, malnutrition and intraspecific strife together accounted equally for 58 percent of mortality (Mech 1977a).

Malnutrition and intraspecific strife also are important causes of mortality and often are related to prey and wolf density. Mech (1977b) found that during a decline in white-tailed deer, malnutrition and intraspecific strife accounted equally for 58 percent of mortality. Malnutrition affected pups primarily, and intraspecific strife mainly involved adult pack members due to increased trespassing in search of deer. During two years of the study, all of the natural mortality in radio-collared wolves was caused by other wolves.

Harvest obviously affects survival directly, but there are also indirect effects. Harvest may increase per capita food availability, thereby increasing survival of the remaining animals. Van Ballenberghe et al. (1975) suggested that natural losses among adults are low in exploited populations, probably due to the presence of relatively few old wolves. As mentioned earlier, the high dispersal rates of heavily harvested populations may result in lower survival rates, due to increased vulnerability to human harvest and intraspecific aggression.

In general, under conditions of low or declining per capita food, natural mortality is high primarily as a consequence of increased malnutrition and intraspecific strife. Mech (1994) also reported that alpha animals were more likely to be killed as a result of intraspecific strife and that 41 percent of wolves killed by conspecifics were within 1.0 kilometer and 91 percent within 3.2 kilometers of their territory's edge. By all indications, wolves run a greater risk of fatal encounters along territory edges.

Moreover, prey animals, including moose (Ballard et al., 1987), musk oxen (Lent, 1978), and white-tailed deer (Frijilink, 1977; Nelson & Mech, 1985), can kill wolves. Although the chances of finding wolves killed and injured by prey are small and prey-caused mortality may be low, nonfatal injuries, particularly from moose, are common. Wolves from areas inhabited by moose were injured more frequently than wolves from non-moose areas. Similarly, Rausch (1967) and Van Ballenberghe (1977) reported high rates of injury based on examination of skeletons and live individuals. Despite these common injuries, most wolves survive, presumably because of their social nature (Rausch, 1967).

Disease is an added mortality factor that has been reported to cause 2 to 14 percent of wolf mortality (Carbyn, 1982; Peterson et al., 1984; Fuller, 1989).

Documentation of disease-related mortality in wild wolves is difficult because often animals dying of disease are not found and carcasses are too decomposed to perform a meaningful necropsy. Wolf social structure probably limits the spread of diseases so that disease-caused mortality may in fact be quite low in wolf populations.

Even though disease-caused mortality is relatively uncommon, many wolves exhibit signs of infection. Zarnke and Ballard (1987) performed the most comprehensive serological study of wolves to date and found that antibodies to rabies, brucellosis, and leptospirosis were rare, while canine parvovirus, infectious canine hepatitis, canine distemper virus, Q fever, and tularemia were relatively common. They concluded that these diseases were enzootic in the population, and usually infection was not caused by contact with domestic animals. All these diseases are capable of causing significant mortality, especially in high-density populations (Zarnke & Ballard, 1987). Yuill (1987) pointed out that parasitic and infectious agents are found in all mammals, in every ecosystem, and can affect population size, distribution, and structure but usually with little obvious effect on their hosts. Death is unusual unless (1) serious illness facilitates transmission, as in rabies; (2) the disease agent does not depend on the host for survival and can complete its life cycle after the host dies; or (3) the pathogen moves through host populations over wide geographic areas, over a long period of time. In any case, coevolution of disease agents and hosts often results in decreased mortality (Yuill, 1987).

Canine distemper virus (CDV) causes a highly contagious and relatively common wolf disease that often is complicated by secondary bacterial infections (Greene & Calpin, 1988); CDV has been determined to be at least partially responsible for wolf population declines in two cases (Peterson et al., 1984; Carbyn, 1982).

Although Carbyn (1982) and Zarnke and Ballard (1987) concluded that CDV was enzootic in their Alberta and south-central Alaska study areas, Stephenson (1982) concluded that it was not, at least in Alaskan wolf populations. Stephenson (1982) attributed a low frequency of CDV antibodies among wolves of various ages to sporadic short-lived introductions of the virus rather than to a continuing enzootic pattern of exposure. If the disease was enzootic, one would expect a wide range of titer levels in each age class, which would indicate continuing exposure to the antigen. It was determined that the low prevalence of CDV antibodies was probably related to the mode of transmission of the virus. CDV usually is transmitted by direct contact rather than by urine or feces, survives less than one day in the environment, and is easily inactivated by sunlight or heat (Greene & Calpin, 1988; Stephenson, 1982). Immunity also is prolonged, and the carrier state is unlikely (Greene & Calpin,

1988). All of these factors probably limit transmission of CDV between wolf packs; it is less likely that the low frequency of CDV titers is caused by high mortality from CDV infections.

Brucellosis and infectious canine hepatitis (ICVH)—the former a contagious bacterial disease, the latter a viral disease that primarily affects the liver of canids—infect wolves but with uncertain consequences. *Brucella suis* biotype 4 is enzootic in caribou in Alaska, and 45 percent of wolves from areas where caribou is the primary prey had positive titers (Zarnke & Ballard, 1987). Stephenson (1982) reported that 95 percent of Alaskan wolves tested for ICHV exposure were positive. They attributed the high level of exposure to ICHV to its mode of transmission. ICHV is environmentally stable and can be shed in the urine of carriers; thus, indirect transmission is much more likely for ICHV than for CDV.

Rates of Increase

Reported annual rates of increase (λ) vary from 0.57 to 2.40 (Rausch, 1967; Ballard et al., 1987; Fuller & Keith, 1980; Fuller, 1989). Keith (1983) found that (λ) was unrelated to wolf density and to an index of ungulate biomass but positively correlated with ungulate-biomass per wolf. Keith (1983) concluded that wolf population dynamics are largely dictated by per capita biomass of the ungulate food resource, as determined by wolf and ungulate densities. Keith (1983) also calculated a maximum rate of increase (1.36) from the highest reproductive and survival rates reported for wolves in the wild. This estimate is in agreement with the rate of increase on Isle Royale during 1952 to 1959 (mean = 1.39), which, because the population was initiated by a few individuals with abundant food, is likely to be the maximum possible increase without immigration. However, Hayes et al. (1991) and Ballard et al. (1987) reported rates of increase of 1.77 to 2.40, respectively. Both of these rates were recorded during population recovery from experimental wolf control. Both control areas were surrounded by naturally regulated wolf populations, so much of these high rates of increase might be attributed to immigration from adjacent areas.

Density

Most often, wolf density appears to be limited primarily by ungulate biomass (Van Ballenberghe et al., 1975; Mech, 1973, 1977a, 1977b; Fuller & Keith, 1980; Packard & Mech, 1980; Keith, 1983; Messier, 1985a, 1987; Peterson & Page, 1988). Even so, social factors control density more directly through intraspecific strife and imposition of limits on the number of breeding females, and indirectly through unequal distribution of food resources among pack members (Packard & Mech, 1980).

Population density can also be limited by harvest (Van Ballenberghe, 1981; Gasaway et al., 1983; Keith, 1983). Mech (1970) concluded that an annual harvest of 50 percent or more was necessary to control wolf population size based on pup-adult ratios, and did not distinguish between harvest and natural mortality. However, a harvest of less than 50 percent could control population size. For example, several studies (Van Ballenberghe, 1981; Gasaway et al., 1983; Keith, 1983; Peterson et al., 1984; Ballard et al., 1987; Fuller, 1989) have found that wolf populations can be limited by harvest levels of 20 to 40%, the lower rate having effects in an area with low ungulate biomass (Gasaway et al., 1983). Peterson et al. (1984) and Fuller (1989) pointed out that harvest effects vary with time and population structure. For instance, if pups composed most of the harvest, the population could withstand an even larger harvest (Fuller, 1989).

Management

Reintroduction Techniques

Wolves can be caught readily in traps, or they may be captured by injecting immobilizing drugs shot from special guns, usually from a helicopter. The 30 wolves captured in Alberta (in January 1995) for transfer to the recovery areas in Idaho and Yellowstone National Park were taken with immobilizing drugs delivered from a helicopter. A veterinarian with extensive experience in capture, immobilization, and subsequent handling of the animals to ensure complete recovery was present. The initial capture operation was successful with the exception that one animal was killed via an improperly delivered dart.

Relocations are generally classified as being either a hard release or soft release. A *hard release* is the immediate and direct release of wolves into a new environment. A *soft release* means that animals are held initially in temporary enclosures where they can adjust to each other and ostensibly recover from the trauma of the capture and transport effort; animals are released at a later date. The Rocky Mountain Wolf Recovery Plan specified a hard release into central Idaho and a soft release into the Yellowstone National Park (Fritts, 1993). Past experience suggests that wolves released into new surroundings abruptly after capture are apt to separate and move independently. This has occurred with the wolves released into central Idaho, where movements have extended up to 160 kilometers (100 miles) from the release area. Of the fourteen wolves alive as of March 1995, two may have paired, while the remaining wolves are alone.

The two release methods each have advantages and disadvantages. While a soft release may encourage wolves to remain together, the required protection and artificial feeding is a substantial undertaking. The expense of constructing enclosures and caring for the wolves, plus the potential conditioning of wolves to humans for food, are added drawbacks of soft release. On the other hand, soft releases improve the chances that wolves will breed and remain in the release area. The soft release used for red wolf restoration in North Carolina has resulted in breeding pairs (Fritts, 1993), although substantial mortality and movements have still occurred.

Releases should occur when: (1) prey is most vulnerable to wolf predation; (2) human activity in the areas is minimal; (3) wolves may be most easily captured and transported; and (4) wolves are in the best condition to endure stress. Prey are likely to be more vulnerable when concentrated by snow at lower elevations of their annual range, most predictably during the late winter or early spring of a deep-snow winter. Because wolves breed in late February and den in early April, the capture of the Idaho and Yellowstone wolves was confined to January. Because breeding was observed in the Yellowstone enclosures and at least one pairing of wolves may have occurred in central Idaho on a major elk winter range, the timing and nature of the January 1995 operation seems to have been appropriate, although its success is still unknown.

Several scenarios may occur regardless of the technique used (Fritts, 1993): (1) adult pairs may separate and move extensively; (2) mortality might occur from expected and unexpected causes; and (3) individuals may exhibit unexpected behavior. In any case, the outcome of reintroductions is hard to predict. At present, three restoration scenarios are being played out in the Northern Rockies: (1) natural recolonization of the region adjacent to Canada; (2) hard release in central Idaho; and (3) soft release in Yellowstone. Each scenario will provide comparisons from which to learn and refine subsequent restoration efforts.

Current Considerations

Management of reintroduced wolf populations, once they are established and reproducing, will be a challenge for wildlife managers. In the Yellowstone region, wolves that reside entirely within Yellowstone National Park will probably not be disturbed by human activity, except if individual wolves become nuisances around campgrounds or settlements. The policy of natural regulation, discussed in Chapter 13, is entirely applicable to these wolves. Wolves that move outside of the park for a portion of the year, or that have home ranges overlapping park boundaries, may be subject to population management for several reasons. First, wolves may cause damage to livestock on surrounding range lands. Wolves learning park boundaries, causing damage outside, and then moving to the security inside may pose especially difficult problems. There are three federal agencies, each with different responsibilities, involved with reintroduced wolf management in and outside of Yellowstone; these are the National Park Service, the Fish and Wildlife Service, and Animal Damage Control. Currently, wildlife agencies in the surrounding states are not involved with wolf management, mainly because state legislatures, not approving of the federal reintroduction project, have vetoed their involvement. However, in the long run, all agencies must be involved if an effective wolf management program is to be developed.

Historically, human-caused mortality, whether legal or illegal, has been the major cause of death for many wolf populations. There is little reason to believe that this will not continue to be the case for wolves that live outside of the park boundaries. Indeed, the only known mortalities of the reintroduced wolves in Idaho and Yellowstone have been attributable to humans. (One wolf from each area was shot; the Yellowstone wolf was outside the park boundary.)

Wolf populations also may need to be managed if their prey base, much of which will be hunted, declines to levels seriously below what habitats will support and do not provide an acceptable available number for the human hunter. Indeed, potential conflicts between hunters and wolves were a major consideration in planning for the reintroductions (Vales & Peek, 1990). These conflicts must involve state wildlife agencies, which are responsible for monitoring big game populations and managing hunter harvest. Wolf population management will require good information on population trends of prey and predator alike, which in turn will require coordination between state and federal agencies in the acquisition of the necessary information. Any wolf management effort will be scrutinized intensively by a wide audience of highly divergent views. However, the alternative of no management might cause ill will and can be expected to precipitate an increase in illegal mortality. The best way to address this issue is to incorporate a contingency management plan for wolf populations that will be implemented as necessary.

An adaptive management format is in order, wherein all interested parties participate in the formulation of a coordinated management plan which states:

1. The existing knowledge.
2. The goals and conditions under which intervention into the predator-prey relationship will be considered.

3. The means by which the intervention will occur.
4. The level of intervention.
5. Follow-up monitoring, coupled with timely dissemination of information to all interested parties.

Hunting seasons may need to be altered as well as wolf population levels. The only means by which this will be equitably accomplished is through provision of well-justified recommendations to an informed and up-to-date lay public, which is committed to the process. A corollary to this might be that goals and objectives will be subject to modification as new information and conditions dictate.

Conclusions

Restoration of the gray wolf into the northwestern states is one of the major events in wildlife management in this century. Leopold (1933) pointed out that game management required incentives in order to be practiced effectively. In 1933, the incentive came largely from sport hunting. With time, incentives have grown to include nonconsumptive uses. For example, nonhunted species have been increasingly valued primarily for aesthetic reasons or because they have become endangered. While the gray wolf has been classified as endangered, it is a species that can significantly interfere with human activities, including hunting and ranching, and with other wildlife. Conversely, the gray wolf is a species that garners great admiration and whose reintroduction was supported by a great majority of the public (Bath, 1991).

Now that wolves have been released into Yellowstone National Park and the central Idaho wilderness, it remains to be seen whether local public support and understanding will continue to increase so that wolf populations can be sustained. A large part of this equation will come from the information base that must be provided by those monitoring the wolf, and how well the information is disseminated to all concerned. The balance of the equation will depend on how effective management of wolf populations will be, if and when management becomes necessary.

By its very presence, the wolf portends to enhance our understanding of the wildlands in which it lives. This, in turn, will focus the attention of a diverse and highly interested public on the wolf and its prey base, a fact that will require extensive information and communication. Just how this all plays out in the long term will determine whether the wolf becomes a permanent resident of the recovery areas.

Literature Cited

Alaska Department Fish & Game. Federal aid in wildlife restoration annual report of survey-inventory activities. Wolf. Volume XVIII Part XV. Proj. S-22-6, Job 14.0, 1988.

Ballard WB. Marrow fat dynamics in moose calves. *J Wild Manage* 1987;51:66–69.

Ballard WB, Franell R, Stephenson RO. Long distance movement by gray wolves, *Canis lupus. Canad. Field-Nat* 1983;97:333.

Ballard WB, Whitman JS, Gardner CL. Ecology of an exploited wolf population in south-central Alaska. *Wildl Monogr* 1987;98.

Bath AJ. Public attitudes in Wyoming, Montana, and Idaho toward wolf restoration in Yellowstone National Park. *Trans N Amer Wildlife Nat Res Conf* 1991;56:91–95.

Berg WE, Chesness RA. Ecology of coyotes in northern Minnesota. In M. Bekoff (ed.), *Coyotes: Biology, Behavior, and Management.* New York: Academic Press, 1978: 229–247.

Boertge RD, Stephenson RO. Effects of ungulate availability on wolf reproductive potential in Alaska. *Canad J Zool* 1992;70:2441–2443.

Boyd DK, Ream RR, Pletscher DH, Fairchild MW. Prey taken by colonizing wolves and hunters in Glacier National Park. *J Wildl Manag* 1994;58:289–295.

Carbyn LN. Wolf predation and behavioural interactions with elk and other ungulates in an area of high prey density. PhD thesis, Toronto: University of Toronto, 1975.

Carbyn LN. Ecology and management of wolves in Riding Mountain National Park, Manitoba. *Canad Wildl Serv Final Rep. No. 10.* Edmonton, 1980.

Carbyn LN. Coyote population fluctuations and spatial distribution in relation to wolf territories in Riding Mountain National Park, Manitoba. *Canad Field-Nat* 1982;96:176–183.

Carbyn LN. Wolf predation on elk in Riding Mountain National Park, Manitoba. *J Wildl Manage* 1983;47:963–976.

Carbyn LN, Paquet P, Meleshko D. Long-term ecological studies on wolves, coyotes, and ungulates in Riding Mountain National Park. *Canad Wildl Serv draft Rep. on File*, Edmonton, 1987.

Frijlink JH. Patterns of wolf pack movements prior to kills as read from tracks in Algonquin Provincial Park, Ontario, Canada. *Bijdrafen tot de dierkunde* 1977;47:131–137.

Fritts SH. Record dispersal by a wolf from Minnesota. *J Mammal* 1983;64:166–167.

Fritts SH. Reintroductions and translocations of wolves in North America. In RS Cook (ed.), *Ecological Issues on Reintroducing Wolves into Yellowstone National Park.* USDI National Park Service Sci. Monogr. NPS/NRYELL/NRSM-93/22. 1993:1–22.

Fritts SH, Mech LD. Dynamics, movements, and feeding ecology of a newly protected wolf population in northwestern Minnesota. *Wildl Monogr* 80, 1981.

Fritts SH, Paul WJ, Mech LD. Can relocated wolves survive? *Wildlife Soc Bull* 1985;13:459–463.

Fuller WA, Novakowski NS. Wolf control operations, Wood Buffalo National Park, 1951–52. *Canad Wildl Serv, Wildl Mgmt Bull* Ser. 1 No. 11. 1955.

Fuller TK. Population dynamics of the wolves in north-central Minnesota. *Wildl Monogr* 1989;105:1–41.

Fuller TK, Keith LB. Wolf population dynamics and prey relationships in northeastern Alberta. *J Wildl Manage* 1980;44:583–602.

Fuller TK, Snow WJ. Estimating winter wolf densities using radiotelemetry data. *Wildl Soc Bull* 1988;16:367–370.

Fuller TK, Berg WE, Radde GL, Lenarz MS, Blair JG. A history and current estimate of wolf distribution and numbers in Minnesota. *Wildl Soc Bull* 1992;20:42–55.

Gasaway WC, Stephenson RO, Davis JL, Shepherd PEK, Burris OE. Interrelationships of wolves, prey, and man in interior Alaska. *Wildl Monogr* 84, 1983.

Gauthier DA, Theberge JB. Wolf predation. In MJ Novak (ed.), *Wild Furbearer Management and Conservation in North America.* North Bay, Ontario: Ontario Trappers, Assn. 1987:199–127.

Gese EM, Mech LD. Dispersal of wolves (*Canis lupus*) in northeastern Minnesota, 1969–1989. *Canad J Zool* 1991;69:2646–2955.

Greene CE, Calpin JP. Viral infections. In RV Morgan (ed.), *Handbook of Small Animal Practice.* New York: Churchill Livingstone Inc. 1988:975–977.

Griffith DB, Scott JM, Carpenter JW, Reed C. Translocation as a species conservation tool: status and strategy. *Science* 1989;245:477–480.

Gunson J. Wolves and elk in Alberta's Brazeau country. *Bugle* 1986;4:29–33.

Haber GC. *Socio-Ecological Dynamics of Wolves and Prey in a Subarctic Ecosystem.* PhD thesis, Vancouver: University of British Columbia, 1977.

Hall ER, Kelson HR. *The Mammals of North America.* New York: Ronald Press, 1959.

Harrington FH, Mech LD, Fritts, SH. Pack size and wolf pup survival; their relationship under varying ecological conditions. *Behav Ecol Sociobiol* 1983;13:19–26.

Holleman DF, Stephenson RD. Prey selection and consumption by Alaskan wolves in winter. *J Wildl Manage* 1981;45:620–628.

Holling CS. Some characteristics of simple types of predation and parasitism. *Can Entomologist* 1959;91:385–398.

Keith LB. Population dynamics of wolves. In LN Carbyn (ed), *Wolves in Canada and Alaska: Their Status, Biology, and Management. Can Wildl Serv, Wildl Manage Bull Ser* 45;1983:66–77.

Kelsall JP. *The Migratory Barren-Ground Caribou of Canada.* Canad. Wildl. Serv. Ottawa: Queen's Printer, 1968.

Kolenosky GB. Wolf predation on wintering deer in east-central Ontario. *J Wildl Manage* 1972;36:357–369.

Lent PC. Musk-ox. In JL Schmidt, DL Gilbert (eds), *Big Game of North America.* Harrisburg, PA: Stackpole Books, 1978: 135–147.

Leopold A. *Game Management.* New York: C. Scribners, 1933.

Lindstrom E. Reliability of placental scar counts in the red fox (*Vulpes vulpes*) with special reference to fading of the scars. *Mammal Review* 198;11:137–149.

Lopez, BH. *Of Wolves and Men.* New York: C. Scribner and Sons, 1978.

Mech LD. *The Wolves of Isle Royale. Fauna of the National Parks of the US.* Fauna Series No. 7. Washington, DC: U.S. Gov. Printing Office, 1966.

Mech LD. *The Wolf: The Ecology and Behavior of an Endangered Species.* Garden City, NY: The Natural History Press, 1970.

Mech, LD. Wolf numbers in the Superior National Forest of Minnesota. *USDA Forest Serv. Res. Paper. NO. NC-97,* 1973.

Mech LD. Wolf pack buffer zones a prey reservoirs. *Science* 1977a;198:320–321.

Mech LD. Productivity mortality, and population trends of wolves in northeastern Minnesota. *J Mammal* 1977b;58:559–574.

Mech LD. Wolf populations in the central Superior National Forest, 1967–1985. *USDA For. Serv. Res. Pap. NC-270,* 1986.

Mech LD. Age, season, distance, direction, and social aspects of wolf dispersal from a Minnesota pack. In BD Chepko-Sade, ZT Halpin (eds), *Mammalian Dispersal Patterns.* Chicago: University of Chicago Press, 1987:55–74.

Mech LD, Frenzel LD, Jr. Ecological studies of the timber wolf in northeastern Minnesota. *USDA For. Serv. Res. Pap. NC-52.* 1971.

Mech LD, Fritts SH, Radde GL, Paul WJ. Wolf distribution and road density in Minnesota. *Wildl Soc Bull* 1988;16:85–87.

Mech LD, Gese EM. Field testing the wildlink capture collar on wolves. *Wildl Soc Bull* 1992;20:221–223.

Messier F. Social organization, spatial distribution, and population density of wolves in relation to moose density. *Canad J Zool* 1985a;63:1068–1077.

Messier F. Solitary living and extraterritorial movements of wolves in relation to social status and prey abundance. *Canad J Zool* 1985b;63:239–245.

Messier F. Physical condition and blood physiology of wolves in relation to moose density. *Canad J Zool* 1987;65:91–95.

Messier F., Crete M. Moose-wolf dynamics and the natural regulation of moose populations. *Oecologia* 1985;65:503–512.

Murie A. *The Wolves of Mt. McKinley.* Fauna Series 5. Washington, DC: U.S. National Park Service, 1944.

Nelson ME, Mech LD. Deer social organization and wolf predation in northeastern Minnesota. *Wildl Monogr* 1981;77:1–53.

Nowak RM. A perspective on the taxonomy of wolves in North America. In LN Carbyn (ed), *Wolves in Canada and Alaska, Their Status, Biology, and Management.* Canad. Wildl. Serv. Rep. Ser. 45. Ottawa. 1983:10–19.

Packard JM, Mech LD. Population regulation in wolves. In AG Klein, MN Cohen (eds), *Biosocial Mechanisms of Population Regulation.* New Haven: Yale University Press, 1980:135–150.

Packard JM, Mech LD. Population regulation in wolves. In FL Bunnell, DS Eastman, JM Peek (eds), *Symposium on Natural Regulation of Wildlife Populations.* Moscow, Id.: University of Idaho For., Wildl., and Range Expt. Stn. Proc. 14. 1983:151–174.

Packard, JM, Mech LD, Seal US. Social influences on reproduction in wolves. In LN Carbyn (ed), *Wolves in Canada and Alaska, Their Status Biology, and Management.* Canad. Wildl. Serv. Rep. Ser. 45. Ottawa. 1983: 78–85.

Parker GR, Luttich S. Characteristics of the wolf (*Canis lupus labradorius* Goldman) in northern Quebec and Labrador. *Arctic* 1986;39:145–149.

Paquet PC, Bragdon S, McCusker S. Cooperative rearing of simultaneous litters in captive wolves. In FH Harrington, PC Paquet (eds), *Wolves of the World; Perspectives on Behavior, Ecology, and Conservation.* Park Ridge, NJ: Noyes Publ. 1982:223–237.

Peterson RO. Wolf ecology and prey relationships on Isle Royale. Fauna Series. 11, Washington, DC: National Park Service, 1977.

Peterson RO, Page RE. The rise and fall of Isle Royale wolves, 1975–1986. *J Mammal* 1988;69:89–99.

Peterson RO, Woolington JD, Bailey TN. Wolves of the Kenai Peninsula, Alaska. *Wildl Monogr* 88, 1984.
Pimlott DH, Shannon JA, Kolenosky GB. The ecology of the timber wolf in Algonquin Prov. Park. Montreal: Ont. Dep. Lands and For. Res. Rep. (Wildl.) 87, 1969.
Potvin F. Wolf movements and population dynamics in Papineau-Labelle reserve, Quebec. *Canad J Zool* 1988;66:1266–1273.
Rabb GB, Woolpy JJ, Ginsburg BE. Social relationships in a group of captive wolves. *Am Zool* 1967;7:305–311.
Rausch RA. Some aspects of the population ecology of wolves, *Alaska Amer Zool* 1967;7:253–265.
Ream RR, Fairchild MW, Boyd DK, Pletcher DH. Population dynamics and home range changes in a colonizing wolf population. In RB Keiter, MS Boyce (eds), *The Greater Yellowstone Ecosystem.* New Haven: Yale University Press, 1991:349–366.
Robinson WL, Smith GJ. Observations on recently killed wolves in upper Michigan. *Wildl Soc Bull* 1977;5:25–26.
Rothman RJ, Mech LD. Scent-marking in lone wolves and newly formed pairs. *Animal Behavior* 1979;27:750–760.
Servheen C, Knight RR. Possible effects of a restored gray wolf population on grizzly bears in the greater Yellowstone area. In RS Cook (ed), Ecological issues on reintroducing wolves into Yellowstone National Park. Washington, DC: USDI National Park Service Sci. Monogr. NPS/NRYELL/NRSM-93/22. 1993:28–37.
Stephenson RO. Serologic survey for canine distemper and infectious canine hepatitis in wolves in Alaska. *J Wildl Diseases* 1982;18:419–424.
Sumanik RS. Wolf ecology in the Kluane Region, Yukon Territory. M.S. Thesis. Houghton, MI: Michigan Technological Univ., 1987.
Thurber JM, Peterson RO. Effects of population density and pack size on the foraging ecology of gray wolves. *J Mammal* 1993;74:879–889.
Thurber JM, Peterson RO, Woolington JD. Coyote coexistence with wolves on the Kenai Peninsula, Alaska. *Canad J Zool* 1992;70:2494–2498.
U.S. Fish and Wildlife Service. Northern Rocky Mountain wolf recovery plan. Denver: U.S. Fish and Wildlife Service, 1987.
Vales DJ, Peek JM. Estimates of the potential interactions between hunter harvest and wolf predation on the Sand Creek Idaho and Gallatin Montana elk populations. In Yellowstone National Park, U.S. Fish and Wildlife Service, University of Idaho, Interagency Grizzly Bear Study Team, (eds), *Wolves for Yellowstone? Report to Congress.* Vol II. Research and analysis. Yellowstone N. P. Wyoming: Yellowstone National Park. 1990.
Vales DJ, Peek, JM. Estimating the relations between hunter harvest and gray wolf predation on the Gallatin, Montana, and Sand Creek Idaho elk populations. In RS Cook (ed), *Ecological issues on reintroducing wolves into Yellowstone National Park.* USDI National Park Service Sci. Monogr. NPS/NRYELL/NRSM-93/22. 1993:118–173.
Van Ballenberghe V. Physical characteristics of timber wolves in Minnesota. In RL Phillips, C. Jonkel (eds), Proc. 1975 Predator Symp. Mont. For. Conservation Exp. Stan. Missoula: University of Montana. 1977: 213–219.
Van Ballenberghe V. Population dynamics of wolves in the Nelchina Basin, southcentral Alaska. In JA Chapman, D Pursley (eds), Worldwide Furbearer Conf. Frostberg, MD Worldwide Furbearer Conf. Inc. 1981:1246–1258.

Van Ballenberghe V. Extraterritorial movements and dispersal of wolves in southcentral Alaska. *J Mammal* 1983a;64;168–171.
Van Ballenberghe V. Rate of increase of white-tail deer on the George reserve: reevaluation. *J Wildlife Manage* 1983b;47:1245–1247.
Van Ballenberghe V, Mech LD. Weights, growth, and survival of timber wolf pups in Minnesota. *J Mammal* 1975;56:44–63.
Van Ballenberghe V, Erickson AW, Byman D. Ecology of the timber wolf in northeastern Minnesota. *Wildl Monogr* 43, 1975.
Weise TF, Robinson WL, Hook RA, Mech LD. An experimental translocation of the eastern timber wolf. *Audubon Conservation Report* 5, 1975.
Yuill TM. Diseases as components of mammalian ecosystems: mayhem and subtlety. *Canad J Zool* 1987;65:1061–1066.
Zarnke RL, Ballard WB. Serologic survey for selected microbial pathogens of wolves in Alaska, 1975–1982. *J Wildl Diseases* 1987;23:77–85.
Zimen E. On the regulation of pack size in wolves. *Z. Tierpsychol* 1976;40:300–341.

IV

Dealing with the Human and Cultural Environments of Parks and Protected Areas

The previous chapters discussed at length the need to protect natural resources, ecological processes, and the landscapes in and around parks and protected areas. In these discussions, the need to understand and protect cultural resources has been largely ignored. Chapter 17 deals with the issue of cultural resources and makes the plea that there should be a balance in resource protection between the natural and cultural resources. It also explains why this balance often has been lacking in the past.

The impact of park establishment and preservation policies on human society has been largely ignored in the preceding chapters. Chapter 18 attempts to correct this oversight by providing a glimpse of how government agencies, conservation groups, and native federations are working together to ensure that the establishment and development of new and existing parks in the territories of Canada are compatible with the cultural context.

Chapter 19 examines the role of science and law in park management. It takes a critical look at what the role of science should be in the decision-making process of the National Park Service. In doing so, it discusses the often-neglected concept of where the burden of proof should lie when seeking to demonstrate potential harm to park resources. Words such as "natural," "unimpaired," and other ambiguous terms commonly used in defining in park policies are also discussed.

17

The Cultural and Natural Resource Management Needs of Parks and Protected Areas: Is There an Appropriate Balance?

Stephanie S. Toothman

Is there an appropriate balance between the cultural and natural management needs of parks and protected areas? Before this question can be addressed, we first need to define what these needs are and what an appropriate balance would be. The following discussion focuses on the units of the national park system managed by the National Park Service (NPS) and draws on the author's professional experience working within the NPS Pacific Northwest Region. As all federal agencies share the same foundation of federal law and policies governing actions affecting the environment and cultural resources, many of these observations are applicable to other agencies responsible for managing protected areas. However, as defined by its own "Organic Act," by its role as leader of the federal preservation program, and by the exceptional significance of the resources it manages, the NPS has special responsibilities. The primary responsibility of the NPS is resource stewardship.

How Do We Define Resource Needs?

All NPS units have been established by Congress to protect and interpret specific, nationally significant resources. These resources usually are identified in the park's establishing legislation and are reflected in the park's name. "National park," "national historical park," "national recreation area," "national historic site," "national battlefield," and "national reserve" are but a sampling of the more than twenty labels presently attached to units of the national park system. While each label evokes an image, whether it be the scenic glories of Yellowstone or Mount Rainier or the somber memories of Gettysburg and Valley Forge, rarely does the image convey the rich diversity of resources that most parks actually contain. Mount Rainier, for example, has more than 100 structures listed on the National Register of Historic Places, including eight national historic landmarks; Gettysburg manages a complex cultural landscape while at the same time interpreting a nationally significant Civil War battlefield and mitigating the impacts of encroaching development at its boundaries.

The first need for appropriate management of resources in the NPS, therefore, is the acknowledgment that most parks contain both natural and cultural resources. In addition, although the legislative mandate that established a given park unit may focus on one type of resource, the laws and policies directing NPS management of natural and cultural resources recognize that this mandate is much broader. Thus, although the establishing legislation of large natural parks (e.g., Olympic or Mount Rainier National Parks) fails to mention cultural resources, this does not excuse these parks from their responsibility to inventory, evaluate, and protect the abundance of prehistoric and historic cultural properties within their boundaries, as required by Section 110 of the National Historic Preservation Act. Similarly, managers of historic sites and parks have a responsibility to inventory, evaluate, and protect the natural resources within park boundaries in accordance with applicable environmental laws and regulations, including the National Environmental Protection Act and the Endangered Species Act. Acceptance of these responsibilities by all park managers and resource management professionals is the second critical need.

Allocation of staff and budgetary resources to manage these resources appropriately and professionally is the third factor in this "needs" equation. Without a strong commitment to identifying and providing the needed professional staff and funding to manage both natural and cultural resources, park managers are not fulfilling their complete responsibilities.

What Is an Appropriate Balance?

The three needs (i.e., acknowledgment of natural and cultural resources, acceptance of responsibility for both resource types, and commitment of staff and finances to manage resources appropriately) provide a framework in which to define what an appropriate balance of the natural and cultural resource management needs of a park or protected area might be. However, to do so we must first define "balance." Many definitions of "balance" emphasize the characteristics of a stable or unchanging system. Given the dynamics of natural and cultural systems and the many factors that influence resource decision making, including politics, budgets, and private property rights issues, using such a definition as a measure of appropriate balance would be unrealistic. Perhaps more viable, both in terms of desirability and achievability, is the following definition: *balance*, a harmonious or satisfying arrangement or proportion of parts or elements.

A park or protected area can be considered to have established an appropriate balance in its resource management program when that program meets the following criteria:

1. A plan exists to identify and evaluate, using professional and nationally recognized criteria and contexts, all natural and cultural resources within the park's boundaries.
2. The need to manage these resources in accordance with laws, regulations and policies, and professionally accepted standards is recognized as the park's primary mission.
3. Staff and budgetary support to manage these resources are sought and allocated according to objective measures of needs and workloads.
4. All park management decisions affecting natural and cultural resources take the significance and requirements of these resources into account as integral parts of the decision-making process.

This description of appropriate balance recognizes that, while most parks have both natural and cultural resources, the nature and significance of these resources vary. This variability requires an unbiased evaluation of what is needed to manage the resources appropriately. Flexibility, based on professional documentation and assessment of the specific needs of each resource type, is essential to define the appropriate balance of staff and budgetary resources devoted to resource management for a particular park. Conversely, guidelines that establish inflexible standards and requirements, that fail to acknowledge the important differences among parks, and that do not allow individual parks to react to changes, both in opportunities and threats to resources, will not work.

Furthermore, achieving an appropriate balance requires that all resource professionals within or supporting a park or protected area recognize that the resources they manage are part of a larger whole, whether it is defined as an ecosystem or a cultural system. Only by working together in inter-disciplinary, mutually supportive, professional partnerships to protect these systems will managers achieve an appropriate balance of resource stewardship.

Historic Problems Affecting the Balance of Natural and Cultural Resource Stewardship

During the past decade, the resource management programs within the NPS have taken some major steps toward achieving an appropriate balance of stewardship for both natural and cultural resources. In order to understand how far the NPS has come and to provide some perspective on future programs, a look back at the state of cultural resource management in the mid-1980s provides some important insights.

The general perspective of most cultural resource specialists in the mid-1980s was that the balance between natural and cultural resource needs was inappropriate (Toothman, 1987). There was a widespread feeling that park management failed to recognize the presence of cultural resources in natural areas, and therefore lacked our first "criterion" for appropriate management. In 1980, amendments to the National Historic Preservation Act converted to law the original Executive Order 11593, which required all federal agencies to inventory, evaluate, and protect cultural resources within their jurisdictions. However, efforts within the NPS to comply with this law were viewed as fragmentary and poorly supported. Cultural resources are not mentioned in the enabling legislation for many natural areas. Reviews of major planning documents and resource management plans prepared prior to the mid-1980s for these natural areas reveal that cultural resources within these areas were rarely considered in depth; many resource management plans did not even contain a cultural resource component. "This failure to recognize the presence of cultural resources in the primary planning documents of these natural areas created a classic 'out of sight, out of mind' situation. When there is no official recognition of cultural resources, then the need to deal with them could go, and has gone, unrecognized" (Toothman, 1987).

The failure to recognize cultural resources within these natural areas was reinforced both by the nature of the resources and the qualifications and interests of staff members. Many natural areas resisted recognizing the importance of historic structures and developments. In part, this antipathy toward old, often difficult-to-maintain structures originated because they were considered to be a drain on park resources and incompatible with current management strategies. Secondly, proposals to list service-designed and service-built structures on the National Register of Historic Places were perceived as inappropriate exercises in self-memorialization (Toothman, 1987).

The significance of other cultural resources in natural areas went unrecognized because no effort was made to identify and understand the historic contexts to which they related. Local and regional themes, unrecognized in park legislation, were represented by a wide range of cultural resources within many park boundaries. The significance of these resources could be understood only within a larger context that included related resources outside park boundaries. Managing these resources in accordance with the requirements of Section 110 of the National Historic Preservation Act required a level of interest, awareness, and knowledge that many staff members of natural areas lacked. This lack of interest and knowledge was frequently compounded by an active antipathy toward cultural resources by professional staff who perceived the presence of cultural resources to be an intrusion on the wilderness values they sought to promote. Disregarding the specific protections for cultural resources provided in the Wilderness Act, many wilderness managers advocated the removal of historic structures in wilderness areas or, at the very least, encouraged a management strategy of "benign neglect" that would achieve the same objective over a longer time period. The conflicts that arose over such proposals created an atmosphere, which still lingers in some areas, of considerable tension and even mistrust between natural and cultural resource professionals.

Finally, efforts to provide some equity for cultural resource management in natural area parks were exacerbated by NPS budget cuts and policies that tended to give funding preference to natural resource management activities. Understandably, natural area managers used their specific legislative mandates to prioritize budgets for resource management needs; if cultural resources were not mentioned in their legislation and administrators did not push the issue, allocation of staff and finances to cultural resources in these areas fell to the bottom of the list of priorities. At the service-wide level, guidelines for allocating the limited Cultural Resources Preservation Fund were such that the money ran out long before cultural resource projects in natural areas could be considered.

Changing Perspectives and Support for Resource Stewardship

While in retrospect the changes in attitudes toward and support for cultural resource stewardship in the NPS over the past decade seem momentous, most of these changes started with small, localized actions. These local decisions, made by individual managers or groups of resource specialists working in the field, in turn heightened momentum to such an extent that it could no longer be ignored by the policymakers who had the power to implement change throughout the NPS. The reaffirmation of resource stewardship (i.e., the integration of cultural and natural resource management) as the primary responsibility of the NPS in the Vail Agenda (Steering Committee, 1992) was the logical culmination of this ground swell of opinions and actions.

It should be acknowledged that within the field of natural resource management there also have been major changes in perspective, including growing support for the movement of cultural resource management toward achieving a parity within the NPS. Many of these changes have helped to bridge the gaps and diminish the frictions that existed between natural and cultural resource professionals, as both move toward the common goal of improving resource stewardship for all resources.

Methods to Ensure Effective Cultural Resource Management

Documenting Cultural Resources

As stated previously, many of the changes in NPS cultural resource management programs started at the field level with resource-specific decisions. The examples that follow are from the Pacific Northwest. They are typical, however, of actions that took place at the park and regional levels throughout the NPS in the mid-to-late 1980s. Finding the means to document the cultural resources that existed in the parks proved to be the key to developing (1) a more widespread awareness that cultural resources do exist and an appreciation that they are important, and (2) an acceptance of the responsibility for managing these resources.

In the Pacific Northwest, the effort to document cultural resources was pursued on several fronts. Cooperative agreements with several universities provided a relatively inexpensive way to begin to inventory field resources and catalog collections, thereby heightening the visibility of these resources. Staff from the University of Oregon and the University of Washington documented sites at Mount Rainier, John Day Fossil Beds, and Olympic. Interagency and intergovernmental agreements provided another source of funding. Seattle City Light (SCL), a publicly owned utility, was the major source of funding for the first comprehensive archeological and historic inventories within the Ross Lake National Recreation Area. The NPS received the funds because SCL needed to satisfy the requirements of Section 106 of the National Historic Preservation Act in order to relicense its hydroelectric project within the recreation area. Finally, volunteers were avidly sought for both rehabilitation work and documentation projects.

Perhaps the most important long-term change, which began with field advocacy, was the redefinition of how the Cultural Resources Preservation Program Fund was allocated. Service-wide competition of all projects, which had pitted the needs of historic structures in natural area backcountry sites against those of Independence Hall and Civil War battlefields, was replaced by an annual base allocation from the fund to all regions. While the percentages on which the allocations were based continued to favor regions where the number of cultural parks was greatest, this change in the allocation formula allowed regions dominated by large natural areas to address their documentation and evaluation needs in a rational manner. Thus, baseline studies of cultural resources in areas such as Craters of the Moon, Mount Rainier, and John Day Fossil Beds could be funded on the basis of regional, rather than national, priorities.

Support for carrying out these baseline studies came from other sources as well. Internally, NPS policy guidelines for the preparation of resource management plans changed, and all parks were made responsible for submitting both a natural and cultural component for resource management plans. Externally, the State Historic Preservation Officers (SHPO) led the preservation community by exerting considerable pressure on the NPS to strengthen its commitment to compliance with the provisions of Section 110 of the National Historic Preservation Act. Criticism of the NPS's propensity to ignore cultural resources not identified in a park's mandate intensified until the agency was compelled, through a series of policies and guidelines, to affirm its commitment to protecting these resources. One such policy was the mandate after 1986 that all structures proposed for removal, regardless of age, be evaluated for eligibility to the National Register through consultation with the appropriate SHPO. This policy was intended to end what some SHPOs perceived to be a developing pattern: some park managers destroyed structures approaching fifty years of age in order to avoid dealing with them as historic structures.

Another management initiative adopted by the NPS to manage historic structures not related to primary park mandates was the historic property leasing program. This program allowed revenues from historic properties leased for commercial uses to remain in the park. Although the program has had mixed results, it has heightened many park managers' awareness of cultural resources and promoted documentation of these resources for the National Register. Completion of National Register documentation for the historic sheep ranch at John Day Fossil Beds, for example, was a direct result of the superintendent's desire to use the program to obtain funds to care for the resource.

In response to field requests to provide guidance for evaluating park structures and landscapes, funding was provided at the national level to develop historic contexts for park-related cultural resources through the National Historic Landmark program. The first of these studies was the 1987 thematic study, "Architecture in the Parks," which finally ended the controversy about the appropriateness of recognizing the significance of NPS architecture. In addition to recognizing the architectural significance of a number of park buildings, including the Ahwahnee at Yosemite, Hopi House at Grand Canyon, and the Administration Building at Mount Rainier, the study firmly established the importance of these structures in defining the philosophy that guided the development of the national park system. The study also provided criteria for evaluating scores of other park structures and discussed their direct relationship to recreational developments in state and local park systems. By the late 1980s, more than 100 structures had been nominated to the National Register at Mount Rainier, 80 at Olympic, 25 at Crater Lake National Parks, and dozens more at other park areas within the Pacific Northwest. The management responsibility for these resources could no longer be denied.

The NPS curatorial programs provided leadership in the late 1980s for another management thrust that has given a tremendous boost to efforts to develop information on cultural resources in all areas. By completing a service-wide analysis of the status of collections held by or for the parks, the chief curator was able to outline for Congress a series of serious deficiencies in collection management programs and provide recommendations for addressing these problems. Congress responded with special initiative money for multiyear programs to address cataloging backlogs and collections security problems. Identification of these problems, and the availability of this funding, heightened park managers' awareness of their collections and provided the means by which to manage them positively.

Following in the curatorial programs' successful footsteps, managers for the service-wide architecture, cultural landscape, history, ethnography, and archeological programs have documented program deficiencies and received special initiative funding for a variety of programs designed to correct these deficiencies. These include the Service-wide Archeological Inventory Program, the upgrading of the List of Classified Structures, and the development of a prototype for a Cultural Landscapes Inventory. Each of these initiatives has had tremendous impact in heightening awareness of the complex array of cultural resources managed in all types of park areas.

Acknowledging Responsibility for Cultural Resource Management

While the actions previously described have been significant in moving the NPS toward addressing the first criterion for achieving a balanced resource management program, there also have been major shifts in NPS perspective regarding the definition of good resource stewardship. Institution of the Resource Management Specialist Trainee program has been a major factor in this shift. Although the program was sponsored by the Natural Resource Management Program and thus focused primarily on natural resource issues, it did expose its trainees to some core cultural resource issues. The resource management specialists graduating from this program have moved into many of the key park resource management positions, bringing to the field a new level of professionalism. Although the cultural resource programs did not participate in this training program as fully as they should have, they still have benefited from having these professionals, with their knowledge of cultural resource laws and policies, in the field where they bring an awareness of the need to identify and properly manage cultural resources.

Other training efforts have contributed to increasing knowledge of cultural resource management responsibilities. Under the team resources training efforts, teams of cultural and natural resource specialists have circulated among the parks presenting programs that focus on the responsibility every NPS employee has for resource stewardship. Most regional cultural resource staffs also have provided basic introductory sessions on cultural resource responsibilities to park staffs on a regular basis. Alternating with these core programs, most regions have sponsored courses on specific topics such as preservation maintenance and the Archeological Resources Protection Act to sensitize all park employees to cultural resource management requirements.

These training efforts have contributed to creating a receptiveness among NPS staff, led by the professional resource management specialists, to adopting a "systems approach" to understanding and developing resource management strategies. While the term "ecosystem management" is normally used to describe this approach, an important conceptual counterpart that has evolved within the cultural resource programs is "cultural landscape analysis." Both concepts emphasize the interdependence of natural and cultural systems and, most importantly, the need to approach resource management from an interdisciplinary perspective, emphasizing integration of research and management objectives. The Ecosystem Management Working Group (1994), in its September 1994 report to the director of the NPS, defined this integrative approach as a "collaborative approach to natural and cultural resource management that integrates scientific knowledge of ecological relationships with resource stewardship practices for the goal of sustainable ecological, cultural, and socioeconomic systems." Elaborating on the relationship of cultural resources to ecosystems, the Working Group further stated:

> Generally, the term ecosystem management is associated with nonhuman biological systems. However, it should not be relegated only to the realm of these natural systems. There is a strong need to include cultural resource considerations in ecosystem management principles. Humans have been influencing ecosystems for thousands of years, and the preservation of cultural traditions and historic, archeological, and other cultural places is a goal valued by the American people. People with traditional association to resources and people who care about cultural resources are valid stakeholders in ecosystems. Natural factors pervade cultural resources, and cultural factors pervade natural resources. Often the natural and cultural factors are intricately connected, as is the case with many national battlefields. Additionally, many cultural areas are becoming increasingly important refuge areas for natural resources. Both natural and cultural resource managers must look at larger areas to manage and interpret the resource fully, and they both must look to the larger natural, social, economic, historical, or cultural patterns and themes of which they are a part. The NPS should reduce the barriers to ecosystem approaches that result from artificially separating cultural and natural resources and strive to replace them with collaborative planning, research, and resource management efforts that reflect the real-world integration of material, human, and natural features. (Ecosystem Management Working Group, 1994)

This statement provides a blueprint for meeting the second criterion for appropriate management—recognition and acceptance of cultural and natural resource management responsibilities by all park managers and resource professionals. At all levels within the NPS, steps have been and are being taken to realize this goal.

Two examples of collaborative research and planning in the Pacific Northwest illustrate the value of this approach. One of the major resource management issues at Olympic National Park has been the proposed removal of non-native mountain goat (*Ovis canadensis*) populations to protect rare and endangered plant communities within the park. Analysis of ethnographic and archeological data provided evidence to refute the statements of goat supporters that there were prehistoric native goat populations on the Olympic peninsula (Schalk, 1993). The second example involves the removal of two historic dams on the Elwha River in order to proceed with restoration of the Elwha ecosystem and its rich fisheries (see Chap. 15). Completion of the studies required to support this proposal has required the collaboration of natural and cultural resource experts working closely with members of the Lower Elwha Klallam tribe. For the tribe, the restoration of the ecosystem and its fishery represents promise of cultural renewal. As for most tribal peoples, the Elwha tribe's cultural system incorporates all of the resources needed to sustain their lifeways, including fish, clean water, and viable forests.

Allocating Staff and Budgetary Resources

Olympic National Park provides a good example of how far the NPS has come in its program to support and understand the needs of cultural resource management (i.e., allocating staff and budgetary resources needed to manage both cultural and natural resources appropriately and professionally). In 1992, Olympic National Park, working closely with the regional cultural resource staff to identify funding for the position, recruited the first cultural resource manager to be hired in any of the Pacific Northwest's major natural areas. This position, designated as chief of cultural resource management, was given parallel status in the park hierarchy to the chief of natural resource management. Part-time and project-funded historian and anthropologist positions were reassigned to report to the chief of cultural resource management, thereby forming the nucleus of a park cultural resources staff. Since then, using a variety of funding sources ranging from special initiatives for archeological survey and professionalism to project funds to support the Elwha study, the division now includes project-funded archeologists and receives support from a newly hired park architect who has responsibilities for the park's historic structures. Most importantly, the usefulness of the cultural resource staff for supporting a wide variety of park issues has not escaped the notice of other superintendents in the region. Both North Cascades and Mount Rainier National Parks have established cultural resource specialist positions and support or propose positions for archeologists, museum technicians, and historians.

Two NPS service-wide staffing initiatives have been crucial in supporting the growth of cultural resource staff in such field areas. The first is the "professionalism" initiative under which the NPS received funding designated to provide new base-funded resource management professionals. The cultural resource specialist positions at Olympic, Mount Rainier, and North Cascades have received funding from this program. Complementing this specific resource management initiative have been base increases to "special needs" parks that also have targeted professional resource specialist positions.

The second, and potentially the most far-reaching, initiative is the Resources Management Assessment Program or R-MAP. The initial objective of this program was to provide "an objective assessment of the base staffing and funding needed to implement a thorough natural resources management and research program [in all parks]" (Natural Resources Management, 1993). Involving both park managers and professional staff at the field and regional levels, the R-MAP task force developed a set of factors for determining needs and allocating natural resource staffing and funding. The project identified 164 factors, beginning with the park's total acreage and ending with the total number of projects listed in the Annual Investigators Reports submitted by the park for the previous three years.

The success of the R-MAP formula for identifying natural resource staffing and funding needs and, most importantly, the interest with which it has been received by the NPS Congressional Oversight Committees, spurred the Western Region's cultural resource professionals to adapt the R-MAP approach to document cultural resource (CR-MAP) needs. CR-MAP was tested in the Western Region and then presented to the NPS Washington office program directors in the fall of 1994. In fiscal year 1995, $75,000 from the NPS Cultural Resource Preservation Fund was allocated to facilitate the application of CR-MAP service-wide. Pilot parks, including both predominantly natural and cultural areas, will be used to demonstrate CR-MAPs utility for all kinds of park areas.

Integrating Resource Concerns into the Planning Process

The most important step toward achieving the goal of integrating resource concerns into planning has been implementation of the requirement that all park resource management plans include both cultural and natural components. As a result, many parks were required to identify and evaluate their cultural resource management needs for the first time; the reverse may be said of many historic areas, which have been required to address natural issues. Most of the funding programs supporting park resource management activities now require that funding requests be supported by project statements in approved resource management plans; this has given these plans high priority among park managers.

An initiative that still is taking shape is the "contract" being developed by the NPS to identify the core responsibilities of every superintendent. This contract includes as one of its key provisions resource stewardship for both natural and cultural resources.

Integrating consideration of resources, both cultural and natural, into a park's decision-making process is the fundamental concept of good resource stewardship. Many parks have adopted project-review processes that are intended to ensure this integration. The transfer of greater responsibility for compliance with the provisions of the National Historic Preservation Act and the National Environmental Policy Act to park superintendents will further reinforce this integration process.

Conclusions

Looking back to the four criteria defining "an appropriate balance" in the resource management programs of parks and protected areas (outlined at the beginning of this chapter), there is tremendous satisfaction in recognizing how much progress the NPS has made during the past decade toward meeting these criteria. Resource stewardship has been confirmed unequivocally as the primary responsibility of the NPS. Concerted efforts have been made to dramatically increase resource data bases for both natural and cultural resources, and the introduction of geographic information systems (GIS) into park planning and management programs promises exciting new opportunities to utilize these data. Reaffirmation of the responsibility for resource stewardship has brought with it renewed commitment to providing the training needed by park managers and staffs alike to understand their legal and professional resource management responsibilities. Furthermore, important initiatives such as R-MAP and CR-MAP are providing the means to identify and defend the need for staff and funds required for effective resource management.

Opportunities for cooperative interaction between cultural and natural resource management specialists are increasing at all levels with the NPS. Cultural resource specialists are being recruited for natural area staffs, and natural resource specialists are joining the staffs of many historic areas. In most NPS support offices, natural and cultural resource professionals are grouped together, along with planning and GIS specialists, to provide more effective support to park programs. Only at the national level does the structural separation between natural and cultural programs persist, relying on program leaders to find ways to work across division lines.

As promising as this is, there are some clouds on the horizon. The most important problem is also the most persistent: the lack of sufficient resources to carry out recommendations for staffing and funding increases. Resource management programs will not escape the cuts that are affecting and will continue to affect all federal programs for the foreseeable future. Thus, it remains to be seen whether good intentions will falter as funds decline and hard decisions or lack of opportunity freeze the system into a status quo that still reflects numerous inequities.

The simple answer to the question of whether an appropriate balance exists is still "no." The institutional bias favoring the natural resources of the traditional national parks still exists throughout the upper-management levels of the NPS and the Department of the Interior. Inequities in staffing at the field level, the result of past allocations and priorities, still exist. More importantly, however, is how close we have come to being able to say "yes, there is an appropriate balance." Given the means to achieve adequate staff and funding resources, the commitment now exists within the NPS to seek such balance as the best means of meeting the responsibility for and setting the highest standards of resource stewardship.

Literature Cited

Ecosystem Management Working Group, Resource Stewardship Team, Vail Agenda. *Ecosystem Management in the National Park Service* (Discussion Draft). National Park Service, September 1994.

Natural Resources Management and Science Task Force, Western Region, National Park Service. *R-Map: Resources Management Assessment Program.* National Park Service, December 1993.

Schalk R. *A Review of the Ethnographic and Archeological Evidence Relating to Mountain Goats in the Olympic Mountains.* Seattle: INFOTEC Research Inc., 1993.

Steering Committee, National Parks for the 21st Century. *The Vail Agenda: Report and Recommendations to the Director of the National Park Service.* Washington, DC: National Park Foundation, 1992.

Toothman SS. Cultural resource management in natural areas of the national park system. *The Public Historian* 1987;9(2):65–76.

18

Hinterlands, Wilderness, and Protected Areas in Northern Canada

D. Scott Slocombe

Unlike the subject areas of many chapters of this book, which deal with the threats to and fragmentation of parks and protected areas in the continental United States, many parts of Canada, and particularly its northern provinces and territories, qualify in the minds of most people as true wilderness. Because the institutional and legislative framework is very different in Canadian provinces versus territories, however, the discussions in this chapter are limited to wilderness and protected areas in Canada's Yukon and Northwest Territories (NWT).

The Yukon covers 483,450 square kilometers and the NWT 3,426,320 square kilometers, both north of 60 degrees north latitude and totaling almost 40 percent of Canada (Canadian Almanac and Directory, 1993). There has been road access to the southern Yukon since the 1940s, and to the northern Yukon and western NWT since the 1960s. The population of the Yukon is about 28,000 while that of the NWT is about 60,000. Aboriginal people make up about 25 percent of the population in the Yukon, 50 percent in the NWT, and a much higher proportion of those residing outside the territorial capitols of Whitehorse and Yellowknife. Both territories have come to the attention of the rest of Canada within only the last 50 to 100 years—first the Yukon after the Klondike Gold Rush of 1898, and both the Yukon and the NWT after World War II due to mineral, military, and transportation developments.

What attention the two territories did receive for most of the twentieth century was due to their resource potential and role in continental defense. Only since the 1960s have the territories gained widespread attention as Native homelands and unique natural environments. This attention has not equated to political power, however. Although both territories now have elected and representative governments, the Canadian Federal Government retains controlling power over almost all land and resources. This contrasts with adjacent northern parts of the Canadian provinces where provincial governments control land and resources. Changes in the territories today derive from the settlement of Aboriginal land claims, devolution of resource and other management responsibilities to the territories, and the growth of tourism and less capital-intensive activities in contrast to traditional nonrenewable extractive resource development (for more general background, see Bone, 1992; Coates & Morrison, 1988; Hamilton, 1994).

The history of protected areas in northern Canada has until recently been a history of national protection efforts, beginning with designation of a few national parks and numerous game sanctuaries in the early twentieth-century, elimination or reduction of most of these sanctuaries and establishment of migratory bird sanctuaries beginning in the late 1950s, followed by the establishment of several national parks in the 1960s and 1970s. All territorial parks and most northern provincial parks were established after 1970. Since 1980, we have seen renewed, diverse efforts to enlarge northern protected areas through territorial and national parks and to expand their role in protecting the essence of Aboriginal culture. These efforts are indicative of the fact that the history of Canada's northern protected areas reflects an ongoing concern with protecting lands that are home to ancient nations and that were long managed through traditional conservation systems. These issues include the fact that Native peoples often have a different perception of the term "protection" and do not see a conflict between wilderness as resource base and wilderness as natural and cultural heritage.

Resource Issues in the Northern Territories

Lands in the Yukon and NWT range from high arctic and alpine tundra to mountains, forests, and taiga, with strong biophysical, as well as cultural, differences from west to east and north to south. Innumerable small and a few vast lakes and rivers are major physical features of the north. The harsh climate and shallow soils combine to create low biological productivity over much of the region. Parks Canada has identified twenty-one terrestrial and eleven marine natural regions in the Canadian North (Figs. 18.1, 18.2). Parks Canada's ultimate goal is to preserve a representative and significant example of each natural region in national parks (cf. Kovacs, 1987).

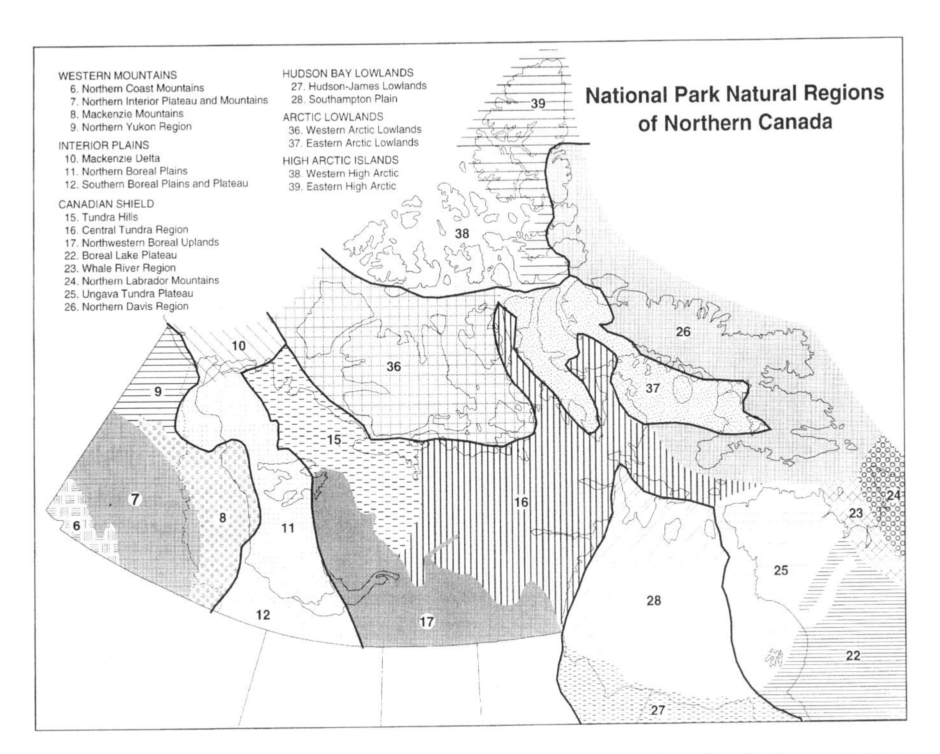

Figure 18.1 National park natural regions of northern Canada (After Canadian Environmental Advisory Council, *A Protected Areas Vision for Canada.* Ottawa: Supply and Services Canada, 1991.)

The territories are also the home of over a dozen distinct aboriginal groups, as well as growing numbers of nonaboriginals. However, historically, efforts to protect northern lands by southern governments and societies have been influenced by priorities that do not necessarily reflect those of northern people. Northern Natives have frequently been negatively affected by wildlife conservation and protected areas initiatives in the past (McCandless, 1985). Historically, these peoples' livelihoods have depended on the land and natural resources through direct subsistence harvesting and, in the last century, on resource development as well. Until recently, withdrawal of lands from large-scale development has been favorably viewed by only a few northern residents, including Native peoples.

Withdrawal of lands for national parks led to the end of mining, hunting, trapping, and guiding by aboriginal and other people. Prior to the 1980s, as in the case of Kluane National Park Reserve (see under Different Approaches to Protection), there was usually no local consultation before or after initial protection. Instead, park establishment plans were dictated by centralized provincial and federal planning offices. The backlash to this procedure was that from

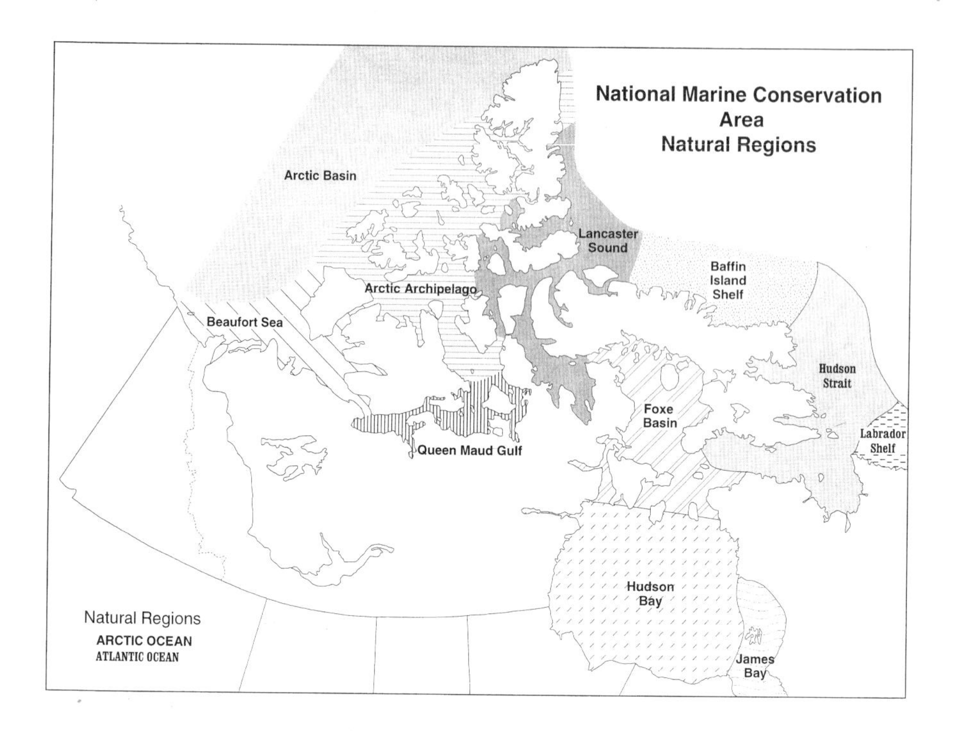

Figure 18.2 National park marine regions of northern Canada (After Parks Canada, Canadian Heritage, *Sea to Sea to Sea: Canada's National Marine Conservation Areas System Plan.* Ottawa: Supply and Services Canada, 1995.)

the 1950s onward the mining industry fought most new protected areas, and from the 1960s onward aboriginal peoples objected to designation of new parks prior to settlement of comprehensive land claims. As a result, by the late 1970s, there was local opposition to all northern national parks proposals. This opposition was ameliorated only by linking new park establishment to settlement of comprehensive land claims beginning in the mid-1980s.

Since the late 1970s, many interests have viewed northern protected areas as final opportunities to protect large and unique areas of wilderness, often of continental significance for wildlife such as grizzly bear (*Ursus arctos*), caribou (*Rangifer tarandus*), and wolves (*Canis lupus*). These interests were often led by southern conservation groups such as World Wildlife Fund (WWF) Canada, the Canadian Arctic Resources Committee (CARC), and the Canadian Parks and Wilderness Society (CPAWS), and were joined by one or two northern groups, notably the Yukon Conservation Society (YCS) founded in 1968. University-based researchers also have had a significant role in the recent development of northern protected areas (e.g., Fenge, 1982; Nelson & Smith, 1987). The Endangered Spaces Campaign (Hummell, 1989) was an early example of efforts led by nongovernment agencies to foster completion of protected areas networks. An example of the success of the above-listed groups is the July 1995 initiation of a program of scientific research and participatory planning to foster completion of the Yukon protected areas network, led by the Yukon chapter of CPAWS, with extensive funding from foundations and other nongovernment organizations.

During the last two decades, there also has been strong growth in adventure and ecotourism, which has increased public acceptance of and illustrated the economic benefits accrued from protected areas (Table 18.1). However, increased tourism also has raised awareness of the issues (and drawbacks) of motorized access as keeping northern tourists in one place long enough to spend money is a challenge. Young fit tourists often head straight for the wilderness, spending little. Older tourists often are in organized tours, or campers and mobile homes, and may spend equally little, particularly if there are no attractions to which they can drive. Moreover, much of what those on organized, and often very expensive, tours spend does not remain in the territories.

Table 18.1 Visitor numbers and economic impact of northern protected areas

	Visitors			
Date	Kluane NPR	Wood Buffalo NP	Auyuittuq NPR	Ivvavik NP
1980	35,208	1980	362	—
1981	53,148	2279	454	—
1982	66,512	2133	342	—
1983	50,709	2275	448	—
1984	60,021	2851	576	—
1985	65,338	2535	470	—
1986	69,431	2745	457	—
1987	81,804	3398	467	—
1988	80,521	5718	408	84
1989	69,000	6600	360	99
1990	75,000	7300	1300	150
1991	67,000	8000	660	93
1992	84,650	7116	298	380
Economic Impact 1987–1988	$7,747,400	$5,413,000	$1,560,000	$471,700
Economic Impact 1992	$7,000,000	$5,700,000	$1,000,000	$440,000

NP = National Park; NPR = National Park Reserve.
Source: Data from Parks Canada (1994), Thompson Economic Consulting (1989), and Stanley & Perron (1994).

Since the 1980s, there has been a growing partnership among Aboriginal groups, conservation organizations, and government agencies to deal with issues of tourism development and wilderness use. This has been made possible in part because of the settlement of comprehensive land claims (cf. Tungavik Federation of Nunavut, 1987) based on traditional land use and occupancy. Such settlements, begun in the late 1970s, are now near conclusion in most of the territories and have usually included protected areas to which Native people have subsistence access and rights to tourism development and to training and employment. However, many issues remain unresolved such as conflicts between those advocating greater resource development and resource extraction and those who wish to preserve protected areas for their environmental and cultural benefits (Slocombe & Nelson, 1992).

History of Protected Areas

Three main themes underlay the development of protected areas in Canada's northern territories. The first and most persistent has been the desire of governments to protect land and wildlife resources in order to support the traditional lifestyles and societies of northern Native peoples. This goal has been complemented in the last forty years or so by an interest in protecting unique and significant geological, ecological, and cultural features of the north. Within the last ten to twenty years, protected areas have been viewed as harbingers of tourism and related economic opportunities.

The earliest protected areas in northern Canada resulted from designation of large parts of the NWT as game preserves, beginning in 1918 when Victoria Island was set aside as a hunting preserve for Native peoples. That reserve was expanded in 1926 as the Arctic Islands Preserve, and still further over the next twenty years. It was then progressively diminished between 1948 and 1966, when it was completely abolished. Other protected areas, including the existing Thelon Game Sanctuary, were established at about the same time to protect wildlife, land, and Native subsistence lifestyles, although most were abolished in the decades following World War II (Foster, 1978; Kovacs 1987; Bregha, 1989). A number of migratory bird sanctuaries were established in the 1950s and 1960s, but most are quite small and have relatively weak provisions for habitat protection. More recently, a National Wildlife Area was established in 1986 at Polar Bear Pass on Bathurst Island.

Wood Buffalo National Park was the first national park in the territories. It was established in 1922 astride the Alberta-NWT border as part of an effort to protect the last of the bison (*Bison bison athabascae*) and the breeding ground of the endangered whooping crane (*Grus americana*) in the marshes of the Peace and Athabasca Rivers. New northern national parks received considerable support from Jean Chretien and Hugh Faulkner in their terms as Ministers of Indian Affairs and Northern Development under then Prime Minister Pierre Trudeau. Between 1968 and 1984, the federal liberal government did much to establish northern national parks and to reserve land for more parks (McNamee, 1992). Justice Thomas Berger's Mackenzie Valley Pipeline Inquiry of 1974–1975 also called for more northern parks, including one on the Yukon North Slope (Berger, 1977) that was later established as part of a land claims settlement.

The Territorial Parks network is small, not least because the federal government owns almost all the land in the territories. Territorial parks systems have been largely campground- and roadside-oriented, with development beginning in the late 1970s. However, with increased recognition of potential economic benefits, both territories now have at least a few relatively large, remote protected areas aimed at protecting particular natural, cultural, or historic resources (Vaughan, 1984; Hamre, 1987; Environment Canada et al., 1992).

During the last fifteen years, there has been growing attention given to integrating northern protected areas with broader planning and development. Thus, the joint Federal, Yukon, and NWT Task Force on Northern Conservation (1984) made establishment of "a comprehensive network of land and/or water areas subject to special protection" a central part of its report, and devoted considerable attention to the issues of what sorts of protected areas, operated how, would be best suited to the north (see also Peterson et al., 1986; Macpherson et al., 1987). A critical goal of northern protected areas was and still is to ensure that they meet the needs of northerners, as well as protect significant resources for larger publics, and avoid the sometimes negative socioeconomic impacts protected area designations have experienced elsewhere in northern Canada and the world (Payne et al., 1992).

Numerous studies, before and since, have identified areas of significance in the north, most comprehensively the International Biological Programme (IBP) (Revel, 1978), Environment Canada (1982), and the NWT government (Ferguson, 1987). Similar inventories, on an ecoregion basis, are under way in the Yukon as comprehensive claims of individual Native groups are settled. IBP sites have long received informal protection from northern and federal governments. Overall, protected areas in northern Canada have a long and varied his-

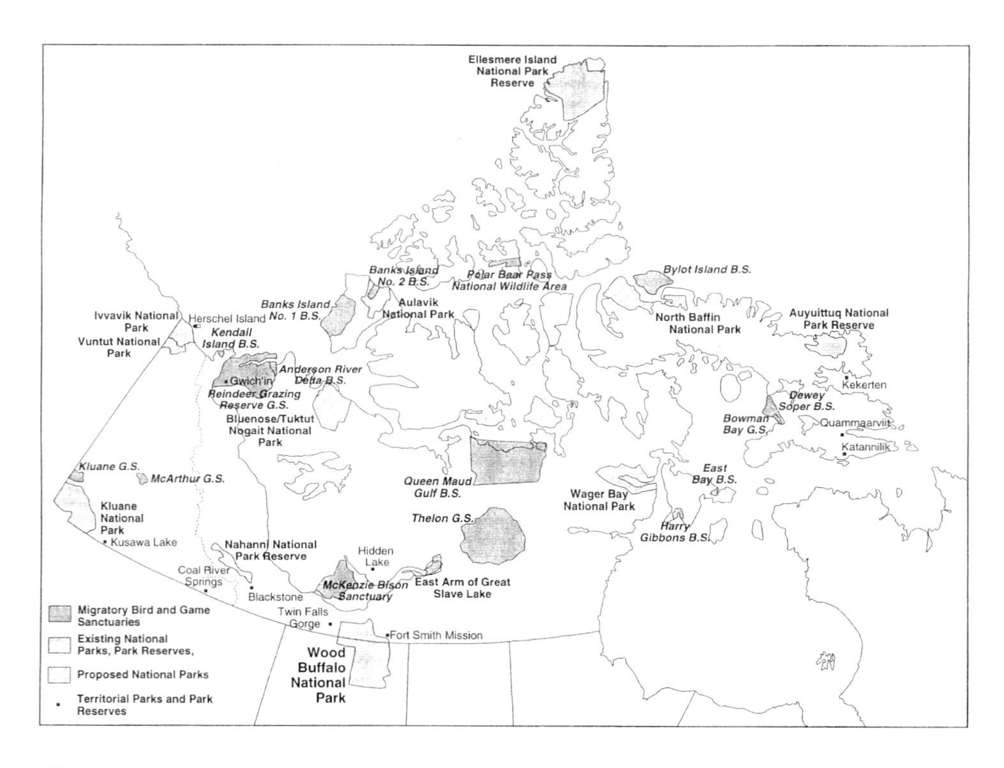

Figure 18.3 Major protected areas of northern Canada (After Environment Canada, GNWT, YTG, *New Parks North Newsletter* (Nos. 1–4). Yellowknife, 1992.)

tory (Fig. 18.3). Some of this diversity in design and purpose, and the innovations that have arisen from it, is explored in the next section.

Different Approaches to Protection

The protected areas of northern Canada exhibit a range of origins and goals. This section briefly examines examples of national and territorial parks, game sanctuaries, and other units as a means to better understand the policy and management context of northern protected areas. These examples illustrate and expand on many of the points made in earlier and succeeding sections (see also Slocombe & Nelson, 1991).

Kluane National Park Reserve is the classic example of a northern protected area developed the "old way." It is one of Canada's most spectacular and significant national parks and includes Canada's highest mountains, the largest non-polar icefield in the world, and significant populations of many big-game and other wildlife species. The threat of hunting by the thousands of servicemen working on construction of the Alaska Highway was the immediate reason the region was set aside in 1942 as a park reserve and then in 1943 as a game sanctuary. Official legislation of Kluane National Park Reserve did not occur until 1976, following protracted efforts and debate and redrawing of the park boundaries to exclude mineral and other resources. Kluane, like other new northern national parks in the 1970s and early 1980s, was designated a national park reserve in recognition of pending land claims and the fact that their final boundaries might change as a result (Theberge, 1978, 1980).

Initial protection of Kluane in the 1940s was enacted without consulting local people and excluded big-game hunting and subsistence activities—a long-remembered cause of local discontent and hardship. The first management plan, completed in 1980, made extravagant promises regarding access to the park, including all-weather roads and even a sightseeing tram up a mountain in very unstable terrain, which led to further local discontent when access development did not happen. The 1990 plan revision, based on extensive public consultation at a time when comprehensive land claims and regional planning were also in progress in the region, was very different and included feasible, practical concessions and access plans. Visitation to the visitor center is high for a northern park, but the number of people actually entering the park is much lower than it could be due to the roughness of terrain and lack of road access (see Fig. 18.1) (Slocombe, 1992). Today, questions over land claims and land selections are an issue only at the local level in the southeast portion of the park; these have resulted in some small boundary changes.

Wood Buffalo National Park is the other long-established northern Canadian national park. It was created in 1922 to protect the last wood bison, and covers 44,670 square kilometers of largely inaccessible forest and wetlands bridging the Alberta-NWT border. A variety of serious environmental problems now threaten the park, including (1) upstream dams that have changed the water regime in its wetlands, (2) upstream pulp and paper mills that are adding pollutants, (3) logging and (4) plans to exterminate the bison as cattle ranching encroaches on the park's southwest borders. As in the case of Yellowstone National Park, agricultural interests have focused attention on brucellosis and tuberculosis endemic in the buffalo herds of Wood Buffalo National Park. A 1990 Federal Environmental Assessment Panel recommendation that all the park's bison be killed was rejected by the park, but the more recent 1993 Northern Buffalo Management Board recommendations for establishment of a buffer zone and extensive research program have still received no government response (McNamee, 1992, 1995). The park is unique as it has a long tradition of Native subsistence harvesting that in its early years was arbitrarily managed. This use was formalized in 1986 through a settlement with the local Cree Band, which established a wildlife advisory board, the majority of whose voting members are Cree. The board has authority to advise on management of a 1.2-million hectare parcel of traditional interest, to determine who receives hunting and trapping permits as well as rights of first refusal on related economic activities in the area of traditional interest, and receives preferential access to park-related training and employment (East, 1991).

Ivvavik (formerly Northern Yukon) National Park was established in 1984, the first national park established as a direct result of the settlement of a comprehensive Native land claim. The park's legislation and plan allow use of the area by native people for subsistence harvesting and ensures that they will benefit from economic development of the park. The park is very remote and visitation varies widely; at most, there are only several hundred visitors per year (Sadler, 1989; East, 1991; Slocombe & den Ouden, 1993). Management planning is conducted jointly between the Native peoples and Parks Canada and effectively includes a larger regional ecosystem. A management plan has recently been developed and approved (Wildlife Management Advisory Committee/North Slope, 1994).

Ivvavik is bordered to the west in Alaska by the Arctic National Wildlife Refuge (ANWR). As a whole, this region is critical habitat for the porcupine caribou herd. There have been efforts to strengthen international cooperation and protection of this region, although current attempts in the United States to open ANWR for oil development may complicate these efforts (Childers, 1994). Ivvavik and regional ecosystem management have been bolstered by the establishment of Vuntut National Park to the south of Ivvavik, through another land claims settlement.

Katannilik Territorial Park was established in 1993 near Lake Harbour on southern Baffin Island. Covering more than 1500 square kilometers, it is the largest territorial natural environment park. The environmental attributes were first identified in the early 1980s through a regional tourism study and were confirmed by another study in the late 1980s. A master planning process was completed in 1991 and, following approval by the Inuit Tapirisat of Canada, land was withdrawn from other uses for park development. Katannilik Park, consistent with the policies of the NWT, seeks to provide high-quality recreational opportunities, maintain the natural and cultural resources that make it significant, and at the same time contribute to the economic base of adjacent communities. Campground and other operational facilities have been completed, and an interpretive facility is scheduled to open in 1996. Promotion of the park is active through regional tourism centers, promotional materials, and the media. Visitation at present is in the low hundreds, and local economic benefits are being closely tracked (Downie & Monteith, 1994).

The Thelon Game Sanctuary was established in 1927 in east-central NWT to protect major wildlife resources. The sanctuary excludes hunting, trapping, and mineral exploration and development. It also contains dramatic scenery centered on the Thelon River and a significant example of the transition from boreal forest to tundra. It is the largest game sanctuary left in the north and is the size of the province of New Brunswick. The 1986 DIAND (Department of Indian Affairs and Northern Development) Northern Mineral Policy raised fears that the sanctuary's boundaries would be changed to accommodate the development of uranium resources and proposed mines in the area. The policy was in fact intended to maximize the area available for mining in the territory. Public opposition to this was quick and strong. Ultimately, the Conservation Advisory Committee established by the Mineral Policy recommended no changes and new, more detailed scientific studies of the sanctuary (Struzik, 1989). The sanctuary currently overlaps the Eastern Arctic Inuit and NWT Dene and Metis comprehensive claims areas. No changes will be made in the status of the area, at least until a sanctuary management plan jointly developed by government and aboriginal interests is approved (Environment Canada et al., 1992).

The Canadian Heritage Rivers program was established in 1984. Canadian Heritage Rivers are designated for their natural values, historical importance, and recreational potential and may include all or part of a river. There are several in northern Canada: the Alsek, Thirty Mile, and Bonnet Plume Rivers in the Yukon; and South Nahanni, Arctic Red, Thelon, and Kazan Rivers in the

NWT (Noel, 1995). Other rivers under consideration or planning include the Tatshenshini. Earlier Heritage River designations were often for rivers in existing protected areas, but recently this has been less common, and comanagement and private lands strategies are developing. The program is unique in its emphasis on community involvement in, and responsibility for, drawing up nomination and management documents. This fits well with current northern approaches to resource management and park development, and the recreational dimension suits many communities' desire to develop and benefit from wilderness and ecotourism.

The Chilkoot Trail, the principal route over the 50 kilometers from Skagway, Alaska, to the headwaters of the Yukon River for prospectors heading to the Klondike during the gold rush of 1898, is an International Historic Site under the management of the Canadian and U.S. National Park Services. It receives very intensive use as tens of thousands of people hike the trail every summer (and a very few ski it in the winter). Another major trail, just being developed for backcountry hikers and cyclists as a NWT territorial park is the Canol Trail. This runs 372 kilometers from Macmillan Pass on the Yukon-NWT border to Norman Wells and follows the access road for a World War II pipeline, which was built to carry oil from Norman Wells, NWT, to Haines, Alaska.

Issues and Approaches, Challenges and Opportunities

The examples outlined previously illustrate the range of opportunities and challenges facing protected areas in northern Canada. Here, key issues are summarized and lessons extracted for protected areas managers facing issues such as regional integration and local participation, aboriginal peoples and comprehensive land claims, resource extraction, tourism development, backcountry access, and demands for economic development.

As for most land and resource management in northern Canada, institutional arrangements are a key issue and question. The federal ownership and control of almost all land and most resources is a continuing annoyance to territorial governments and citizens. Federal protected areas often have been a particularly visible manifestation of this outside power. The development of territorial parks systems, which have included development of participatory management and advisory and comanagement boards at most existing and virtually all new protected areas, is one response to this situation. It is equally true that the diversity of agencies and institutions with some authority over some or all protected areas complicates their designation and management (Nelson & Jessen, 1984; Nelson & Grigoriew, 1987; Nelson & Smith, 1987). Cooperative, multijurisdictional approaches often required or fostered by comprehensive land claims processes are focusing and clarifying some of these problems.

Integration of protected areas into broader land and resource planning has been a goal for at least twenty years. Achieving this goal is closely linked to the general success of integrated regional and/or environmental planning in the north. Progress slowed for a while with the end of the federal northern land use planning program in the early 1990s but is gaining momentum as territorial governments and land claims processes take over land use and resource development planning. Indeed, the cooperative development of new protected areas involving local, territorial, Aboriginal, and federal interests is probably one of the areas of resource and environmental planning where such integrated, multistakeholder planning is taking place. In an allied vein, it is significant that the agenda of northern protected areas has increasingly been taken over by northerners, both practically and conceptually (cf. Peepre & Jickling, 1994).

At the local level, comanagement is an increasingly common approach to involving and representing diverse user and ownership groups. Comanagement still is evolving and has been most successful in the Yukon North Slope and Mackenzie Delta area as a result of the Inuvialuit Final Agreement. At its best, comanagement allows use of local knowledge that is often important to understanding land and wildlife dynamics in particular places, improving consideration of local needs and goals, and fostering local support for conservation and management measures that may be needed (Berg et al., 1993; Bailey, 1994; MacLachlan, 1994).

Development of new resource knowledge and understanding has long been a goal for northern protected areas and land and resource planning. A few areas, such as the Yukon North Slope, Lancaster Sound, northern Quebec, and the Kluane region have been intensively studied, but much of northern Canada has not. At one level, basic biophysical and socioeconomic information is needed for land and resource planning. At another level, and nearly everywhere, the integration of traditional knowledge and scientific knowledge is a key priority.

The progress of land claims in northern Canada during the last decade has significantly changed the land and resource management regime. These changes have not come quickly but to some extent have been paralleled by changes in Alaska and northern Australia (Gardner & Nelson, 1980, 1981; see also Chap. 9). The details of specific management institutions and planning processes are still being developed at local and regional levels. In general, settlements include new protected areas and several land management, development assessment, and wildlife management boards that usually operate at local or regional levels, with broad coordination at the territorial level. The main issue this raises for protected areas is the impact of involving local people on the levels of development and subsistence hunting in protected areas, including national parks. This concern has aroused strong opposition from some environmentalists and scientists. There is also, of course, an argument to be made that traditional subsistence activities are a natural, even coevolved, part of the ecosystems in many wilderness areas (Chipeniuk, 1989). Efforts to minimize the disruption of park visitors by subsistence activities are common and, so far at least, few visitors appear to have been upset.

The other immediate effect of comprehensive claims settlements has been, and will continue to be for some years yet, the establishment of new national parks and other protected areas. The sheer number of new parks is astonishing: three have been established in the last few years and perhaps another ten or so will be in the next five or ten years. This increase is stressing the resources of Parks Canada, which must manage the new parks, as efforts to add money to their budget, quite separate from the agency that agrees to land claims, have not always been successful or enduring.

However, even with all these new terrestrial parks, development of marine parks in northern Canada will continue to be a challenge. Despite a national marine parks policy in 1986, an Arctic Marine Conservation Strategy in 1987, and the Arctic Environmental Strategy in 1991, work on arctic marine protected areas (or indeed marine protected areas anywhere in Canada) has barely begun. Legislation for national marine conservation areas is expected in the near future, which could facilitate the creation of the first northern marine protected areas (Environment Canada et al., 1992; Beckmann, 1994; Parks Canada, 1995).

Nonrenewable resource development remains a significant source of economic opportunities in northern Canada and, for many people, a primary potential source of government revenues to help support greater territorial independence from Ottawa. After a number of relatively quiet years, mineral development is on the increase in the Yukon and the much-publicized diamond area of the central NWT. Mineral interests have always been the strongest opponents of protected areas in northern Canada and that does not appear to be changing quickly. Park and mining tensions may have been eased slightly with the recent announcement of a settlement between the British Columbia government and Royal Oak Mines over cancellation of the Windy Craggy Mine and the establishment of the Tatshenshini-Alsek Provincial Wilderness Park on the British Columbia-Yukon border (Cernetig, 1995).

The steady growth in tourism in the territories during the last decade has had a progressive effect on perceptions of and interest in protected areas. Through the 1980s, the potential economic benefits of major protected areas from Kluane to Auyuittuq became clearer to local and territorial interests. New protected areas often became more acceptable and the territorial governments became much more interested in establishing their own, often tourism- and recreation-oriented, protected areas. Access and development of facilities have remained issues in many parks, for example Kluane, but the growth of back-country use and spending associated with it is even leading to greater acceptance of wilderness parks and reducing demands for large-scale infrastructural development.

The changes in the nature and management of protected areas in Canada's territories during the last fifty years reflect many factors. Perhaps the most significant changes have been in federal-territorial relations, in Native–non-Native relations and land claims, in the size and make-up of northern populations, and in the interest of southern Canadians in northern Canada. A progressive devolution of some resource and wildlife management responsibilities and controls over economic development and political power to the territorial governments have improved local and territorial abilities to promote and capitalize on protected areas development. As a result, the move to settle comprehensive land claims and involve Native peoples in resource and protected areas management has provided a base for cooperative management and use of resources and protected areas.

The influx of southern professionals to the territories in the last twenty years created new constituencies in the politics of issues such as social services and environmental protection. During the same time period, southern Canadians and organizations became more aware of the resources of the territories and more concerned about their protection, thus providing, in particular, a constituency for federal protection of the northern environment.

The wide distrust of protected areas that developed in the late 1960s and 1970s has since been tempered by long-term efforts to build trust among local and Native peoples, by new cooperative efforts giving power to those groups, by initiatives by territorial governments, by growing evidence of the economic benefit of protected areas through tourism, and by the development of a conservation community in the north.

Conclusions

Protected areas in northern Canada have gone through several phases. At the very least, they have evolved from tracts imposed by paternalistic external forces, through protection sometimes arbitrarily imposed for the good of all Canadians, to more and more collaboratively developed areas ostensibly designed both to protect natural resources and cultural sites as well as to support sustainable traditional lifestyles of local peoples. Now the prime area of evolution is in multi-party, participatory mechanisms for research, planning, and management of protected areas. Not only are different levels of government involved but also Native peoples and even nongovernment organizations.

The size of many of Canada's northern protected areas presents a unique opportunity and challenge. Many of the national parks are big enough to protect whole watersheds and ecosystems, but this sometimes makes minimizing the disruption of local peoples' lives difficult. As big as northern protected areas often are, managers still must deal with the potential impacts of activities on adjacent lands. The evolution of northern protected areas is continuing and faces continuing challenges, not least from large-scale nonrenewable resource development and from the potential success of tourism development.

In northern parks, at least, Canada seems to have learned that local people need to be involved in park planning and management. The future challenge will be integrating protected areas with local and regional goals through regional land and resource planning exercises, under whatever guise and to build on the many grounds or interests for establishing parks that have emerged in recent years. Financing management may be a difficult challenge in the future as the days of rapid growth of tourism and of land claims initiatives and funding are probably nearing an end. Creative approaches and greater territorially based initiatives will be needed.

Reconciling multiple uses with conservation will simultaneously challenge managers. Subsistence, recreation, and backcountry use will all compete with conservation considerations and each other. Multiple use is not necessarily a bad thing, particularly if it can reduce conflict while increasing overall protection levels. It can be approached in several ways that are already evident in the territories. First, a range of protected area types must be established, often under different management regimes. Second, different uses must be permitted within individual parks, facilitated by the large size and diversity of most northern parks, but necessarily constrained by biophysical hazards and sensitivity.

The final conservation challenge is completing a representative system of terrestrial and marine protected areas. This is especially true for marine and tundra areas. A greater diversity of protected area types could also be beneficial. Most sites are currently in IUCN-World Conservation Union categories II and IV (national parks and managed nature reserves/wildlife sanctuaries), with a very few category I and III areas (strict nature/scientific reserves and national monument/natural landmarks) (see Table 1.1) (Conservation of Arctic Flora and Fauna, 1994). There is hope, through the actions of nongovernment organizations, land claims organizations, and territorial and federal governments, that this situation will change in the near future.

Literature Cited

Bailey J. Managing Protected Areas in the North: What We Know, What We Need to Learn about Co-Management. In Peepre JS, Jickling B. (eds), Northern protected areas and wilderness. Whitehorse: Canadian Parks and Wilderness Society and Yukon College. 1994:91–97.

Beckmann L. Marine conservation in the Canadian arctic. *Northern Perspect,* 1994;22(2/3):33–39.

Berg L, Fenge T, Dearden P. The Role of Aboriginal Peoples in National Park Designation, Planning and Management in Canada. In: P Dearden, R Rollins (eds), *Parks and Protected Areas in Canada.* Toronto: Oxford University Press, 1993:225–255.

Berger, T. *Northern Frontier, Northern Homeland: The Report of the Mackenzie Valley Pipeline Inquiry. Supply & Services.* Ottawa: Environment Canada, 1977.

Bone, RM. *The Geography of the Canadian North.* Toronto: Oxford University Press, 1992.

Bregha F. Conservation in the Yukon and the Northwest Territories. In M. Hummel (ed), *Endangered Spaces: The Future for Canada's Wilderness.* Toronto: Key Porter Books, 1989:211–225.

Canadian Environmental Advisory Council. *A Protected Areas Vision for Canada.* Ottawa: Supply and Services, Canada, 1991.

Cernetig M. B. C. Royal Oak sign $167-million mining deal. *Toronto Globe and Mail,* August 19, 1995. P. A3.

Childers R. Culture and conservation on the Arctic borderlands. *Northern Perspect,* 1994; 22(2/3):40–43.

Chipeniuk R. The vacant niche: An argument for the recreation of a hunter-gatherer component in the ecosystems of northern national parks. *Environments* 1989;20(1):50–59.

Coates KS, Morrison WR. *Land of the Midnight Sun: A History of the Yukon.* Edmonton: Hurtig, 1988.

Conservation of Arctic Flora and Fauna. The State of Protected Areas in the Circumpolar Arctic, 1994. *CAFF Habitat Conservation Report 1.* Trondheim: Norwegian Directorate for Nature Conservation, 1994.

Downie BK, Monteith D. Economic impacts of park development and operation: Katannilik Territorial Park, Lake Harbour, NWT. *Northern Perspect* 1994;22(2/3):7–17.

East KM. Joint Management of Canada's Northern National Parks. In PC West, SR Brechin (eds), *Resident Peoples and National Parks.* Tucson: University of Arizona Press, 1991:333–345.

Environment Canada. *Canada's Special Places in the North: An Environment Canada Perspective for the '80's.* Ottawa: Supply and Services Canada, 1982.

Environment Canada, Government Northwest Territories, Yukon Territorial Government. *New Parks North Newsletter* (Nos. 1–4). Yellowknife, 1992.

Fenge T. Towards comprehensive conservation of environmentally significant areas in the Northwest Territories of Canada. *Environ Cons* 1982;9:305–313.

Ferguson RS. *Wildlife Areas of Special Interest to the Department of Renewable Resources.* Yellowknife: Department of Renewable Resources, 1987.

Foster J. *Working for Wildlife: The Beginning of Preservation in Canada.* Toronto: University of Toronto Press, 1978.

Gardner JE, Nelson JG. Comparing national parks and related reserve policy in hinterland areas: Alaska, northern Canada, and northern Australia. *Environ Cons* 1980;7:43–50.
Gardner JE, Nelson JG. National parks and native peoples in northern Canada, Alaska, and northern Australia. *Environ Cons* 1981;8:207–215.
Hamilton JD. *Arctic Revolution: Social Change in the Northwest Territories, 1935–1994.* Toronto: Dundurn Press, 1994.
Hamre GM. Conservation in the Canadian Arctic with Special Emphasis on Territorial Parks. In JG Nelson, R Needham, L Norton (eds), *Arctic Heritage: Proceedings of a Symposium.* Ottawa: Association of Canadian Universities for Northern Studies, 1987:557–566.
Hummel M (ed). *Endangered Spaces: The Future for Canada's Wilderness.* Toronto: Key Porter Books, 1989.
Kovacs TJ. National Overview for Canada on National Parks and Protected Areas in the Arctic. In JG Nelson, R Needham, L Norton (eds), *Arctic Heritage: Proceedings of a Symposium.* Ottawa: Association of Canadian Universities for Northern Studies, 1987:530–556.
MacLachlan L. Co-management of wildlife in northern aboriginal comprehensive land-claim agreements. *Northern Perspect,* 1994;22(2/3):21–27.
MacPherson NM, Peterson EB, Peterson NM. Yukon protected areas: A values framework and preliminary database. Whitehorse: Department of Renewable Resources; Ottawa: Environment Canada, 1987.
McCandless RG. *Yukon wildlife: A social history.* Edmonton: University of Alberta Press, 1985.
McNamee K. Wood Buffalo National Park: Threats and possible solutions for a World Heritage Site. Fourth World Congress on National Parks and Protected Areas, San Jose, Costa Rica, 1992.
McNamee K. All's quiet as the buffalo roam. *Nature Alert* 1995;5(2):1.
Nelson JG, Grigoriew P. Institutional arrangements for individual environmentally significant areas: The case of Aishihik, Yukon. *Environ Cons* 1987;14:347–356.
Nelson JG, Jessen S. *Planning and Managing Environmentally Significant Areas in the Northwest Territories.* Ottawa: Canada Arctic Resources Committee, 1984.
Nelson JG, Smith PGR. Institutional arrangements for a system of environmentally significant areas: The case of the east Beaufort Sea area, Canada. *Environ Cons* 1987;14:207–218.
Noel L. Over the watershed: Celebrating ten years of Canada's Heritage Rivers. *Explore* 1995;71:48–55.
Parks Canada. Parks Canada Visitor Use Statistics: 1992–1993. Ottawa: Parks Canada, Co-ordination Branch, Program Management Directorate, 1994.
Parks Canada, Canadian Heritage. *Sea to Sea to Sea: Canada's National Marine Conservation Areas System Plan.* Ottawa: Supply and Services Canada, 1995.
Payne RJ, Rollins R, Tamm S, Nelson C. Managing Social Impacts of Parks and Protected Areas in Northern Canada. In JMH Willison et al (eds), *Science and the Management of Protected Areas.* Amsterdam: Elsevier, 1992:513–518.
Peepre JS, Jickling B (eds). *Northern Protected Areas and Wilderness.* Whitehorse: Canadian Parks and Wilderness Society and Yukon College, 1994.
Peterson EB, Allison LM, Peterson NM. *Evaluation of the "Protected Area" Concept for the Northwest Territories* (2 Vols). Yellowknife: Department of Renewable Resources; Ottawa: Department of Indian Affairs and Northern Development, 1986.

Revel RD. The International Biological Programme in the Subarctic and Arctic regions of Canada. In RF Keith, JB Wright (eds), *Northern Transitions* (Vol 2). Ottawa: Canadian Arctic Resources Committee, 1978:237–250.

Sadler BA. National parks, wilderness preservation, and native peoples in Northern Canada. *Nat Resource J* 1989;29:185–204.

Slocombe DS. The Kluane/Wrangell–St. Elias National Parks, Yukon and Alaska: Seeking Sustainability Through Biosphere Reserves. *Mount Research Dev*, 1992;12(1):87–96.

Slocombe DS, den Ouden S. Ecosystem Management in the Ivvavik (Northern Yukon) National Park Region. In WE Brown, SD Veirs Jr (eds), *Partners in Stewardship: Proceedings of the 7th Conference on Research and Resource Management in Parks and on Public Lands.* Hancock: George Wright Society, 1993:221–227.

Slocombe DS, Nelson JG. Canada and the Arctic. In R Burton (ed), *Nature's Last Strongholds* (Vol 3 of *Illustrated Encyclopedia of World Geography*). New York: Oxford University Press, 1991:44–51.

Slocombe DS, Nelson JG. Management issues in hinterland national parks: A human ecological approach. *Nat Areas J* 1992;12:206–215.

Stanley D, Perron L. The economic impact of northern national parks (reserves) and historic sites. *Northern Perspect* 1994;22(2/3):3–6.

Struzik E. The wildlife bank: Will the government make a withdrawal from the NWT's Thelon Game Sanctuary? *Nature Can* 1989;18(2):23–29.

Task Force on Northern Conservation. *Report of the Task Force on Northern Conservation.* Ottawa: Indian Affairs and Northern Development; Whitehorse: Yukon Department of Renewable Resources; Yellowknife: Department of Renewable Resources, 1984.

Theberge JB. Kluane National Park. In EB Peterson, JB Wright (eds), *Northern Transitions* (Vol 1). Ottawa: Canadian Arctic Resources Committee, 1978:153–189.

Theberge JB. *Kluane: Pinnacle of the Yukon.* Toronto: Doubleday Canada, 1980.

Thompson Economic Consulting Services, Scace and Associates. *Visitor Profile and Economic Impact Statement of Northern National Parks (Reserves) and Historic Sites: Summary report.* Ottawa: Canadian Parks Service, Socioeconomic Branch, 1989.

Tungavik Federation of Nunavut. Land Claims, National Parks, Protected Areas and Renewable Resource Economy. In JG Nelson, R Needham, L Norton (eds), *Arctic Heritage: Proceedings of a Symposium.* Ottawa: Association of Canadian Universities for Northern Studies, 1987:285–297.

Vaughan A. The History and Future of Parks in the Northwest Territories: A Government of the Northwest Territories Perspective. In Canadian Arctic Resources Committee (ed), *National and Regional Interests in the North.* Ottawa: Canadian Arctic Resources Committee, 1984:319–335.

Wildlife Management Advisory Committee/North Slope (WMAC/NS). *Ivvavik National Park Management Plan.* Parks Canada, Ottawa: Canadian Heritage, 1994.

19

The Role of Science and Law in the Protection of National Park Resources

John Lemons
Kirk Junker

Since their inception, management of national parks has been controversial because of conflicting views regarding how to preserve their resources while still allowing for their use and enjoyment. A number of national park scholars have analyzed the historical, legal, public policy, scientific, and ethical dimensions of park management with a view toward resolving conflicts in a manner that allows for appropriate use while still providing protection of the resources. For example, Runte (1979) has shown how historical attitudes of both the public and park managers have influenced current park management. Sax (1980), Lemons and Stout (1984), and Keiter and Boyce (1991) have analyzed the legislative mandates governing management of parks in an attempt to clarify their management and policy implications. Leopold et al. (1963), Robbins (1963), the National Park Service (NPS, 1980a), Lemons (1986a), and Wright (1992) have assessed the status of science in national parks as a basis for management decisions. Finally, the value-laden and ethical dimensions of national park management also have been analyzed (Sax, 1980; Lemons, 1987a; Rolston, 1988; Grumbine, 1994). Perhaps predictably, almost all studies have recommended significant improvements in the application of the disciplinary research to the resolution of management problems.

With respect to the application of science in the protection of park resources, scientists have attempted to assess the biological and ecological status of resources and have made recommendations to improve scientific capabilities in order to enhance their protection. In part, improvements in scientific capabilities are assumed to enhance protection of park resources because they will provide additional and more accurate information to managers for use in decision making. In addition, improvements in science will be of benefit when the NPS is called on to justify decisions based on science because, as the agency responsible for rule-making, it has the burden of proof for showing environmental impact or risk from existing or proposed activities. This burden of proof requires that the NPS demonstrate with reasonable certainty that there is a scientific need for protection of resources and that recommended solutions to problems have a reasonable chance of success.

Clearly, improved scientific capabilities are important to the understanding of park resources and problems confronting them. Nevertheless, we suggest that further recommendations for more scientific information about park resources must be supplemented by a critical examination of the role of science in NPS decision making and an assessment of how burdens of proof function and who should have the burden of proof to demonstrate potential harm to resources. That is, should those who seek to protect park resources from harm from human activities or projects be required to demonstrate the potential for harm with reasonable certainty, or should those whose actions might potentially harm the resources be required to demonstrate that their actions will not cause harm?

We raise these questions because we believe they are understudied and bear directly on the protection of park resources for two reasons, which we will discuss subsequently. The first reason, the burden for demonstrating environmental impact or risk from existing or proposed activities, stems from two powerful prescriptions. The first prescription is based on standard scientific norms governing acceptability of scientific evidence, namely, to accept conclusions at a 95 percent confidence level and minimize type-I error, which leads to the acceptance of false positive results. In natural resource problems, a type-I error would be to conclude that harm to resources will result from existing or proposed human activities when, in fact, no harm will result. The second prescription is based on rules of evidence and procedure in legal proceedings and agency administrative decision making that generally place the burden of proof on the moving party (i.e., the NPS) and thereby force the NPS to support its requirements with scientific certainty. While these two types of prescriptions might reduce speculative conclusions, they are inconsistent with a precautionary approach regarding the protection of natural resources. A precautionary approach would be to minimize type-II error, which errs on the side of a desire not to impose unnecessary costs of environmental protection or restrictions on resource use. The second reason these questions affect park resources is that pervasive uncertainty confounds scientists' understanding of parks' natural resources. This uncertainty makes it extremely difficult to fulfill the above-mentioned scientific and legal prescriptions. In our opinion, most discussions about how to enhance protection of park resources have not focused on these issues.

In the remainder of this chapter we discuss the roles of science and law as they relate to the ability to fulfill burden-of-proof requirements to demonstrate harm to park resources from human activities. Specifically, we discuss: (1) the ambiguity of national park legislation, (2) the interface between scientific capabilities and law, (3) alternative roles for the use of science in national park management, (4) the meaning of prescriptive national park legislation, and (5) how prescriptive national park legislation fails to protect resources given scientific uncertainty about the resources.

Ambiguity in Meaning of National Park Legislation

Legally speaking, NPS legislative mandates to protect parks and to ensure that their use is consistent with the parks' fundamental purpose must be used as the basis to resolve park problems. Statutes enacted by Congress must be taken by the NPS as expressing public desires and representing the national interest (Mantell, 1979; Lemons & Stout, 1984).

The most relevant legislation that guides NPS policy is the National Park Service Organic Act (Ch. 408, 39 Stat. 535 [1916], as amended, 16 U.S.C. §§ 1, 1(a)-1, 3, and 20). (The Act of October 31, 1983, P.L. 98–141, § 1, 97 Stat. 909 provides that this act currently is known as the Public Lands and National Parks Act of 1983.) Section 1 states that the purpose of parks is ". . . to conserve the scenery and the natural and historic objects and the wildlife therein and to provide for the enjoyment of the same in such manner and by such means as will leave them unimpaired for the enjoyment of future generations." Section 1(a)-1 reaffirms the fundamental purpose of Section 1 and provides that NPS decisions are to be consistent with this fundamental purpose. Section 3 authorizes the Secretary of the Interior to undertake certain specific actions to protect park resources, including the creation of penal sanctions for violations of rules and regulations promulgated by the Secretary governing the use and management of parks. Section 20 is a statement of congressional findings and purpose regarding park concessions, accommodations, facilities, and services. This section provides for the establishment of concession facilities and visitor services provided they are consistent with the fundamental purpose of national parks as described in Section 1. Importantly, Congress stated in Section 20 that the "... preservation of park values requires that such public accommodations, facilities and services as have to be provided within these areas should be provided only under carefully controlled safeguards against unregulated and indiscriminate use, so that the heavy visitation will not unduly impair these values and so that development of such facilities can best be limited to locations where the least damage to park values will be caused." All of the aforementioned sections appear to mandate that park administrators make decisions that are consistent with the legislative language contained therein. In fact, the National Park Service Organic Act explicitly directs the Department of the Interior, through the NPS, to promote and regulate the use of parks by such manner and such measures as conform to the fundamental purpose of said parks (16 U.S.C. §§ 1, 2–4).

However, despite this apparent mandate, the exact meaning of the National Park Service Organic Act as a basis for decision making about park resources is questionable. The traditional interpretation of its mandate is that parks are to be managed for two conflicting purposes: (1) maintenance of park resources in an unimpaired condition and (2) provision for their use and enjoyment by the public. The NPS has not been entirely explicit or consistent on how it interprets these conflicting purposes. At times it has stated that park resources should be managed (1) according to a strict policy of preservation (NPS, 1980a), (2) to balance equally the goals of preservation and use (NPS, 1978; Mantell, 1979), and (3) to maximize use and enjoyment of parks (Shabecoff, 1981). The National Park Service Organic Act also is ambiguous because it uses terms that it does not define, such as "natural" and "unimpaired" in Sections 1 and 1(a)-1, and "park values" in Section 20. These words and terms are normative because they serve as promulgated standards for park resources. That is, according to the National Park Service Organic Act, park resources should be "natural," "unimpaired," and be managed to promote "park values." Despite the normative nature of the words and terms, the National Park Service Organic Act does not define their meaning or provide guidelines that can be used to indicate whether park resources conform to these standards. This leaves the terms open to interpretation in each use and subject to politics and lobbying. Unfortunately, there are few judicial interpretations of the National Park Service Organic Act to indicate the extent to which parks should be managed for preservation versus visitor use and enjoyment (Lemons & Stout, 1984). Consequently, NPS administrators have considerable discretion in how they balance protection and use of park resources. Nevertheless, despite this discretion they still must be able to demonstrate that there is a reasonable basis for their decisions and that they are consistent with the legislative mandates regarding protection of park resources.

Complicating the implementation of NPS legislation and its efficacy is the fact that many threats to park resources originate from sources external to their boundaries (Keiter & Boyce, 1991). Examples of external threats include transboundary air and water pollution and conflicts with other agencies or private landholders regarding management of wildlife that cross park boundaries. Because external threats to park resources arise in jurisdictions outside of parks, with few exceptions they are not regulated by legislation designed to promote the protection of park resources. Consequently, NPS administrators have little influence in determining whether or to what extent such threats are mitigated in order to protect park resources.

Interface of Science and National Park Legislation

In attempting to mitigate the numerous types of threats to park resources and to provide for both their use and protection, decision makers use scientific information as one of the bases for assessing whether their decisions are consistent with the fundamental mandates of NPS legislation—namely, that resources will be preserved in their natural state, that they will be unimpaired, and that the park values will be preserved. Despite the use of science as a basis for decisions that conform with the fundamental purpose of parks, both theoretical and informational uncertainty limits the ability of decision makers to make such decisions.

One limitation stems from the number and types of threats to park resources. It is a common perception that park resources are in danger of change from threats originating both internally and externally to park boundaries (NPS, 1980a; Lemons, 1986a, 1987a; Keiter & Boyce, 1991; Wright, 1992). Over 70 different kinds of threats in the categories of aesthetics, air pollution, physical removal of resources, encroachment by exotic species, visitor physical impacts, water quality pollution and water quantity changes, and park operations and planning of facilities have been identified, and over 2000 specific internal threats and over 2400 external threats have been identified for biological, physical, aesthetic, cultural, and park operations resources. The sheer number and types of threats have overwhelmed the organizational capabilities of the NPS to deal with them effectively; they also have created demands for the application of methods and techniques of the ecological and environmental sciences for problem solving that are not likely to be met anytime soon. Consequently, a consensus exists that a lack of adequate understanding and documentation of most threats to park resources constrains the ability of park managers to make more informed decisions.

Another limitation stems from the inability of science to measure and monitor park resources in a manner that will enable decision makers to know with reasonable certainty whether their actions are consistent with the meanings of the normative language contained in the National Park Service Organic Act. For example, the term "natural" often is used in a sense that implies freedom from human influence, especially that of modern humans (Devall & Sessions, 1984; Smith & Theberge, 1986). Alternatively, some people argue that modern human use and influence should not necessarily be excluded from natural areas if such use is compatible and harmonious with the ecosystem (Callicott, 1991). Both concepts of "natural" have been used by NPS managers. For instance, the primary goal for many of the Alaskan parks, Isle Royal National Park, and the large wilderness areas of other parks is to manage them in a condition relatively free from human impact. On the other hand, management of some parks (e.g., Acadia National Park) and some areas within parks (e.g., Yosemite Valley) permits human alteration of the landscape (Lemons, 1987a).

Confusion also exists with respect to the understanding of the word "natural," even when it is being used in a sense that implies freedom from human influence, especially that of modern humans. For example, as a guide to management decisions for park ecosystems, Parsons et al. (1986) define "natural" as ". . . the unimpeded interaction of native ecosystem processes and structural elements." Alternatively, Bonnicksen and Stone (1985) define a natural ecosystem as one that ". . . portrays, to the extent feasible, either the same scene that was observed by the first European visitor to the area or the scene that would have existed today, or at some time in the future, if European settlers had not interfered with natural processes."

In addition to these problems, the ability of the ecological sciences to specify a meaning for "natural" also has been questioned for several other reasons. First, "natural" often is defined as an existing condition prior to human perturbation of ecosystems. Jorling (1976) and Shrader-Frechette and McCoy (1993) argue that this definition ignores the fact that humans are a part of nature and therefore need to be included in the definition of "natural." Second, given the historical and current impacts of humans on ecosystems, there are probably few or no ecosystems in existence that have not been or are not affected by humans. Third, it is difficult or impossible to know with reasonable certainty what ecological conditions existed in ecosystems prior to human influence. Fourth, given global consequences of modern humans it is improbable to maintain ecosystems unperturbed by humans. Consequently, ecology cannot provide an unambiguous or noncontroversial definition of "natural." Ambiguity about the meaning of "natural" has permitted wide oscillations in NPS policy regarding the acceptability of phenomena such as nonhuman-caused fires, floods, and fluctuations of animal populations as well as of human intrusions in park ecosystems.

Likewise, although the legislation states that park resources should be maintained in an "unimpaired" condition, there are no guidelines or criteria that have been promulgated that define what constitutes "unimpaired" or how this condition should be assessed. Schaeffer et al. (1988) and Cairns et al. (1992) have proposed ecosystem structural and functional attributes that theoretically can serve as indicators of the conditions of the biological and ecological resources. Examples of attributes at the ecosystem level include (1) habitat for desired diversity and reproduction of desired organisms, (2) phenotypic and genotypic diversity among organisms, (3) relationships between food chains and desired biota, (4) relationships between nutrient pools and cycling and desired organisms, (5) relationships between energy flux and maintenance of trophic structure, (6) homeostatic mechanisms and damping of undesirable oscillations, and (7) the capacity to decompose, transfer, chelate, or bind anthropogenic chemicals and radionuclides. Parameters necessary to assess these ecosystem attributes include (1) individual measures such as fitness, disease, mutation, reproduction, physiology, acclimation, and individual behavior; (2) measures for populations such as intraspecific behaviors, epidemiology, genotypic variation, phenotypic variation, reproduction, physiology, and adaptation; (3) system-level factors such as interspecific behavior, decomposition, production, recovery, resilience, resistance, connectivity, indicator species, keystone species, successional patterns, guild theory, species, diversity, and vegetative diversity; and (4) abiotic elements such as nutrient and mineral retention, leaching, physiography, structural diversity, and chemical composition.

Kay (1992) and Karr (1992) have proposed that the following be used by decision makers to determine whether ecosystems were unimpaired: (1) the ability of ecosystems to maintain optimum operations under conditions as free as possible from human intervention; (2) the ability of ecosystems to cope with changes in environmental conditions (i.e., stress); and (3) the ability of ecosystems to continue the process of self-organization on an ongoing basis; i.e., the ability to continue to evolve, develop, and proceed with the birth, death, and renewal cycle. By "optimum conditions" presumably it is meant the situation in which the external environmental fluctuations that tend to disorganize ecosystems by making them less effective at dissipating solar energy and the organizing thermodynamic forces that make ecosystems more effective at dissipating solar energy are balanced. By this line of reasoning, activities that interfere with these three abilities of ecosystems would be said to be impaired.

Ecologists disagree about the utility of the types of indicators used in assessing environmental impacts or so-called ecosystem health of parks, and they also disagree about whether structural or functional attributes should be emphasized. Further, there is no a priori way to determine the criteria that should be used to assess whether parks' biological or ecological resources are impaired. The selection of one criteria over another may reflect management goals to monitor, assess, or protect particular resources, or it may reflect a manager's values concerning the importance of respective biological or ecological attributes.

Despite the fact that legislation refers to the preservation of "park values," nowhere does it define the values or indicate how decisions regarding conflicts between values are to be resolved. Such definitions and decisions are left up to NPS decision makers. Rolston (1988) has provided the most comprehensive list and analysis of values often attributed to park resources: (1) market or commercial values, (2) life support values, (3) recreational values, (4) scientific values, (5) genetic and biodiversity values, (6) aesthetic values, (7) cultural symbolization values, (8) historical values, (9) character-building values, (10) therapeutic values, (11) sacramental values, (12) aspirational or option values, and (13) intrinsic values. Generally speaking, market or commercial values often conflict with most of the other values. However, conflicts between the various nonmarket values also can exist, such as when recreational values conflict with aesthetic or intrinsic values. Although it is not the role of science to determine what values should have precedence over others, scientific information can be used to provide information on whether certain values are being threatened by activities.

Alternative Concepts of the Use of Science in National Park Management

Numerous studies focusing on the role of science in protecting park resources have provided specific examples and analyses of the contributions of scientific knowledge to the understanding of park resources as well as in identifying areas where more scientific knowledge is needed for informed management decisions (Leopold et al., 1963; Robbins, 1963; NPS, 1980a; Lemons, 1986a; Keiter & Boyce, 1991; Wright, 1992). In these studies, few examples of specific resources problems were provided in which scientific information was conclusive. Consequently, the studies recommended that scientific research in national parks be improved. Generally speaking, the studies have analyzed the types and extent of problems requiring scientific information for their solution, levels of financial and administrative support for scientific research, organizational structures to support scientific research, the quality and quantity of scientific researchers, and NPS attitudes and philosophies regarding levels of support for scientific research.

Some researchers recommend that the use of science in national park decision making be based on classical scientific methods and techniques that they believe will yield robust scientific knowledge and minimize speculation (Murphy, 1990; Drew, 1994). Peters (1991) recommends that the primary way to improve the utility of the ecological sciences in decision making is to judge every ecological theory on the basis of its ability to predict. These recommendations are based on assumptions that (1) there should be a strong emphasis on quantitative data acquisition, (2) more and better data will solve problems, and (3) the scientific method as commonly understood has great ability to discern facts about the natural world. Consequently, the necessary goal of science in protection of natural resources is said to include (1) formulating hypotheses and conducting observations to test them, (2) acquiring quantitative data, (3) developing an understanding of processes and linkages among variables, and (4) developing reasonably certain predictions (Bella et al., 1994). According to this line of reasoning, scientific methods and techniques should be based ideally on hypothesis-deduction methods or other methods yielding "strong inferences" in order to increase the likelihood of accurate results and predictions. Further, when interpreting the results of ecological or environmental studies, many researchers maintain that type-I rather than type-II error be minimized because it is the most conservative course of action in situations of uncertainty and because minimization of type-I error conforms to traditional scientific practices that seek to reduce the chances of accepting false positive results (Simberloff, 1987). This means that scientists would apply tests of statistical significance that reject an experiment's or study's results if the probability of their being due to chance is greater than, for example, five percent. By minimizing type-I error, it is said that science can play a central role in informing debates about problems of protecting park resources because of the likelihood that decisions based on speculative thinking will be reduced (Miller, 1993).

Of course, not all ecological studies involve the use of hypothesis-deduction methods; other forms of scientific explanation are inductive and do not test for inferences. Shrader-Frechette and McCoy (1993) have critically analyzed numerous methods and techniques of ecology, including hypothesis-deduction methods but also various other approaches used in site-specific studies, statistical models, mechanistic models, comparative studies, theoretical modeling, historical studies, and case studies. Their conclusion is that the methods and techniques of ecology are of limited use in providing decision makers with descriptive scientific conclusions for conservation issues because they have not yielded sufficient information for the development of general theories capable of serving as a basis for precise or reasonably certain predictions useful for conservation decisions. All of the methods and techniques are limited by both informational and theoretical uncertainty, and they all involve value-laden judgments, assumptions, inferences, and evaluations that serve to increase uncertainty. Consequently, Shrader-Frechette and McCoy (1993) conclude that ecology is more of a science of case studies than a science of generalizable laws and that its primary value is with its heuristic rather than predictive capabilities. Their conclusions are consistent with those of other researchers who have critically analyzed the extent to which the methods and techniques of science are capable of yielding reasonably certain information appropriate to serve as a basis for management and public policy decisions.

For example, Sagoff (1988) argues that the role of ecology should be to identify ecological indicators that might allow scientists to diagnose perturbations in species or ecosystems early enough so that mitigation measures could be implemented. This type of diagnosis does not depend on knowing generalizable laws and basing predictions on them; rather, it involves the integration of diverse information to make a general argument for one rather than another interpretation of the causes or consequences of ecological impacts. Recently, Bella et al. (1994) have argued that the role of the ecological sciences in problems of environmental change ought to be in the identification of indicators of ecological change rather than in the prediction of the consequences of human activities with reasonable certainty. Lemons (1986b) has analyzed the different meanings of "stress" as applied to organisms, populations, species, and ecosystems and has concluded that the theoretical differences among the different meanings are so great that when combined with informational uncertainty concerning the assessment and evaluation of the causes of stress and their effects, little basis for reasonably certain predictions exists. Lubchenco et al. (1991) have identified numerous scientific uncertainties regarding protection of biodiversity and have proposed a research agenda to obtain more information about it. In their report, they acknowledge the limited role scientists can play in making reasonably accurate predictions about the effects of human interventions in ecological systems. Cairns et al. (1992) note that in theory both structural and functional attributes of ecosystems can be used as a basis for ecological predictions but that practically speaking there is a significant lack of knowledge about them. Finally, the Committee for the National Institute for the Environment has developed a comprehensive proposal to reform environmental research in the United States in order to provide more suitable information for decision making (Committee for the National Institute for the Environment, 1994).

A number of researchers have recommended that the role of science in natural resources problems such as national park management should reflect the limited predictive capabilities of ecology (Shrader-Frechette & McCoy, 1993;

Miller, 1993). Consequently, so-called holistic or "post-normal" scientific approaches have been proposed for use in natural resources decision making (Lemons, 1995). Proponents of post-normal approaches reject the belief that (1) problem solving should reflect an emphasis on data acquisition per se, (2) more data will necessarily solve problems, and (3) the scientific method is objective and value neutral. Consequently, the post-normal science approach emphasizes (1) adequate formulation of problems so that data will contribute to public policy goals; (2) that most results from scientific studies will not yield reasonably certain predictions about future consequences of human activities and that many problems of protecting park resources therefore should be considered to be "trans-science" problems requiring research directed toward useful indicators of change rather than precise predictions; and (3) the need to evaluate and interpret the logical assumptions underlying the empirical beliefs of scientists with a view toward ascertaining more fully the validity of scientific claims and their implications. While post-normal science is not easy to characterize, it seeks a broad and integrated view of problems and places more emphasis on professional judgment and intuition and is less bound by analytically derived empirical facts. Proponents of the application of post-normal science to protection of park resources maintain that its claims are more amenable for practical public-policy purposes than the claims of predictive science approaches (Miller, 1993; Bella et al., 1994). Post-normal approaches are based more on "retroduction" and conceptual analysis than are hypothesis-deduction methods, and by necessity they emphasize explanation and heuristic understanding of the complexities of nature rather than predictions. One result of basing decision making on post-normal science is that type-II error would be minimized.

Clearly, there is a need to improve the capabilities of science and its use in order to enhance the protection of park resources. Although there is a debate about the role science should play in decision making, because there are different types of problems (e.g., ranging from determining the effects of air pollution on trees to determining the role exotic species play in altering park ecosystems) that need to be understood and resolved, there is no singular scientific approach or methodology that can be applied to their understanding or resolution. However, in our opinion, efforts to enhance the protection of park resources should not be based on the call to improve scientific capabilities solely or even primarily. Instead, we suggest in the following sections that the enhanced protection of resources will require assessing the prescriptive mandates of NPS legislation and the implications of pervasive scientific uncertainty in fulfilling burden-of-proof requirements to demonstrate harm to park resources. Because of the inherent difficulty of meeting burden-of-proof requirements with less-than-certain scientific evidence, we also propose that the enhanced protection of park resources likely will require a more precautionary legislative approach that allows protection of resources when there is a reasonable basis for concern but where scientific conclusions are uncertain about the consequences of human activities.

General Meaning of Prescriptive Legislation

In order to provide for the management of national parks, Congress delegated legal authority to NPS managers who use their judgment, informed by scientific and legal experts, to decide such questions as how many people shall be allowed to visit parks, where in the parks visitors shall be allowed to go, when parks may be visited, what types of uses will be allowed of parks, who may conduct business in parks, what facilities may be built in parks, and whether and how attempts to restore park resources will be made. On a more limited basis, park managers also participate in the sharing of information or decision making about external threats to parks and possible measures to use in their mitigation (NPS, 1980a; Keiter & Boyce, 1991). In some cases, decisions about these types of questions are made so as to determine a case-specific outcome, such as whether to control alien mountain goats in Olympic National Park. More often, decisions are made within the context of broader public policies, which results in a more general outcome not specific to any particular case; an example is the development of the General Management Plan for Yosemite National Park (NPS, 1980b). Theoretically, NPS statutes prescribe policies in order to guide decisions in specific cases. However, many of the legislative statutes can be read at times to conflict and they fail to define terms and therefore can be considered to be sufficiently ambiguous so as to allow wide decision-making discretion to park managers.

The nature of prescriptive legislation is such that for a party to be able to sue to enforce the legislative mandate, that party must prove that a manager has failed to carry out the prescribed action. In practice, this is usually accomplished by the suing party by proving that acts occurred, which by their very nature, contradict or contravene the prescription. That is to say, evidence of the failure to conduct prescribed activity consists of positive acts that are contrary to the prescription. Evidence does not typically consist of some failure to act altogether or some nonact. This effectively places the burden of proof on those wishing to curtail acts where prescriptive legislation operates.

In the case of the National Park Service Organic Act, this burdensome burden of proof is made even more difficult because acts that a party may try to use to demonstrate a manager's failure to carry out the prescriptive mandate of "use" may in defense be characterized as "preservation," and thereby comply with the prescription, and those acts demonstrated to fail as "preservation," may in defense be characterized as "use," and thereby comply with the prescription. Consequently, a party must prove that certain undesired acts by a manager are neither preservation or use before that party can prevail in litigation relying on the ostensibly prescriptive mandate of the National Park Service Organic Act. We wish here to examine this problem more closely.

What is it to say that NPS laws are prescriptive? Typically, prescription is understood to mean that certain acts or behaviors are being recommended, suggested, or required. If behaviors or acts are prescribed in a law, then they are understood to be required under the threat of sanction. In this sense, prescription is understood dialectically in that it is opposed to proscription, which is to suggest against or to forbid certain behavior or acts. Section 1 of the National Park Service Organic Act (16 U.S.C. §§ 1) is ostensibly prescriptive insofar as it states that ". . . the [National Park] service thus established shall promote and regulate the use of Federal areas known as national parks . . . by such means and measures as conform to the fundamental purpose of the said parks, . . . which purpose is to conserve the scenery and the natural and historic objects and the wildlife therein and to provide for the enjoyment of the same in such manner and by such means as will leave them unimpaired for the enjoyment of future generations." The important question that this prescription raises is just what is being prescribed, use or preservation? As mentioned previously, for a variety of reasons, this question has not been answered by the courts in interpreting the apparent conflicting prescriptions of the statutes (Sax, 1980; Lemons & Stout, 1984), nor has it been definitively answered by comparing this statute with other statutes (Lemons, 1987b), nor by arguments from reason, authority, or emotion outside the statute's language (Mantell, 1979; Lemons, 1987a).

If one simply looks to the text of the National Park Service Organic Act for direction on how to administer the national parks, one finds simple and explicit direction to promote the use and preservation of the parks. But, because this apparently simple and explicit prescription can function in practice to support contradictory goals (use and preservation), it functions as no helpful prescription at all. Instead, administrators find that they must look outside the explicit (but contradictory) prescription of the National Park Service Organic Act to other implicit prescriptions if they are to develop a lasting and consistent program that can do more than act as the "king's seal" on whatever program is proposed, depending on the politics and lobbying at hand.

Therefore, we suggest that if the National Park Service Organic Act is actually prescriptive, its powerful prescriptions stem from factors other than the explicit, but confusing, prescriptions of the statute. The foremost among these external and yet implicit prescriptions is the placement of the burden of proof. The burden of proof is placed implicitly on those seeking to enhance protection of park resources both in the mechanics of the operations of law and science. This burden of proof effectively stifles such protection efforts. Once the burden of proof is set, like placing the magician's rabbit in the hat, the outcome of the deliberative process over use and preservation is also unwittingly set. But before explaining why this may be so, we will examine the way in which the explicit language of this ostensibly prescriptive statute operates in practice. We do this because at the practice level we find that the statute functions proscriptively.

Explicit Prescriptions of the Organic Act Fail

When is law proscriptive? When a statute explicitly forbids activities, it functions proscriptively. Decision makers who wish to act legally take the statutes as a prohibition of explicitly identified activities and assume that other activities are either prescribed or allowed unless they are known to constitute an abuse of discretionary power.

Unlike NPS legislation that ostensibly is prescriptive, most statutory environmental law is proscriptive and therefore influences the consideration of the legislation in legal forums. For example, when a statute such as the National Park Service Organic Act explicitly creates a prescriptive mandate for a park manager, such as when it requires that ". . . the service shall . . . ," a failure to act in such a way as to carry out the statute's mandate is far more difficult for citizens to prove and therefore to sanction through the courts than is proving an act that is contrary to the prescription, especially under conditions of scientific uncertainty. In effect, one can attempt to prove that a park manager has failed to act per his or her mandate by demonstrating what he or she has done instead, which does not fulfill the mandate. Therefore, the burden of proof becomes one of proving a contradictory act rather than proving a nonact. Even when statutes appear to be prescriptive, they operate proscriptively when an individual's actions are tested against the prescriptions; it is rare that evidence is presented of an individual's inaction in the face of a statutorily mandated action. In summary, one proves that an administrator has failed to carry out a legislative mandate by presenting a tribunal with evidence of the administrator's acts that are contrary to the mandate, not by presenting evidence of nonacts in the face of a mandate. As a consequence, most laws are not written prescriptively but proscriptively precisely because of, in part, the difficulty involved in proving a nonact. Because of this difficulty, the National Park Service

Organic Act functions in a proscriptive manner even though its explicit language ostensibly is prescriptive.

In the context of public policy, the fact that the National Park Service Organic Act functions proscriptively in practice has implications for the role of science in decision making. Given the ambiguous and potentially conflicting interpretation of NPS legislation, how may we say that science informs park decision makers? When scientific information is reasonably certain (if ever), ideally we would expect park managers to apply that information to park problems and to generate a definitive and "correct" solution to the problems. Practically speaking, this expectation is not met often because science is seldom capable of providing reasonably certain information or descriptions to immediate problems. In theory, waiting for more scientific information (data) may increase the level of certainty (but it may not). However, while waiting for more data, the status quo (i.e., the immediate problem) continues. Waiting for more data (called more certainty) is a decision to allow current problematic or unacceptable threats to park resources to continue. Decision makers would like to avoid making a decision without sufficient data so as to avoid the appearance of being wrong or subjective. However, waiting for more data is tantamount to an assumption that deteriorating resource conditions will go on hold until the next quanta of data arrive. In part, few activities have been explicitly identified either by the courts or by decision makers as being prohibited because of the pervasive scientific uncertainty that exists about park resources and the conflicting interpretations of NPS statutes. It is in this sense that the statutes can be said to be proscriptive at the practice level, because it is only those few activities informed by scientific certainty that are capable of being prohibited; remaining activities are assumed to be either prescribed or allowed.

Because, in practice, the statutes function in a proscriptive manner, scientific uncertainty makes it extremely difficult for decision makers to demonstrate harm to park resources and therefore to identify activities that should be prohibited or restricted in order to protect the resources more fully. In addition, pervasive scientific uncertainty creates certain tensions between the practices of science and regulatory goals and approaches. When scientists are concerned about the search for valid conclusions in normal scientific endeavors, statements about cause and effect relative to the consequences of human activities ideally are based on, for example, the standard norm to accept scientific findings at the 95 percent confidence level. However, government decision makers cannot wait until all desired scientific information is available prior to deciding on regulatory approaches. Unlike scientific investigations in which conclusions may be delayed until more scientific information is available, decision makers are expected to act in a timely manner, often on the basis of weak data and scientific theory. When decision makers promulgate regulations on this basis, those opposed to the regulations can criticize them on the grounds that they are making speculative conclusions.

Because decisions about park resources must be made in the face of pervasive scientific uncertainty, legal rules on the use of scientific evidence in court proceedings may determine when NPS law or other laws that might be used to protect park resources may be enforced or implemented. If rules of evidence restrict the use of scientific evidence in court proceedings to that which is highly certain, enforcing or implementing the law may be impossible in matters where certain scientific evidence is unavailable. Therefore, ruling on the use of scientific evidence in legal proceedings must be understood to be an important public policy choice about when law may be enforced or implemented.

Regardless of when during the accumulations of data a park manager makes a decision, the legal check on that decision is the same—the abuse-of-discretion scope of review. This review considers when a manager made his or her decision. If knowing a particular scientific fact or information can be shown to require a manager to decide differently than he or she did, the manager has not abused his or her discretion if the particular fact or information was only known or knowable after the discretionary decision was made. In other words, park managers are not responsible for knowing all that is knowable at any given time. They are held to the same standards as that which exists for judges in courts of general jurisdiction in that they need only to make decisions based on the facts or information presented or available.

The rules of the use of scientific evidence in legal proceedings differ depending on the type of proceeding (Brown, 1995). In actions for the recovery of damages, the U.S. Supreme Court in *Daubert v United States* (1993) announced the following four-pronged analysis to assist courts in determining whether evidence is relevant and reliable:

1. Scientific methods used by experts to derive an opinion must be capable of being tested and capable of being shown to be false.
2. Publication and peer-review of scientific methods used by experts to derive an opinion strengthen the admissibility of evidence but nonpublication does not impart inadmissibility.
3. Admissibility of evidence is strengthened by the use of methods that have a known and low error rate.
4. Admissibility and acceptance of expert opinion will be enhanced if the opinion is based on methods that have been accepted generally within the scientific community (the court did make clear that this analysis was not a definitive checklist, however).

Under *Daubert,* evidence that establishes a reasonable basis for concern about harm but does not conclusively establish causation is not admissible.

Laws dealing with environmental matters avoid many of the problems of admissibility in actions by giving the government the administrative power to (1) take legal action if it determines that an activity creates a threat to the environment, and (2) create standards and regulations that can be enforced (Brown, 1995). Through such a grant of power the government avoids the problem of having to show causation as an established fact. Instead, the government might only have to prove that an environmental threat might exist. Further, environmental harm might be presumed if the government can show a violation of the standards or regulations. Environmental law also can avoid some admissibility problems of uncertain scientific evidence by limiting judicial review of administrative actions to the record created by the administrative agency. In cases where judicial review is from the administrative record, the court does not call witnesses nor admit evidence but simply reviews the record of public comment about the proposed action prepared by the administrative agency (but the record can, and does, include public hearings and consultation with scientific experts). Because no witnesses or evidence is heard in such court proceedings, there are fewer problems of admissibility of scientific evidence. In record-review matters, expert administrators may apply expertise to draw conclusions from suspected but not completely substantiated evidence, relationships between facts, from trends among facts, from theoretical projections of imperfect data, and from probative preliminary data not yet certifiable as fact. However, for such conclusions to be upheld on court challenge, administrators must be able to demonstrate a step-by-step proof of cause and effect (*Ethyl Corporation v EPA,* 1976). In record-review matters, the agency's actions are afforded deference and must be upheld if they are based on relevant factors and are not a clear error of judgment. Consequently, in court review of an agency's administrative record, the burden of proof is on those challenging administrative decisions. If such decisions were made under conditions of scientific uncertainty, challengers may have great difficulty in meeting the burden of proof.

In addition to the problems of admissibility of uncertain scientific evidence described previously, other proscriptive manifestations of science, law, and business make it difficult to fulfill burden-of-proof arguments to enhance protection of park resources. In order to obtain more accurate and credible results, scientists often will define problems in a narrow manner by isolating and studying selected variables under controlled conditions. This approach leads to formulating environmental problems in particular ways, but it presents what almost is an intractable problem due to the fact that it attempts to understand complex systems by isolating a few variables so that they can be studied under narrowly controlled and simplified conditions. This approach becomes problematic because when dealing with more complex systems (such as park ecosystems), the understanding of the interactions among variables that determine the way in which individual variables express themselves is not able to be discerned. Consequently, because of the pervasive scientific uncertainty inherent in complex environmental problems, establishing cause-and-effect relationships between the activities and impacts at the 95 percent confidence level is precluded in many instances. Further, by simplifying and reducing parts of complex environmental problems to a more manageable scale, scientists often study a scientific problem that may be very different from the more complex environmental problem from which it stems. For example, in attempting to provide answers to problems of aerial spraying of pesticides in New Brunswick forests in Canada, researchers utilized strict scientific research guidelines and norms in order to attempt to discern whether there was a cause-and-effect relationship between pesticide exposure from spraying and human health effects. While theoretically this problem is amenable to scientific investigation, the complexities involved in understanding ecological phenomena as well as human responses to the pesticides precluded a definitive answer to the problem. More importantly, the research approach served to focus attention away from more fundamental questions concerning the misdirection and redesign of resource policy that focused on whether and under what conditions people should have been exposed to underdetermined risks of pesticide spraying without their consent (Miller, 1993).

The act of a scientist positing a hypothesis and deducing and testing principles therefrom or favoring minimization of type-I rather than type-II error is analogous to the setting of the burden of proof in law. So long as logically deduced principles do not contradict each other, a scientist's hypothesis is considered valid. The inherent and implicit power or right that a scientist has to posit a hypothesis is not a scientific process, but it is analogous to placing the burden of proof in law because it sets the stage for identifying and describing certain facts about nature but not others. Those who may wish to contest the validity of principles logically deduced from a tested hypothesis generally are assumed to have the burden to prove a competing hypothesis. Further, when scientists tend to favor minimization of type-I error as being the most conservative course of action under conditions of scientific uncertainty, they are establishing normative rules of conduct; namely, that an experiment's results cannot be assumed to be valid if the probability of their being due to chance is greater than five percent. With respect to determining whether harm exists to a particular park resource due to an activity, the effect of such a rule is to act as if it is better to minimize type-I error (rejecting a false null hypothesis that there is no harm to a park resource) than it is to increase the chances of concluding a finding of harm to a park resource when there is none. Such a rule increases the chance that a scientist will make credible scientific conclusions, but it also increases the chance of reaching a conclusion that there is no harm to park resources when, in fact, there is. The standard of proof that should be required of regulatory decisions is a public policy and ethical question, not a scientific one.

Shrader-Frechette (1994) analyzes the reasons why there should be an ethical preference for minimizing type-II error in preservation issues (thereby increasing the risk of type-I error) and why the burden of proof for demonstration of no adverse environmental harm from development or human activities should be placed on those calling for such development or activities. She bases her conclusion on the following:

1. Minimizing the chance of not rejecting false null hypotheses with important public policy consequences is reasonable on the grounds of protecting the present and the future public.
2. The proponents of development or activities that potentially threaten environmental resources typically receive more benefits from the development or activities than do members of the public and, consequently, minimizing type-II error would result in a more equal distribution of benefits and risks.
3. Natural resources typically need more risk protection than do promoters of development or human activities because the advocates for protection usually have fewer financial and scientific resources than developers or promoters of activities that potentially can harm the resources.
4. The public ought to have rights to protection against decisions that could impose incompensable damages to natural resources.
5. Public sovereignty justifies letting the public decide the fate of development and human activities that potentially threaten natural resources.
6. Minimizing type-II error would allow enhanced protection of nonhuman species that typically receive inadequate consideration in decision making based on cost-benefit methods.

In the United States, the notion of stating the law, stating environmental problems for public policy purposes, and assigning the burden of proof under the law places great emphasis on personal liberties and individual freedoms. This emphasis is consistent with most statutes because individuals and corporations are permitted and encouraged to proceed with whatever activities they wish unless or until someone else proves that such activities are a public harm. When such harm is proven, one response is for Congress to enact legislation to mitigate against it; a second is for a manager to take mitigation actions under his or her agency's statutes; and a third is litigation under existing statutes. However, as we have seen, the demonstration of harm with reasonable certainty is difficult under conditions of scientific uncertainty. Moreover, the scientific framing of natural resource problems as established by scientists determines the possible field of responses from which Congress or an agency such as the NPS has to choose when making decisions about activities that might cause harm to the resources.

Business also has adopted a mantle of individual liberty and generally asserts the right to pursue its interests until market forces or the demonstration of public harm from its activities curtail them. This assertion of the right to act also functions to place the burden of proof. Here it is placed on those who would curtail business activity, as when the NPS seeks to limit logging on external lands or on park concessions.

For example, in order to create an adequate business environment in parks to provide for visitor services and accommodations, Congress supplemented the statement of national park purpose (i.e., the National Park Service Organic Act, 16 U.S.C. § 1) with the National Park System Concessions Policy Act (The Act of October 9, 1965, P.L. 89–249, 79 Stat. 969, 16 U.S.C. § 20). In enacting this statute, Congress intended to establish a financial policy for concessions and to provide the proper atmosphere for private investment to meet the demands of increased park use (Aspinall, 1965). Section 20(a) states that the interest of the NPS is to promote private as opposed to public concessioners. Section 20(b) guarantees the concessioner a property right in the form of a possessory interest and a chance to realize controlled profits. Sections 20(c) and 20(d) grant monopoly rights and preferential rights for renewal of contracts to existing concessioners. Park concessioners argue that these sections were intended to encourage development and use of concession facilities (U.S. Congress, 1976). Their rationale is that the operative sections of law, such as 16 U.S.C. §§ 20(c) and 20(d), are needed only to protect large investments, which implies that substantial commercial developments are permitted. Concessioners also argue that the guarantee of a possessory interest mandates continuation of existing facilities in parks, as substantial (and perhaps unavailable) compensatory funds would be required to terminate a concession contract. In practice, these provisions establish a burden of proof on those who might wish to demonstrate harm to park resources resulting from activities or facilities of park concessions (Lemons, 1987b).

In typical environmental litigation, the government acts as an advocate of the public's interests in opposing private interests that benefit the private operator but harm the public. This situation is different with respect to national parks, however. The government is not the clear public advocate opposing private interests. Instead, the government must act as a manager of varying interests. What complicates this situation is the fact that these interests do not always identify themselves as competing, because it is the park manager who makes them competitors by his or her interpretation of the purpose of the enterprise being considered (i.e., whether it is use or preservation).

Because national parks are public lands, the notion of exploiting the land for private gains determined primarily by market forces is mostly inoperative. Instead, park concessions are permitted to conduct business by the discretion of decision makers acting under the mandates of NPS legislation. Because no concession would admit to operating with an openly hostile attitude toward protection of park resources, probusiness antienvironment positions do not have much of an identifiable voice in park management decisions. Consequently, the managers must make decisions to carry forward the mandate of NPS statutes without having parties who clearly identify themselves as having opposing interests. This situation, combined with the fact that most park decision makers have adopted a middle-of-the-road approach so that decisions reflect, in their view, a balance between protection and use of resources, means that the more politically difficult decisions to enhance protection of resources have been avoided (Lemons & Stout, 1984; Lemons, 1987a).

In effect, our analysis suggests that three factors work in tandem to constrain the enhanced protection of park resources: (1) scientific uncertainty about park resources is pervasive; (2) although the National Park Service Organic Act ostensibly is prescriptive, it functions in a proscriptive manner; and (3) traditional practices and beliefs in law, science, and business place on those seeking enhanced protection of park resources a burden of proof that is not likely to be met in many instances.

Conclusions

Many people believe that national park resources are being threatened by a variety of activities. Some scholars of national parks have examined national park legislation with a view toward seeking solutions to problems, while others have made recommendations to improve scientific capabilities in order to provide more information to decision makers. While these are important recommendations, we believe that they need to be supplemented by (1) an assessment of the difficulties posed by scientific uncertainty in fulfilling burden-of-proof requirements to demonstrate harm to park resources and (2) a reassignment of the burden to those who seek to conduct activities that might cause harm so that they might prove their activities will not do so.

Various types of burden-of-proof requirements are entrenched in prevailing law, methods of science, and business. Given the ambiguities surrounding the meaning of NPS statutes, pervasive scientific uncertainty and science's emphasis on minimizing type-I error, and an emphasis on beliefs and practices that assert business and individual liberties until the demonstration of harm, it seems that those promoting the enhanced protection of park resources will continue to have difficulty in meeting burden-of-proof requirements imposed by law, science, and business, whether formal or informal. While recommendations to improve scientific capabilities to enhance protection of park resources have merit, simply calling for more support for research or for more data will not eliminate uncertainty about the effects of human activities on the resources, because uncertainty is inherent in the nature of scientific practices and, in particular, in the field of ecology. Consequently, legal burden-of-proof requirements are not likely to be met by those seeking to demonstrate harm to park resources. Allowing uncertainty to delay park management decisions is to make a tacit decision to allow and thereby promote the status quo; therefore, no decision is in fact still a decision.

In our opinion, the question of how uncertainty regarding the protection of park resources translates into decision making should be addressed through legislative mandates. One suggestion would be to establish legislation that clearly adopts precautionary principles that will better deal with uncertainty. In other words, the burden of proof could be reassigned whereby certain activities in parks that might cause harm would not be allowed to continue or go forward until those advocating it could prove, with the same degree of scientific certainty now required of those who would seek curtailment, that the activities will not cause harm. Extrapolated to law and business and focused on the national parks, the adoption of the precautionary principle in interpreting the National Park Service Organic Act would require that rather than allow an agent to go forward with a legal, science, or business position that may potentially harm a park resource until opponents can scientifically prove harm by the necessary quantified reasonable degree of scientific certainty, the burden of proof would be placed on the agent, using the same standard to form a clear position regarding NPS statutes. The recommendation to adopt such a precautionary principle and shift the burden of proof to those undertaking activities that might cause harm to park resources (i.e., to prove that the activities will not cause harm) is consistent with precautionary principles contained in Agenda 21 and adopted by governments of the world at the 1992 Earth Summit Conference (Johnson, 1993; Junker, 1994). It also is consistent with conclusions that it is better to minimize type-II error in conservation decisions (Shrader-Frechette & McCoy, 1993; Lemons & Brown, 1995; Lemons, 1996). In other words, from an ethical perspective, it is more prudent to accept the higher risk of an erroneous conclusion that activities will cause harm than it is to accept the lower risk of a false null hypothesis that no harm will result from activities that potentially threaten some of our nation's most important natural resources—our national parks.

Literature Cited

Aspinall L. *111 Congressional Board* 23,636. 1965.

Bella DA, Jacobs R, Hiram L. Ecological indicators of global climate change: A research framework. *Environ Manage* 1994;18:489–500.

Bonnicksen TM, Stone EC. Restoring naturalness to national parks. *Environ Manage* 1985;9:479–486.

Brown DA. The Role of Law in Sustainable Development and Environmental Protection Decisionmaking. In J Lemons, DA Brown (eds), *Sustainable Development: Science, Ethics, and Public Policy.* Dordrecht: Kluwer Academic, 1995:64–76.

Cairns J Jr, Niederlehner BR, Orvos DR. *Predicting Ecosystem Risk.* Princeton: Princeton Scientific Publishing, 1992.

Callicott JB. The wilderness idea revisited: The sustainable development alternative. *Environ Prof* 1991;13:235–248.

Committee for the National Institute for the Environment. A proposal to create a National Institute for the Environment (NIE). *Environ Prof* 1994;16:93–191.

Daubert v United States. 113 Sup. Ct. 2786 (1993).

Devall B, Sessions G. The development of natural resources and the integrity of nature. *Environ Ethics* 1984;6:293–322.

Drew GS. The scientific method revisited. *Cons Biol* 1994;8:596–597.

Ethyl Corporation v EPA. 541 F. 2d 1 (1976).

Grumbine ER. What is ecosystem management? *Cons Biol* 1994;8:27–38.

Johnson SP (ed). *The Earth Summit: The United Nations Conference on Environment and Development* (UNCED). London: Graham & Trotman/Martinus Nijhoff, 1993.

Jorling TC. Incorporating ecological principles into public policy. *Environ Policy Law* 1976;2:140–146.

Junker K. Understanding the rhetorical nature of science in the implementation of Agenda 21. *Environ Prof* 1994;16:349–355.

Karr JR. Ecological Integrity: Protecting Earth's Life Support Systems. In R Costanza, BG Norton, BD Haskell (eds), *Ecosystem Health.* Washington, DC: Island Press, 1992:223–238.

Kay J. A non-equilibrium thermodynamics framework for discussing ecosystem integrity. *Environ Manage* 1992;15:483–495.

Keiter RB, Boyce MS (eds). *The Greater Yellowstone Ecosystem.* New Haven, CT: Yale University Press, 1991.

Lemons J. Research in the national parks. *Environ Prof* 1986a;7:127–137.

Lemons J. Ecological stress phenomena and holistic environmental ethics—A viewpoint. *Int J Environ Stud* 1986b;27:9–30.
Lemons J. United States' national park management: Values, policy, and possible hints for others. *Environ Cons* 1987a;14:329–340.
Lemons J. Title 16 United States Code §55 and its implications for management of concession facilities in Yosemite National Park. *Environ Manage* 1987b;11:461–472.
Lemons J. Ecological Integrity and National Parks. In L Westra, J Lemons (eds), *Perspectives on Ecological Integrity.* Dordrecht: Kluwer Academic, 1995:177–201.
Lemons J (ed). *Scientific Uncertainty and Environmental Problem-Solving.* Cambridge, MA: Blackwell Science, 1996.
Lemons J, Brown DA (eds). *Sustainable Development: Science, Ethics, and Public Policy.* Dordrecht: Kluwer Academic, 1995.
Lemons J, Stout D. A reinterpretation of national park legislation. *Environ Law* 1984;15:41–65.
Leopold AS, Cain SA, Cottam LM, Gabrielson IN, Kimball TL. Wildlife management in the national parks. *Trans N Am Wildl Conf* 1963;28:28–45.
Lubchenco J, et al. The sustainable biosphere initiative: An ecological research agenda. *Ecol* 1991;72:371–412.
Mantell M. Preservation and use: Concessions in national parks. *Ecol Law Quart* 1979;8:1–54.
Miller A. The role of analytical science in natural resource decision making. *Environ Manage* 1993;17:563–574.
Murphy D. Conservation biology and scientific method. *Cons Biol* 1990;4:203–204.
National Park Service. *Draft Environmental Statement, General Management Plan, Yosemite National Park, California.* Washington, DC: U.S. Department of the Interior, 1978.
National Park Service. *State of the Parks Report—1980. A report to Congress.* Washington, DC: U.S. Department of the Interior, 1980a.
National Park Service. *Yosemite General Management Plan.* Washington, DC: U.S. Department of Interior, 1980b.
Parsons DJ, Graber DM, Agee JK, Wagtendonk JWV. Natural fire management in national parks. *Environ Manage* 1986;10:21–40.
Peters RH. *A Critique for Ecology.* Cambridge: Cambridge University Press, 1991.
Robbins WJ. *A Report by the Advisory Committee to the National Park Service on Research.* Washington, DC: National Academy of Sciences, 1963.
Rolston H III. *Environmental Ethics, Duties to and Values in the Natural World.* Philadelphia: Temple University Press, 1988.
Runte A. *National Parks: The American Experience.* Lincoln, NE: University of Nebraska Press, 1979.
Sagoff M. Ethics, ecology, and the environment: Integrating science and law. *Tenn Law Rev* 1988;56:78–229.
Sax J. Mountains Without Handrails: Reflections on the National Parks. Ann Arbor, MI: The University of Michigan Press, 1980.
Schaeffer DJ, Herricks EE, Kerster HW. Ecosystem health I: Measuring ecosystem health. *Environ Manage* 1988;12:445–455.
Shabecoff M. Administration seeks greater role for entrepreneurs at federal parks. *New York Times* March 29, 1981. §1, at 1, col. 1.

Shrader-Frechette K. *Ethics of Scientific Research.* Lanham, MD: Rowman & Littlefield, 1994.
Shrader-Frechette K, McCoy E. *Method in Ecology.* Cambridge: Cambridge University Press, 1993.
Simberloff D. Simplification, danger, and ethics in conservation biology. *Bull Ecol Soc Am* 1987;68:156–157.
Smith PRG, Theberge JB. A review of criteria for evaluating natural areas. *Environ Manage* 1986;10:715–734.
U.S. Congress. 94th Congress, 2d sess, HR Report No. 869. 1976.
Wright RG. *Wildlife Research and Management in the National Parks.* Urbana, IL: University of Illinois Press, 1992.

V

Parks as Baselines

This section examines from an applied perspective the role of parks as baselines.

Chapter 20 examines the specific links connecting scientific research and park management. It recognizes that these links are often delicate, not well defined, and frequently based on tradition. It further recognizes the complexity of the research needed to deal with today's complex environmental problems and emphasizes, in particular, the value of long-term research. The chapter describes the characteristics and qualities of a complete integrated scientific research program, which, in general, are applicable to all parks and protected areas.

The value of long-term research is undeniable, and parks often offer the best venue for pursuing a program of long-term ecological research. Many examples could have been chosen to illustrate the construct and value of long-term research programs. The program at Crater Lake National Park, Oregon, was chosen for several reasons: Certainly, the park is a world-class scenic and scientific wonder; the limnologic focus of its long-term monitoring is also unique. In addition, as a result of public and scientific concerns over the continued maintenance of the environmental quality of Crater Lake, the monitoring program has influenced management and public perceptions directly. The results of the Crater Lake program are described in Chapter 21.

20

Research in Parks and Protected Areas: Forging the Link Between Science and Management

David L. Peterson

When large fires swept through the Yellowstone National Park region in 1988, even the casual observer was aware of some obvious impacts on the landscape and vegetation. The publicity and drama of this event were magnified primarily because the fires occurred in one of North America's most sacrosanct protected areas. As a result, along with criticism of NPS fire management policies (see Chap. 14), the media and the scientific community were able to communicate some of the basic scientific principles involved in this large disturbance. Unfortunately, the importance of parks and protected areas in other natural resource issues appears to be less apparent to the general public. For example, survival of the whooping crane (*Grus americana*), whose total population was less than 40 birds in the 1960s, is due largely to research and management activities in Aransas National Wildlife Refuge (Texas) and Wood Buffalo National Park (Alberta and Northwest Territories, Canada), which provide the crane with winter and summer habitat, respectively. Much of our understanding about the biology and behavior of the timber wolf (*Canis lupus*) is based on research conducted in Isle Royale National Park (Michigan). Both of these examples are high-profile issues of public interest (i.e., endangered species) and illustrate the relevance of research in parks and protected areas.

It was not until the post–Earth Day social climate of the 1970s that environmental issues and ecological concepts gained a foothold in the minds of the general public in North America. As Faber (1977) observed, "The brook trout and spotted salamanders of Adirondack Mountain lakes are becoming vanishing species because of deadly rain and snow containing corrosive acids that kill them." Twelve years later, Egan (1989) reported that "[t]he 30-year practice of cutting vast stands of ancient trees on national forests in the Pacific Northwest will likely be sharply reduced by a government finding that such logging is wiping out a rare owl." These sensationalistic statements from the popular media may or may not be true, but they are at least partially based on scientific observations. It is the role of research to clarify where these statements fall along the gradient between fact and fiction. Acidic deposition and protection of endangered species are complex issues—biologically and socially—and parks and protected areas can play a critical role in understanding these issues.

Gone are the days of the *Forstmeister,* when local resource managers dictated management activities according to straightforward guidelines. Natural resource issues are increasingly complex from administrative, political, legal and social perspectives (Harmon, 1994). Most ecological phenomena are spatially and temporally complex (Heal, 1991; Fig. 20.1) and in many cases unpredictable (Gleick, 1990). The emerging paradigm of "ecosystem management" discussed earlier (see Chaps. 2–5) implies staggering information needs, complex manage-

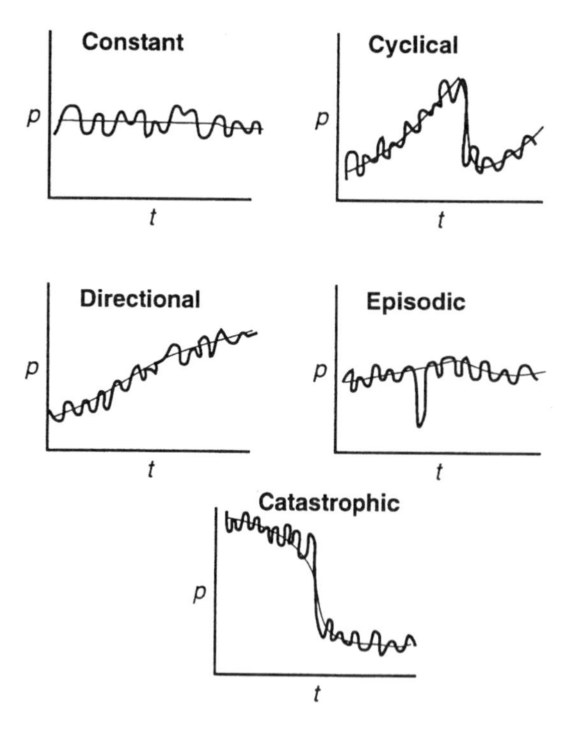

Figure 20.1 General long-term trends in ecosystems that must be considered with respect to research programs and management decisions. Response of ecosystem parameters (p; e.g., productivity, organic turnover, etc.) is indicated over time (t). (From OW Heal, The Role of Study Sites in Long-term Ecological Research: A UK Experience. In PG Risser (ed), *Long-term Ecological Research: An International Perspective.* West Sussex, UK: John Wiley and Sons, 1991:23–44. Reprinted with permission of John Wiley and Sons.)

ment analyses, increasing compliance responsibilities, and demands that managers undertake innovative and often risky actions.

One of the best weapons for addressing increasingly complex management problems is good scientific information. This requires good research. Furthermore, it requires a tight linkage between research and management. The science-management relationship often has been a delicate one, perhaps because the relationship itself has not been well defined. Underwood (1995) suggests four categories of research that define this relationship (Fig. 20.2): (1) *available and directed,* in which existing information is directly applicable to problems; (2) *applied and environmental,* in which managerial hypotheses are tested

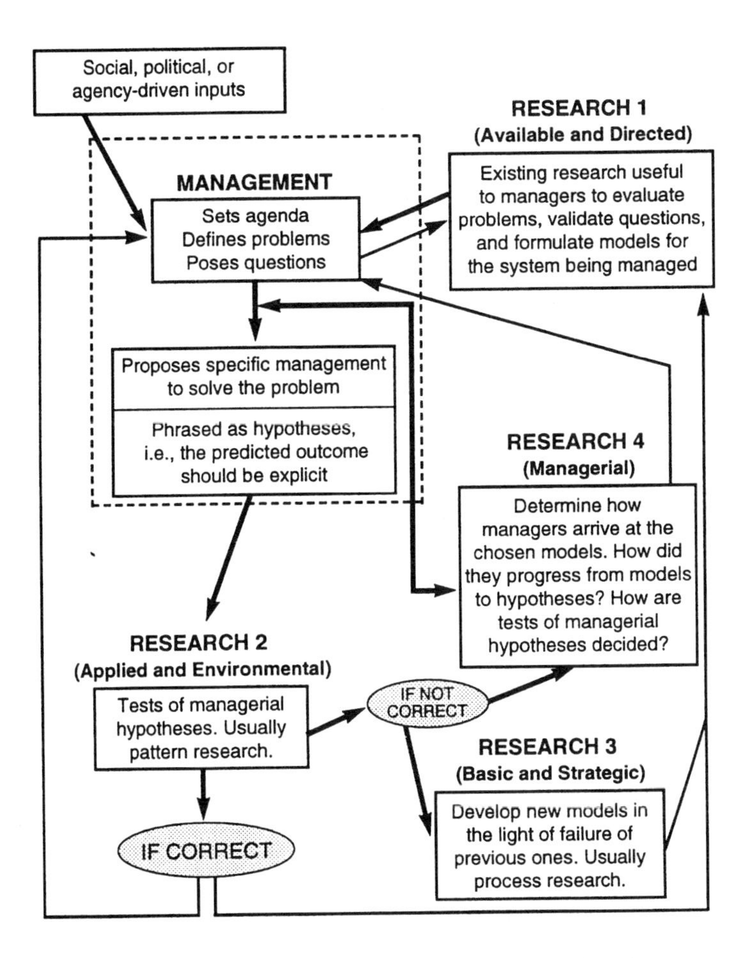

Figure 20.2 Representation of relationships between research and resource management decision making. Arrows with heavy lines indicate direct links; arrows with light lines indicate feedback from one type of research to another. (Adapted from AJ Underwood, Ecological research and (and research into) environmental management. *Ecol Appl* 1995;5:232–247.

(pattern research); (3) *basic and strategic,* in which new models and approaches are developed (process research); and (4) *managerial,* in which the managerial process itself is evaluated. In each category, there are clear feedbacks between research and management, as well as links between different research components. Regardless of the specific components of this conceptual model, nothing is more important to developing a strong science-management relationship than a clear statement of objectives.

The Scope of Research in Parks and Protected Areas

Defining Objectives

The degree to which research is integrated in park resource management varies widely, and the specific role of research as a component of the broader stewardship of natural resources is generally poorly defined. For example, the National Park Service (NPS) has no legislative authority to conduct research on its own lands even though the agency has long had a small contingent of research scientists. In 1994, these scientists were transferred to the newly created National Biological Service, and the NPS must now depend on that agency for scientific expertise.

The value of parks as venues for research has long been recognized in the scientific community (Wright & Hayward, 1985; Parsons, 1989). Parks provide examples of ecosystems with a reasonable level of ecological integrity and functionality and their legal protection makes them suitable for long-term studies (Franklin et al., 1972). Such research contributes to integrated science-based resource management by:

1. Providing information on baseline conditions of natural resources as a component of a broader program of long-term monitoring.
2. Improving our understanding of functional characteristics and processes in "natural" systems.
3. Interfacing with educational, interpretive, and training programs of park agencies and the general public.

The lack of basic information on natural resources derived from scientific research and monitoring is a constant source of frustration to most park managers who strive to make informed management decisions. A critical first step in any research program is to clearly define objectives (Hinds, 1984; Silsbee & Peterson, 1991). Objectives should be developed cooperatively between scientists and re-

Table 20.1 Summary of objectives of research information, appropriate ecosystems for study, and potential users

Objective	Appropriate ecosystem	User
Inform internal decision maker	Ecosystems involved in specific management decisions	Park manager
Influence external decision maker	Ecosystems threatened by outside activities	Decision makers and managers of lands adjacent to park
Satisfy legal requirement	Depends on legal requirement	Park administrators, regulatory agencies, legislative bodies
Provide a better understanding of resources	Broad spectrum but mainly ecosystems subject to potential change	Scientists, various park managers
Provide background information	Broad spectrum	Scientists, park personnel, visitors
Provide early warning of global or regional problems	Ecosystems sensitive to change	Park decision makers, regulatory agencies, external decision makers
Provide data for comparison with areas having less protection	Ecosystems comparable to areas outside park	Decision makers and managers of lands adjacent to parks

source managers to ensure research will produce a mutually acceptable "product." A diversity of objectives is possible, and there can be multiple objectives for the same project. Specific objectives have implications for the ecosystems selected for study and for the users of the scientific information (Table 20.1).

"Natural" Processes and Landscape Issues: Looking Outside and Inside Parks

Parks and protected areas are increasingly viewed as "islands" within landscapes containing a variety of resource management objectives. While the "island" analogy may be a realistic biotic perspective, it is limiting for both scientists and resource managers. Parks and protected areas clearly are not isolated from surrounding lands with less protected status; there are continuous flows of materials (e.g., water, atmospheric gases) and organisms (e.g., large mammals, anadromous fish, human visitors) across political boundaries. Furthermore, these areas are sources of biota as well as refuges (Peterson et al., 1996); biotic exchange is critical at both short- (e.g., seasonal migration) and long- (e.g., postglacial plant establishment) temporal scales.

Understanding the "natural" ecological processes of parks is critical for management of their landscapes and ecosystems (Rowe, 1984). For example,

90 percent of all low-elevation forests in the western (contiguous) United States have been harvested during the last century; most of the remaining uncut forests are in parks or wilderness areas. Quantification of productivity, stand dynamics, and disturbance patterns can be used to draw inferences regarding the size and configuration of forests needed to retain functional integrity of ecological processes. Quantification of habitat requirements for salmon (*Oncorhynchus* spp.) in rivers that run from inland mountains to the Pacific Ocean helps to define the viability of populations spawning in stream channels that cross both undisturbed and clearcut lands. These kinds of "landscape" issues require interagency cooperation in conducting research, interpreting data, and integrating results into management activities.

Managing the physical and biological complexity of landscapes requires an understanding of ecological processes at a broad range of spatial and temporal scales (Magnuson et al., 1991). Long-term research is needed to quantify variation in space and time and to account for the complexity of natural systems (e.g., infrequent but large-scale disturbances such as fire) (Heal, 1991). Parks and protected areas are ideally suited for long-term research programs.

The value of long-term research is illustrated by the Long-Term Ecological Research (LTER) program, which has been funded by the U.S. National Science Foundation since 1976 (Callahan, 1984). This program includes 18 sites in a wide range of ecosystems. Long-term measurement of physical and biological variables provides a better understanding of individual systems. Sevilleta National Wildlife Refuge, New Mexico, is one of the LTER sites. Research in this protected area is providing information on conifer woodlands, shrub-steppe systems, shortgrass prairie, and low-elevation desert. Much of the information is relevant to ecosystems found throughout the southwestern United States, indicating that investment at one site has potential payoffs for natural resource management at many other sites, including millions of hectares of public land.

Scientific Programs and Social Values

Recognizing that most environmental issues are driven by social and cultural values provides an important context for the development of research on natural resources. For example, decreased visibility at Grand Canyon National Park, Arizona, is caused at least partially by emissions from fossil fuel combustion. Grizzly bears (*Ursus horribilis*) at Glacier National Park, Montana, are a management problem only because they have occasional encounters with people and livestock. Hurricanes in Everglades National Park, Florida, pose no threat to the existence of ecosystems but damage buildings and infrastructure used by people.

Air quality became a high-profile issue in the 1970s, with damage to forests and aquatic systems being widely reported in the scientific literature and the popular media. While many of the original claims regarding linkages between pollutants and impacts have never been proven, there is now a much greater awareness of the large-scale impacts of air pollution. Wildland areas previously thought to be "pristine" are indeed susceptible, even in areas remote from urban pollutant sources.

The potential threat of air pollution was recognized by the resource management staff of Sequoia and Kings Canyon National Parks in California in the early 1980s. These parks, located in the southern Sierra Nevada, receive elevated levels of ambient ozone throughout the summer as a result of pollutant transport from the San Francisco Bay region and Central Valley. The parks, with the assistance of the Air Quality Division of the NPS, developed a multidisciplinary research program involving federal agencies and university scientists to assess the impacts of ozone on natural resources in the parks. This highly successful program demonstrated that there were indeed serious impacts to coniferous tree species (Peterson et al., 1991), including potential injury to the parks' namesake species, giant sequoia (*Sequoiadendron giganteum*) (Grulke et al., 1989).

The commitment of park management to science-based resource management was a key to the success of the program. In addition, the rigorous data demonstrating the impacts could not have been achieved without considerable destructive sampling and highly visible research facilities (e.g., open-top fumigation chambers, scaffolds allowing access to tree canopies). This type of research would be viewed as unacceptable in many parks. Aesthetic values, managerial philosophy, and visitor perceptions can impact judgments on what types of research are appropriate in parks (Sellars, 1989). If the role of parks and protected areas in long-term research and environmental issues becomes more prominent, there may be greater emphasis on intrusive research activities. The appropriateness of these activities is subjective but should be discussed openly among managers, scientists, and the public.

Developing a Research Program

More often than not, park research consists of a series of individual projects without programmatic themes or long-term planning. This is because most projects are directed by scientists working independently with funding from diverse sources. Much of this research produces useful scientific information on natural resources without impacting park budgets or personnel. However, there is an opportunity for these projects to have greater impact if they are integrated in the context of a larger programmatic structure. A formal research program allows park managers to identify information needs and establish research priorities. It also allows an individual park to set the course for scientific activities on its own lands.

Setting Priorities

Prior to developing priorities for research, park resource managers need to examine managerial objectives on their lands. A resource management plan is the most common basis of this examination and suggests what kinds of information are needed to guide management activities. This assessment should not be done in isolation: an open forum involving managers, scientists, and various stakeholders (e.g., local residents, interest groups, other agencies) is especially helpful.

Identifying priorities can be a difficult process because of divergent viewpoints and the general complexity of natural resource issues. It would not be unusual for a park to consider over 100 issues (from its management plan), use five or more evaluative criteria, and have ten people participate in the assessment. Simply keeping track of the decision-points involved in a typical priority assessment quickly becomes overwhelming. Nevertheless, many organizations still deal with complex decision making by using the BOGSAT (bunch of guys/gals sitting around a table) method (Peterson et al., 1994). Informal decision making may in fact produce good decisions, but those decisions may be difficult to document or defend because of the lack of quantitative process.

There are some straightforward quantitative tools that can assist with the development of priorities. One such tool, the Analytic Hierarchy Process (AHP) (Saaty, 1980), has been used to support priority setting in a variety of managerial situations. It is particularly relevant for parks because it allows for consideration of multiple objectives and viewpoints (Schmoldt et al., 1994). The AHP uses hierarchies to structure decision making, and applies judgment measures and formal mathematics to express and quantify individual preferences. It can be operated with minimal training on a personal computer and can be used interactively with many participants. As shown in Figure 20.3, a complex decision-tree is reduced to a series of individual judgments, which are subsequently quantified and ranked. These rankings express the priorities developed by the group, and provide quantitative support for the research (or other) program.

Achievement of scientific priorities is constrained by budgets, personnel limitations, and political influences. The ideal program will rarely be attained. Budget and personnel criteria can be integrated with the AHP and linear programming techniques to further refine priorities in a more realistic context (Schmoldt et al.,

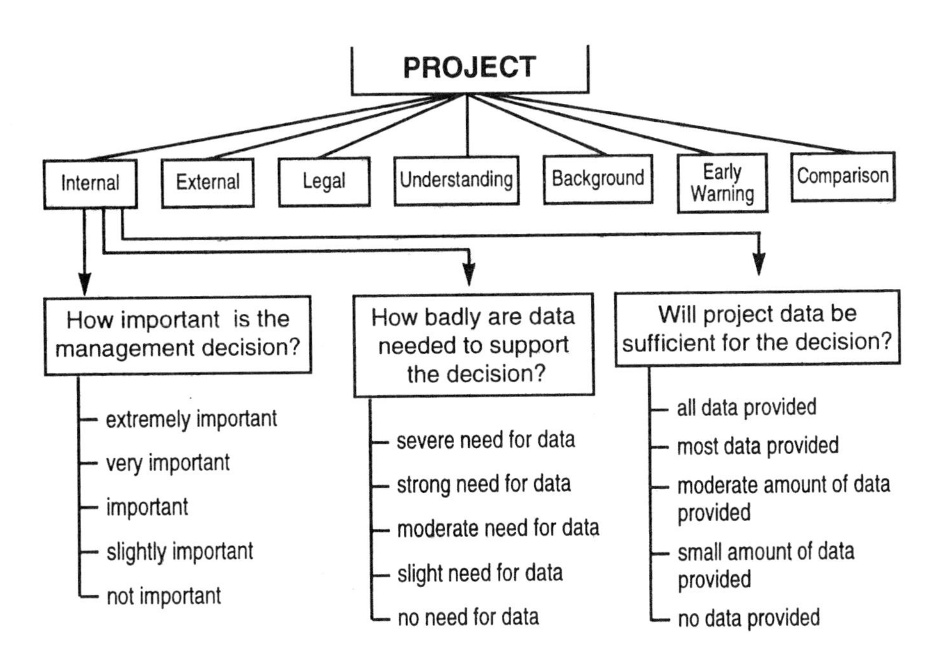

Figure 20.3 The Analytic Hierarchy Process (AHP) can be used to assist resource managers and scientists in setting priorities and making decisions. In this case, a portion of a decision-tree is shown in which the relative importance of seven objectives (from Table 20.1) is determined for a particular project. Each of the objectives has similar sets of questions and criteria used for ranking.

1994). Although an expensive, high-priority research project may never be implemented, it may be possible to implement less expensive, lower priority projects. Again, priorities can be formally quantified to provide support for decision making. Political constraints are another matter. It is the responsibility of resource managers and cooperating scientists to "market" the proposed research program convincingly and demonstrate strong linkages between science and management.

Program Development and Coordination

Parks and protected areas often have surprisingly little oversight of individual research projects on their lands. This must change. Scientists must be able to demonstrate the value of their research from both managerial and scientific perspectives. Proposed projects require formal review by park staff, with an emphasis on scientific responsibility and accountability. This should be a cooperative process, with scientists coordinating projects with park staff at all stages.

A set of consistent guidelines can greatly assist park managers in administering a research program and evaluating individual projects. Scientists should be required to submit formal research proposals or work plans that clearly articulate their proposed activities. In addition to scientific objectives and methods, these proposals should indicate how the project will (1) address managerial objectives, including potential applications; (2) enhance the park research program; (3) coordinate with other research projects in the park; (4) comply with park regulations and guidelines; and (5) ensure high-quality data (through quality-assurance protocols). These proposal elements will raise awareness of researchers of the special requirements of working in parks. In addition, managers can ensure greater rigor in the scientific quality of research projects by including scientists in a peer review of proposals.

Individual research projects can be particularly effective if they lead to long-term monitoring efforts or contribute to an existing monitoring program (Silsbee & Peterson, 1993; Peterson et al., 1995). Research can often define the initial scope (or severity) of a problem (e.g., measuring fecal coliform bacteria concentrations in several streams in a particular year), while monitoring can assess variation in resource condition over time (e.g., measuring fecal coliform bacteria concentrations over a ten-year period). Monitoring is essential for demonstrating long-term trends (see Chap. 21) and for identifying departures from the natural range of variation (e.g., the effects of roads on ungulate populations). Long-term research and monitoring can be synonymous in some cases, and the distinction is not critical, as long as high-quality information is being provided to the park.

Incorporating Traditional and Local Knowledge

Cultural research often depends on local sources of information, including oral histories and interviews. Oral tradition is particularly important in studies of indigenous cultures because it represents a "library" of knowledge that has been accrued over many generations; elders are important "librarians" of this knowledge. Traditional ecological knowledge and the observations of local residents can be extremely valuable in understanding natural resources and developing research projects (Bielawski, 1992; Harmon, 1994).

The value of traditional ecological knowledge has been demonstrated for several parks in Alaska and northern Canada. For example, interviews with community elders have provided valuable information on wildlife distribution and subsistence resource use in Kobuk Valley National Park, Noatak National Preserve, Cape Krusenstern National Monument, Lake Clark National Park and Preserve, and Denali National Park, Alaska. In the Bering Land Bridge National Preserve region of Alaska, local native residents are actively involved in various biological and geological research projects (Schaaf,

1995). Local communities and traditional ecological knowledge are included at all levels of planning and research in the highly successful Mackenzie Basin Impact Study (Northwest Territories, Canada), a large multidisciplinary study of the potential impacts of climatic change on natural resources and human populations (Cohen, 1995). Local residents often have many years of observational knowledge or informal records about natural resources. Researchers and managers need to consider the value of traditional knowledge and look for opportunities to involve local residents in park planning and scientific activities.

Quality Assurance and Data Management: The Key to Scientific Credibility

Credible research programs need to ensure the quality of data being collected. Quality assurance (QA) provides accuracy, precision, continuity, and consistency in data collection and analysis. Thorough QA documentation of project objectives, study design, and methods should be required for all research projects. A strong QA program is the cornerstone of credibility for a research program and will ensure that data will stand up to scrutiny in a court of law if necessary.

It is the responsibility of individual researchers to provide QA information for their projects. This is typically done through a QA plan (Peterson et al., 1995) that contains a description explaining and justifying the overall approach to the project, including (1) a clear statement of objectives, including how the data will be used; (2) a general description and justification of study methods; (3) population of inference and geographic area; (4) statistical approach; and (5) related studies and data sets.

A QA plan also includes a methods manual that describes site-selection criteria, methods and detailed sampling protocols, methods for calibrating and maintaining instruments, methods of documenting precision and accuracy of measurements, and data analysis and reporting methods. Evaluating the adequacy of QA plans, similar to that of the proposal itself, is best done by park managers in cooperation with scientists as peer reviewers.

Sound data management is another important component of a QA program. Appropriate data management protocols ensure that data are stored and transferred accurately and are secured from loss or damage. Data structures and format should be documented in sufficient detail so that someone not involved in the original project can interpret their meaning and evaluate precision and accuracy. A data management plan should be submitted with each QA plan; it specifies the disposition of all data and the means to access them.

Although QA and data management procedures may seem burdensome, they ensure that a park has a technically sound, credible, and defensible program. If appropriate guidelines are instituted at the beginning of each project, they become a routine part of operations. Managers should encourage scientists to include QA and data management costs in budget and personnel requirements for each project.

A New Role for Research in Parks and Protected Areas

Managers Who Know Science, Scientists Who Know Management

The presumed gap between science and resource management is a constant source of discussion at conferences and workshops (e.g., McAninch & Strayer, 1989). Scientists are often criticized for not being sufficiently informed about management needs and potential applications of scientific data. Managers are often criticized for not having adequate scientific training to make informed decisions about natural resource activities. However, the evolution of ecosystem management seems to have motivated a recent movement toward science-based management. Many resource managers have embraced this concept and made significant efforts to improve the scientific basis of management decisions.

One reason for the long-standing gap between researchers and resource managers has been contrasting goals and objectives (Huenneke, 1995; Table 20.2). Job-performance standards and reward systems are powerful motivators for professional expectations and achievements. Unless the importance of science-based management and protection of natural resources is recognized—and required—by all employers, significant changes in actual job performance cannot realistically be expected. There is a need for scientists who understand and appreciate the value of park and public lands management, and who are willing to work on applied problems. We also need resource managers who have better scientific backgrounds and technical training that can be integrated into planning and on-the-ground applications. The distinction between management-oriented scientists and science-oriented managers may decrease in the future if there is a greater commitment to science-based management accompanied by a shift in professional objectives and expectations (Table 20.3).

Integrated Programs for Parks, Landscapes, and Regions

Consider the complex political landscape of the Olympic Peninsula in northwestern Washington (Fig. 20.4). At the heart of this one-million hectare region

Table 20.2 Goals and objectives of academic scientists and resource managers

Job component	Academic scientist	Resource manager
Motivation	Questions driven by theory, basic science	Questions driven by need to address specific problem
Goals	Publish in high-quality journals; compete for research funds	Provide data to manager to guide management; derive guidelines for action
Financial considerations	Acquire grant funds	Accomplish goals as cost-effectively as possible
Staffing	Train students in modern techniques; find students jobs; recruit and support students	Cost-effective workers
Time frame/work schedule	Conform to class/academic schedules; projects chosen to fit schedule	Work quickly to obtain data; long planning range of agency budget process
Service	To state and public; idealistic goals	Explicit responsibilities to agency; idealistic goals

Source: Adapted from LF Huenneke, Involving academic scientists in conservation research: Perspectives of a plant ecologist. *Ecol Appl* 1995;5:209–214.

is Olympic National Park, which has a strong focus on resource preservation. The park is surrounded by national forest and state, tribal, and private land, for which commercial timber harvest has been the primary objective for the past 50 to 100 years. Ecosystems do not respect political boundaries, and fragmentation of the Olympic Peninsula landscape by logging, urbanization, and agriculture have had significant impacts on the functionality of many ecosystems. A recent shift toward protection of wildlife, fisheries, and other components of biological diversity outside the park has produced a greater awareness that agencies and institutions must work together to achieve the societal goals of a healthy environment and economic well being.

Resource managers on the Olympic Peninsula are now sharing data in compatible geographic information system (GIS) formats. This facilitates communication and administration of interboundary issues. Research scientists in different agencies and institutions are also communicating their objectives and results with each other and with resource managers, in some cases through formal coordinating groups.

Complex landownership patterns are found throughout North America. Clean water, clean air, healthy wildlife and fish populations, and economically viable local communities are resource management objectives—and in some cases legal requirements—for most agencies and institutions. Processes and organisms that cross boundaries can be protected only through cooperation and

Table 20.3 Desirable goals and objectives of scientists and resource managers working in parks and protected areas

Job component	Management-oriented scientist	Science-oriented resource manager
Motivation	Questions driven by critical gaps in knowledge about resources and by management problems	Questions driven by need for science-based long-term solutions to management problems
Goals	Publish results in formats relevant for resource managers; develop cooperative interdisciplinary research programs	Work with scientists to collect data that will guide management; develop guidelines based on high-quality scientific information
Financial considerations	Develop cost-effective research programs with multiple agencies; work toward long-term funding	Prioritize management objectives to optimize allocation of limited funds; initiate cost sharing
Staffing	Train students and technicians in applied research; recruit students with interest in resource management	Recruit and train employees with scientific backgrounds; provide training for new concepts
Time frame/work schedule	Develop a program of long- and short-term projects; adapt schedules to work with management personnel in the field; ensure timely data	Work closely with scientists to facilitate time frame and relevant data; incorporate analytical and scientific concepts in long-term agency planning
Service	Develop research and scientific expertise relevant to resource management agencies and general public	Develop a managerial ethic that values scientific principles; provide timely solutions to management problems; protect natural resources

communication in both scientific and management activities. Research programs and projects designed for protected areas will have greater effectiveness at the landscape scale by incorporating the interests and objectives of outside landowners. By including a broader range of stakeholders at the beginning of a project, scientists will have the opportunity for wider applicability of their findings. Furthermore, scientists must ensure that their results are communicated to a wide audience of potential users, preferably through means other than just academic journal articles.

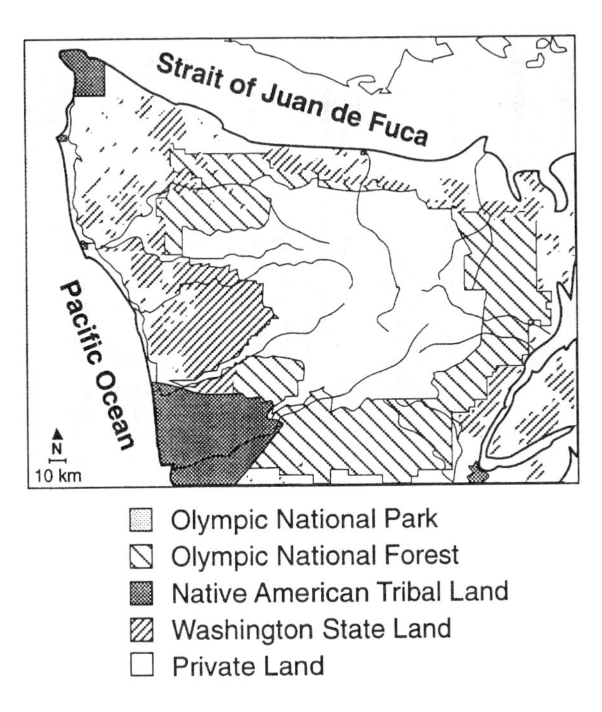

Figure 20.4 The complex land ownership of the Olympic Peninsula, Washington, illustrates the need for interdisciplinary, cooperative research efforts to support natural resource management.

Parks as the Core of Scientific Programs

The concept of biosphere reserves, as originally proposed by the United Nations Educational, Scientific, and Cultural Organization (UNESCO, 1970; Price, 1995), provides a model of how parks can serve as the focus of regional research programs. Biosphere reserves are ideally designed such that there is a core protected area (such as a park) with little or no resource exploitation. The core is surrounded by a buffer zone in which some resource exploitation is permitted. The buffer zone is surrounded by a transition area in which there is a variety of land-use patterns, human settlements, and an emphasis on sustainable economic development. This situation is well-represented by the Olympic Peninsula (see Fig. 20.4), and, in fact, Olympic National Park (but not surrounding lands) is designated as a biosphere reserve.

The core area of biosphere reserves is designated for preservation of natural resources, but it is also suitable for limited research and monitoring activities, provided they are not overly intrusive or destructive. Regardless of whether a park is designated as an official biosphere reserve, it is a site where natural systems and processes can be studied without disturbance by humans. Research in the buffer zone and transition area can be developed in coordination with park research, so that various ecosystem variables (e.g., productivity, carbon storage, erosion rates) can be compared. Data on natural processes in the core area can be used to interpret ecosystem responses in adjacent nonprotected areas and to assess the impacts of human management activities (e.g., timber harvest). Research in all zones should be coordinated as much as possible. This requires cooperation among agencies and institutions and may take considerable effort, but the result is a better understanding of landscape processes and a greater potential for applications in resource management.

Inclusion of the "transition zone" in a broad research program emphasizes the importance of including social, economic, and cultural factors in biological studies (Pedevillano & Wright, 1987; Machlis, 1992). Social elements should be integrated with ecosystem-scale research and management to the greatest extent possible (Agee & Johnson, 1988). For example, if scientists are studying how to enhance grizzly bear populations, what will be the impact of increased bear encounters on park visitors? Will the bears impact livestock on adjacent private land? Including local residents and stakeholders during the development of the research program may have the benefit of increased support for the program (Cohen, 1995) and reduced antagonism and litigation. Broad participation in the design and conduct of a research project helps to maintain a focus on the original objectives and potential applications of the data.

Acknowledgments

I thank James K. Agee, Darryll R. Johnson, and Miriam R. Peterson for helpful comments on the chapter.

Literature Cited

Agee JK, Johnson DR (eds). *Ecosystem Management for Parks and Wilderness.* Seattle: University of Washington Press, 1988.

Bielawski E. Inuit indigenous knowledge and science in the Arctic. *Northern Perspect* 1992;20:5–8.

Callahan JT. Long-term ecological research. *BioScience* 1984;34:363–367.

Cohen SJ. An Interdisciplinary Assessment of Climate Change in Northern Ecosystems: The Mackenzie Basin Impact Study. In DL Peterson, DR Johnson (eds), *Human Ecology and Climate Change: People and Resources in the Far North.* Washington, DC: Taylor and Francis, 1995.

Egan T. U.S. stand on owl seen saving trees in west. *The New York Times* April 27, 1989. Sec. A:18.

Faber H. Deadly rain imperils 2 Adirondacks species. *The New York Times* March 28, 1977. Sec. 1:31.

Franklin JF, Jenkins RE, Romancier RM. Research natural areas: Contributors to environmental quality programs. *J Environ Qual* 1972;1:133–139.

Gleick J. *Chaos: Making a New Science*. New York: Penguin, 1990.

Grulke N, Miller P, Wilborn R, Hahn S. Photosynthetic Response of Giant Sequoia Seedlings and Rooted Branchlets of Mature Foliage to Ozone Fumigation. In RK Olson, A Lefohn (eds), *Effects of Air Pollution on Western Forests* (Transactions Series 16). Pittsburgh: Air and Waste Management Association, 1989:429–441.

Harmon D (ed). *Coordinating Research and Management to Enhance Protected Areas*. Cambridge, UK: IUCN (World Conservation Union) Publication Services, 1994.

Heal OW. The Role of Study Sites in Long-term Ecological Research: A UK Experience. In PG Risser (ed), *Long-term Ecological Research: An International Perspective*. West Sussex, UK: John Wiley and Sons, 1991:23–44.

Hinds TW. Towards monitoring of long-term trends in terrestrial ecosystems. *Environ Cons* 1984;11:11–18.

Huenneke LF. Involving academic scientists in conservation research: Perspectives of a plant ecologist. *Ecol Appl* 1995;5:209–214.

Machlis GE. The contribution of sociology to biodiversity research and management. *Biol Cons* 1992;62:161–170.

Magnuson JJ, et al. Expanding the Temporal and Spatial Scales of Ecological Research and Comparison of Divergent Ecosystems: Roles for LTER in the United States. In PG Risser (ed), *Long-term Ecological Research: An International Perspective*. West Sussex, UK: John Wiley and Sons, 1991:45–70.

McAninch JB, Strayer DL. What Are the Tradeoffs Between the Immediacy of Management Needs and the Longer Process of Scientific Discovery? In GE Likens (ed), *Long-term Studies in Ecology*. New York: Springer-Verlag, 1989:203–205.

Parsons DJ. Evaluating National Parks as Sites for Long-term Studies. In GE Likens (ed), *Long-term Studies in Ecology*. New York: Springer-Verlag, 1989:171–173.

Pedevillano C, Wright RG. The influence of visitors on mountain goat activities in Glacier National Park, Montana. *Biol Cons* 1987;39:1–11.

Peterson DL, Arbaugh MJ, Robinson LJ. Regional growth changes in ozone-stressed ponderosa pine (*Pinus ponderosa*) in the Sierra Nevada, California, USA. *The Holocene* 1991;1:50–61.

Peterson DL, Schmoldt DL, Silsbee DG. A case study of resources management planning based on multiple objectives and projects. *Environ Manage* 1994;18:729–742.

Peterson DL, Silsbee DG, Schmoldt DL. A planning approach for developing inventory and monitoring programs in the National Park Service. *National Park Service Natural Resources Report* (NPS/NRUW/NRR-95/00). Denver: Government Printing Office, 1995.

Peterson DL, Schreiner EG, Buckingham NM. Gradients, vegetation, and climate: Spatial and temporal dynamics in mountains. *J Biogeog* 1996.

Price MF. Biosphere Reserves: A Flexible Framework for Regional Cooperation in an Era of Change. In DL Peterson, DR Johnson (eds), *Human Ecology and Climate Change: People and Resources in the Far North*. Washington, DC: Taylor and Francis, 1995.

Rowe JS. Forestland Classification: Limitations of the Use of Vegetation. In JG Bockheim (ed), *Proceedings of the Symposium on Forest Land Classification: Experience, Problems, and Perspectives.* Madison, WI: University of Wisconsin, Department of Soil Science, 1984:132–147.

Saaty TL. *The Analytic Hierarchy Process.* New York: McGraw-Hill, 1980.

Schaaf J. Understanding Northern Environments and Human Populations Through Co-operative Research: A Case Study in Beringia. In DL Peterson, DR Johnson (eds), *Human Ecology and Climate Change: People and Resources in the Far North.* Washington, DC: Taylor and Francis, 1995.

Schmoldt DL, Peterson DL, Silsbee DG. Developing inventory and monitoring programs based on multiple objectives. *Environ Manage* 1994;18:707–727.

Sellars RW. Science or scenery? A conflict of values in national parks. *Wilderness* 1989; Summer:29–38.

Silsbee DG, Peterson DL. Designing and implementing comprehensive long-term inventory and monitoring programs for National Park System lands. *Natural Resources Report* (NPS/NRUW/NRR-91/04). Denver: National Park Service, 1991.

Silsbee DG, Peterson DL. Planning for implementation of long-term resource monitoring programs. *Environ Monit Assess* 1993;26:177–185.

Underwood AJ. Ecological research and (and research into) environmental management. *Ecol Appl* 1995;5:232–247.

United Nations Educational, Scientific, and Cultural Organization (UNESCO). Plan for a Long-term Intergovernmental and Interdisciplinary Program on Man and the Biosphere (Document 16C/78). 16th Session, General Conference. Paris: UNESCO, 1970.

Wright RG, Hayward P. National parks as research areas, with a focus on Glacier National Park, Montana. *Bull Ecol Soc Am* 1985;66:354–357.

21

Exploring the Dynamics of Crater Lake, Crater Lake National Park

Gary L. Larson

Crater Lake National Park was established by Congress in 1902, twelve years before it established the National Park Service (NPS). Crater Lake was recognized to be a primary resource of the park, especially for its deep-blue color and extreme water clarity. Considerable concern was voiced, therefore, when results from limnological studies conducted between 1978 and 1981 suggested that lake clarity had declined relative to the results of intermittent studies conducted earlier. In 1982, the NPS convened a panel of limnologists to evaluate the existing limnological data for Crater Lake. The panel concluded that the data base was insufficient to determine if lake clarity had changed and recommended a limnological study of the lake. A study was initiated by the NPS in the summer of 1982. In the fall of 1982, Congress passed Public Law 97-250, which authorized the Secretary of the Interior to promptly initiate studies on the status and trends of changes of lake water quality for ten years and to implement immediately such actions as necessary to ensure retention of the lake's pristine water quality.

Broad program goals were to: (1) develop a reliable data base for future use; (2) develop a better understanding of physical, chemical, and biological characteristics and processes of the lake; (3) establish a long-term monitoring program to examine limnological characteristics of the lake through time; and (4) investigate the possibility of long-term changes in lake conditions and, if such changes were found and determined to be human related, identify the causal factor(s) and recommend ways of mitigating the change(s).

The ten-year limnological study of Crater Lake began in the summer of 1983. The study was reviewed periodically by a panel of professional aquatic ecologists. After the 1985 field season, enough preliminary work was accomplished to develop two conceptual models to guide additional monitoring and research (Larson et al., 1993a). These models were used to develop a set of working objectives and a format for the monitoring program.

The ten-year limnological program was completed in 1992 and the results were published in 1993 (Larson et al., 1993b). The results of the numerous studies clearly demonstrated that some components of the lake system exhibited little interannual change, whereas other components exhibited considerable long-term "cyclic" changes. A key challenge to the researchers was to separate the natural dynamics of the lake system from any changes caused by human activities. Nonetheless, the lake was considered to be pristine, with the exception of impacts from introduced fish.

This chapter demonstrates the value of long-term monitoring in our attempts to understand the natural dynamics of key components of lake systems. Variables of the Crater Lake system that do not show much interannual change are compared to those exhibiting considerable interannual variation.

Study Area

The lake basin was formed by catastrophic collapse of the sides of volcanic Mount Mazama following a violent eruption about 6850 years ago (Bacon & Lanphere, 1990). At the present time, the subcircular caldera is between 8 and 10 kilometers in diameter. The rim of the caldera averages about 2100 meters in elevation, although some mountain peaks extend to about 2500 meters.

Crater Lake covers the floor of the caldera. The lake occupies about 78 percent of the catchment area (flat map) of the caldera. The average long-term elevation of the lake is 1882 meters. At this elevation, the lake has a surface area of 53.2 square kilometers, a maximum depth of 589 meters, a mean depth of 325 meters, and a shoreline length of 31 kilometers (Fig. 21.1; Byrne, 1965; Phillips & Van Denburg, 1968). Over 40 small permanent and ephemeral inlet springs and streams drain into the lake, but there is no surface outlet.

Methods

Field studies of the lake were conducted from late June through September from 1983 to 1985 because the lake was not accessible during other months of the year.

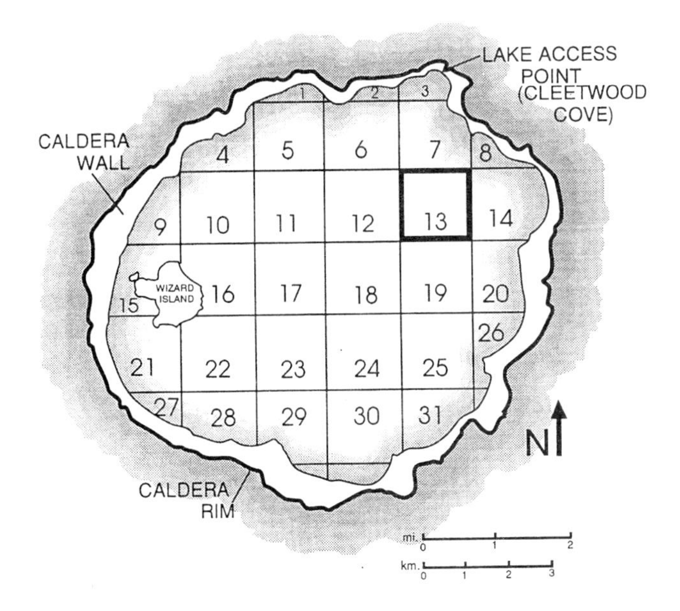

Figure 21.1 Grid system of Crater Lake stations established by Hoffman (1969). Station 13 was the main monitoring site. (After GL Larson, CD McIntire, RW Jacobs [eds], Crater Lake Limnological Studies Final Report. *National Park Service Technical Report* [NPS/PNROSU/NRTR-93/03] 1993b.)

Research vessels were flown to the lake by helicopter or were winched down the side of the caldera in early summer and removed from the lake by helicopter in fall. A boat house was built on Wizard Island in 1985. From that time on research vessels (an 8-meter-long double-pontoon boat and a 5.5-meter-long Boston Whaler) were stored in the boat house, allowing sampling during winter and spring.

Monitoring was primarily conducted in the deepest basin of the lake because much of the historical data had been collected at this site (Station 13; see Fig. 21.1). This site also was similar limnologically to the conditions at the second deepest basin in the lake, which was located in the southern portion of the lake. Physical, chemical, and biological sampling techniques and laboratory methods are described in Larson et al. (1993b).

Results

Lake Level Fluctuation

The surface elevation of Crater Lake has been recorded at nearly annual intervals since 1892. Prior to 1961, lake elevation measurements in summer and

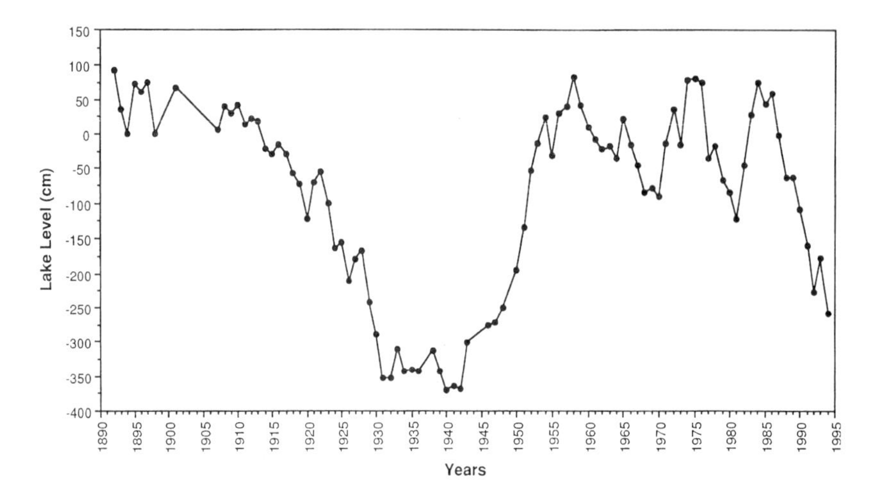

Figure 21.2 Variation in the surface elevation (cm) of Crater Lake from the long-term average elevation of 1882 meters on September 30 between 1892 and 1994. The long-term average elevation is represented as 0 on the Y-axis.

early fall were recorded using a staff gauge and by determining the distance between the lake surface and a brass pin that was hammered into the side of a cliff along the shore of the lake. From 1961 to present, lake level measurements have been recorded automatically by a stage-height recorder operated by the United States Geological Survey.

During the period from 1892 to 1913, the lake level was higher than the long-term average (1882 meters). From 1914 to 1942, the level dropped about 4.6 meters. The lake surface increased in elevation in succeeding years and returned to the level of the late 1890s by 1958. Between 1958 and 1980, the level fluctuated about 0.8 meter above and below the long-term average at approximately ten-year intervals. After 1981, however, the lake level dropped precipitously, and by 1994 the level was nearly as low at it was in the early 1940s (Fig. 21.2).

Changes in the lake level are directly related to changes in the water budget of the lake (Redmond 1993). During colder periods of the year when stream runoff is low, changes in lake level are controlled primarily by precipitation and evaporation at the surface of the lake. Redmond's results suggest, in fact, that the highest rates of evaporation at Crater Lake occur on the coldest winter days without precipitation because the lake is seldom covered by ice and snow in winter. Redmond (1993) also shows that the amount of precipitation entering the caldera closely correspond to the Southern Oscillation Index (SOI), one phase of the phenomenon known as El Niño. For example, negative SOI values are correlated to reduced precipitation inputs to Crater Lake between the months of October and March.

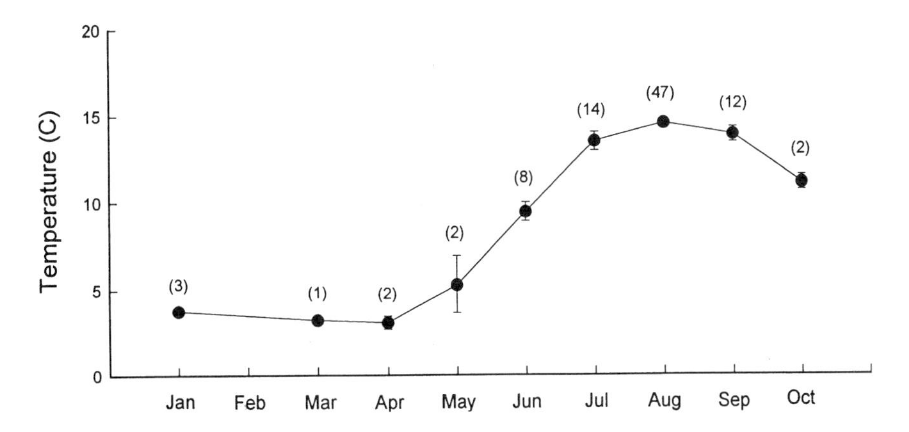

Figure 21.3 Mean monthly temperatures (+/− standard error) of Crater Lake at a depth of 1 meter between 1982 and 1990. (After GL Larson, CD McIntire, RW Jacobs [eds], Crater Lake Limnological Studies Final Report. *National Park Service Technical Report* [NPS/PNROSU/NRTR-93/03] 1993b.)

Declining lake levels between 1991 and 1994 exposed rock bars along the western and southwestern edges of the lake. The bars were formed from extensive avalanche slides in these areas of the caldera. Approximately 30 ponds (avalanche pits) were found on the bars; these water resources had not been seen in the caldera for approximately 50 years.

Near-Surface Water Temperature

Crater Lake is unique among lakes in the Cascade Mountain Range of Oregon in that its large volume of water retains enough heat to keep the lake surface from freezing in winter. However, ice has been observed on the lake for short periods of time in the winters of 1949, 1985, and 1986. During these three periods the weather was described as clear and extremely cold, while the lake surface was calm. Near-surface water temperatures exhibit considerable seasonal changes, with the highest temperatures occurring in August (Fig. 21.3). However, no obvious long-term changes in near-surface temperatures have been observed during the period from 1895 to 1990 (Fig. 21.4).

Secchi Disk Clarity

The depth at which a Secchi disk disappears is a function of the contrast between the light reflected from the disk and the scattering and absorption of light in the water column between the disk and the surface of a lake. Particles are the

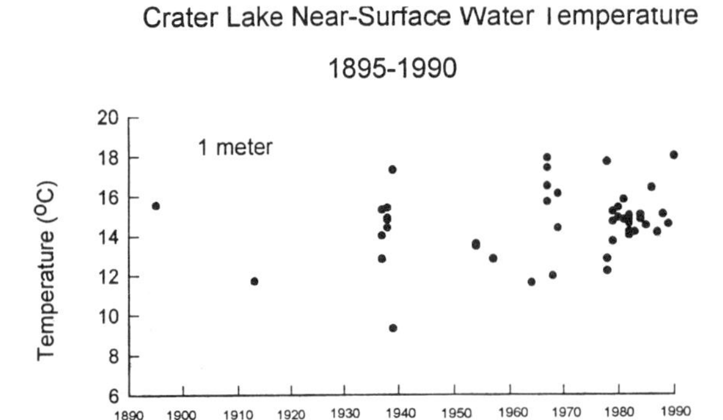

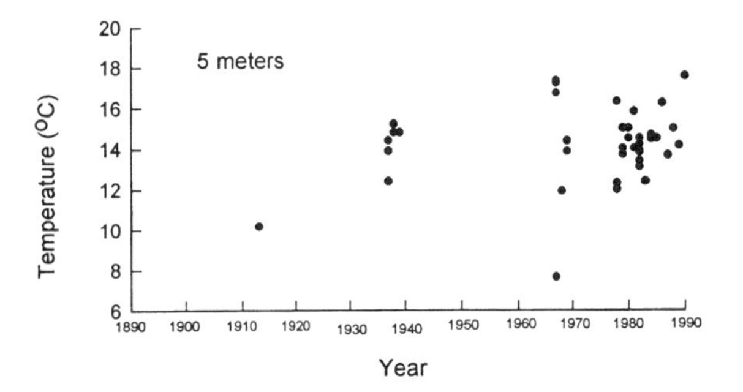

Figure 21.4 Water temperatures in August at depths of 1 meter and 5 meters below the water surface of Crater Lake between 1895 and 1990. (After GL Larson, CD McIntire, RW Jacobs [eds], Crater Lake Limnological Studies Final Report. *National Park Service Technical Report* [NPS/PNROSU/NRTR-93/03] 1993b.)

key to altering this optical contrast in lakes with low concentrations of dissolved and colored substances, as is the case of Crater Lake. The relationship between Secchi disk clarity and particle density is hyperbolic. Thus, a small change in particle density in very clear lakes like Crater Lake will correspond to large increases or decreases in Secchi disk clarity. However, it should be noted that the hyperbolic relationship is actually a family of curves depending on the optical properties of the particle community. Furthermore, Secchi disk clarity readings are influenced by time of day, sky conditions, and lake surface conditions. Under optimal conditions, the sky is clear of clouds at midday and the lake surface is flat and calm.

In Crater Lake, seasonal changes in Secchi disk clarity showed some general patterns, but the patterns were difficult to predict from year to year (Larson & Hurley, 1993). Secchi disk clarity from 1978 to 1991 appeared to be low during the snow avalanche and snowmelt runoff periods of late winter and early spring (Fig. 21.5). Clarity increased as snowmelt runoff subsided and reached its maximum in June and early July. As the lake surface increased in temperature, clarity decreased and was at a minimum in August. Clarity increased in fall owing to dilution of the particle density caused by deepening of the epi-

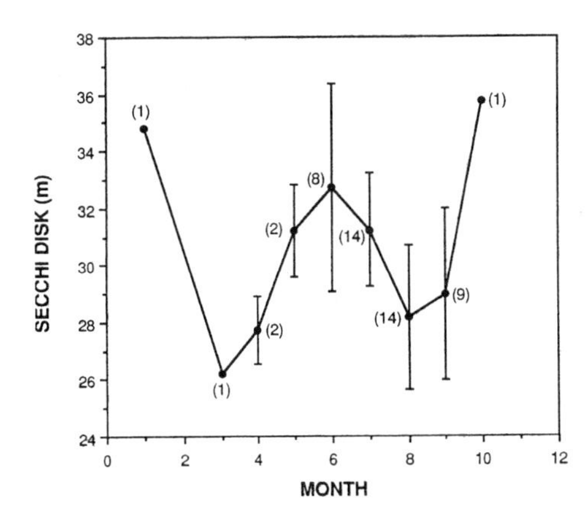

Figure 21.5 Mean Crater Lake Secchi disk readings (+/− 1 standard deviation) per month from 1978 to 1991. Sample sizes are indicated in parentheses. (After GL Larson, M Hurley, Clarity of Crater Lake Measured with a Secchi Disk. In GL Larson, CD McIntire, RW Jacobs [eds], Crater Lake Limnological Studies Final Report. *National Park Service Technical Report* [NPS/PNROSU/NRTR-93/03] 1993:301–316.)

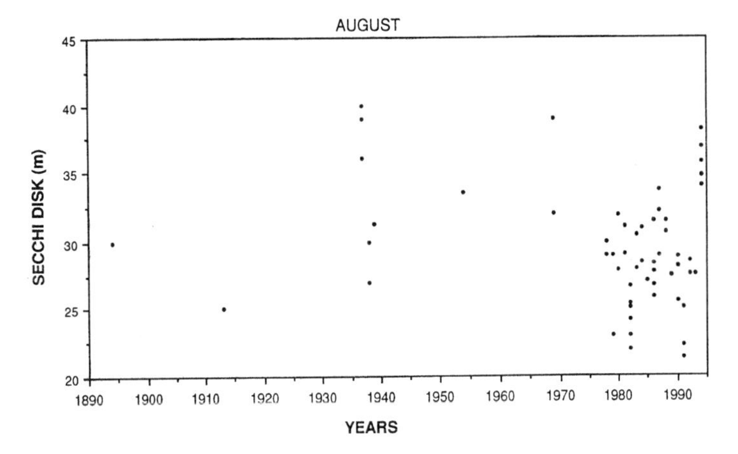

Figure 21.6 Mean daily Crater Lake Secchi disk readings in August from 1896 to 1991. (After GL Larson, M Hurley, Clarity of Crater Lake Measured with a Secchi Disk. In GL Larson, CD McIntire, RW Jacobs [eds], Crater Lake Limnological Studies Final Report. *National Park Service Technical Report* [NPS/PNROSU/NRTR-93/03] 1993:301–316.)

limnion; however, fall storms often were associated with increased densities of abiotic particles from disturbed sediments along the shoreline and from stream runoff, which may explain reduced Secchi disk clarity in the fall of some years (Larson & Hurley, 1993).

Historically, the deepest Secchi disk readings were recorded in August, a fact contrary to the observations made between 1978 and 1993 (Fig. 21.6). The deepest August Secchi disk readings were recorded in 1937 and 1969 and were about 25 percent deeper than those recorded in August from 1978 to 1993. However, the readings in August of 1994 were comparable to those recorded in 1937 and 1969 (see Fig. 21.6), and one reading was a record for the lake at 40.8 meters (20-centimeter disk). Given the sensitivity of the Secchi disk in clear lakes relative to small changes in particle densities and lake surface conditions, it seems likely that the conditions required to obtain readings at 39 meters or more are rare rather than typical of Crater Lake. Arthur Hasler (University of Wisconsin, personal communication, 1985) suggested that a lack of rain during the 1937 field season was responsible for reduced erosion and thus the unusual lake clarity that he recorded. Comparable conditions existed during the 1994 field season. Farner (1939) commented that shallow Secchi disk readings in 1938 were, in his view, due to mud slides in the caldera. Similar qualitative observations of mud flows and turbidity from caldera streams were made during the period from 1982 through 1993, which appeared to correspond to periods of reduced Secchi disk depth. Thus, it seems reasonable to conclude that the Secchi disk clarity of Crater Lake is dynamic relative to both short-term and long-term environmental changes.

Water Chemistry

Water chemistry characteristics were assessed as part of the long-term limnological study. Some of the variables increased in value with lake depth (i.e., total alkalinity, conductivity, nitrate-N, and orthophosphorus-P). Nitrate-N was below detectable concentrations in the upper 200 meters of the water column. Other variables decreased in value with increased lake depth (i.e., pH, total Kjeldahl-N). Total phosphorus was fairly uniform in concentration throughout the water column. In all cases, however, no long-term changes in the values of these water quality variables were detected (Larson & McIntire, 1993).

Total Chlorophyll

The low productivity of Crater Lake is reflected in low concentrations of total chlorophyll. In winter, chlorophyll is uniform in concentration to the depth of mixing of the water column (maximum depth of about 200 meters) by wind energy. During summer and early fall when the lake is thermally stratified, a chlorophyll maximum develops at a depth of 100 to 140 meters. Seasonal patterns are inconsistent in that concentrations were low in winter and high in summer in some years, whereas the reverse was true in other years. Long-term temporal patterns in concentrations of total chlorophyll are shown in Figure 21.7 (McIntire et al., 1993).

Seasonal and long-term patterns of chlorophyll concentrations in Crater Lake probably are controlled in part by changes in nutrient supply, particularly nitrogen. Bulk atmospheric precipitation provides over 90 percent of the ni-

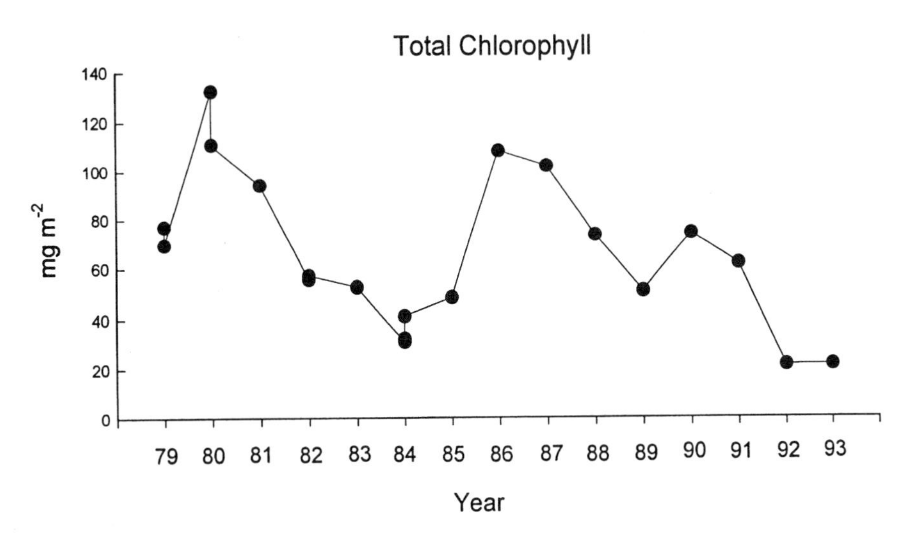

Figure 21.7 Cyclic behavior of total chlorophyll in Crater Lake in August between 1979 and 1990.

trogen entering the lake each year (Nelson et al., 1993). In winter, nutrient inputs can be mixed to a depth of about 200 meters. Also, cold water in the upper portion of the lake can sink into the deep lake under certain climatic conditions, which in turn causes water in the deep lake to upwell in winter and early spring (McManus, 1992). Although the occurrence and extent of the upwelling varies annually, the input of nitrogen from the deep lake into the euphotic zone is several orders of magnitude greater than inputs from bulk precipitation and stream runoff (Dymond et al., 1993). The upwelling of nitrogen-rich water from the deep lake appears to be an important process regulating the long-term patterns in chlorophyll concentrations in Crater Lake.

Zooplankton

Changes in total zooplankton densities and biomasses, integrated to a depth of 200 meters (the maximum depth distribution of the zooplankton), are illustrated in Figure 21.8. Zooplankton densities were highest in 1985 and 1986, whereas zooplankton biomass was highest in July 1985. In some years zooplankton biomass increased during the year; in other years it decreased. The pattern of zooplankton density, however, did not always correspond to the pattern for biomass, indicating annual variations in species composition (Larson et al., 1993c).

The integrated densities (to a depth of 200 meters) of the two dominant crustacean species show that one, *Bosmina longisrostris*, followed a pattern

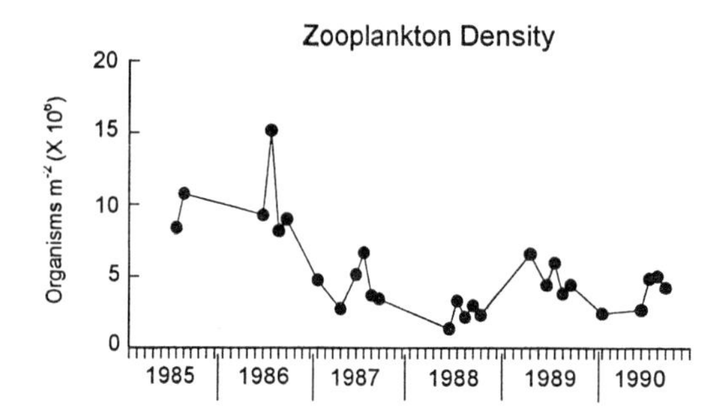

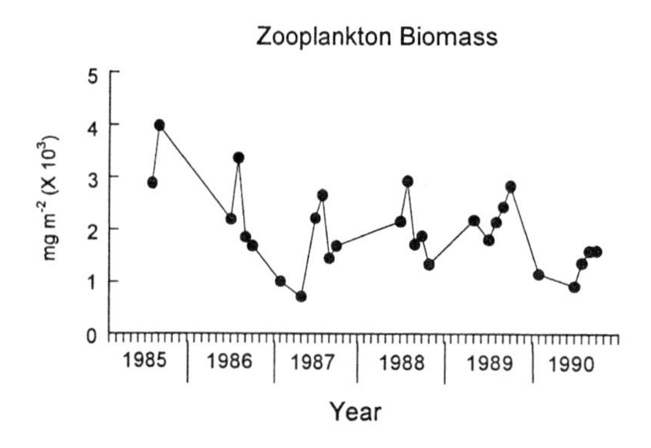

Figure 21.8 Temporal variation in total zooplankton density (**A**) and biomass (**B**) in Crater Lake between July 1985 and September 1990. (After GL Larson, CD McIntire, R Truitt, M Buktenica, E Karnaugh-Thomas, Zooplankton. In GL Larson, CD McIntire, RW Jacobs [eds], Crater Lake Limnological Studies Final Report. *National Park Service Technical Report* [NPS/PNROSU/NRTR-93/03] 1993c:457–558.)

similar to that for total zooplankton biomass, whereas the density of the second, *Daphnia pulicaria,* did not begin to increase until the summer of 1987, and it was only abundant in the water column in 1988 and 1989.

Patterns of integrated density for rotifers were variable. *Keratella quadrata* and *Kellicottia longispina* had density maxima in 1985 and were rare between 1987 and 1990. *Keratella cochlearis* and *Polyarthra dolichoptera* were at maximum densities in 1986; however, the latter species exhibited a more uniform density than the former throughout the study. Density maxima for *Synchaeta oblonga* occurred in 1987, 1989, and 1990. *Filinia terminalis* reached density maxima in 1985, 1986, and 1989.

Early studies of the zooplankton of Crater Lake were limited in scope but suggested that the assemblage probably consisted of one species of *Daphnia,* one species of *Bosmina,* and several rotifer species (Karnaugh, 1988). Specimens of *Daphnia* collected by Evermann (1897), Kemmerer et al. (1924), Brode (1938), Hasler (1938), Hasler and Farner (1942), and Hoffman (1969) were identified as *Daphnia pulex.* Recent examination of specimens, however, suggests that all of the *Daphnia* collected prior to 1985 were actually *D. pulicaria* (Karnaugh, 1988). In addition, specimens of *Bosmina* collected in 1913 by Kemmerer et al. (1924) and identified as *Bosmina longispina* were probably

B. longistrostris (Karnaugh, 1988). Little information was available about rotifer populations in Crater Lake prior to 1985.

Indications are that the zooplankton assemblage in Crater Lake may undergo long-term changes in density and biomass. Most taxa found during the present study were abundant in 1985. The relatively high densities of *D. pulicaria* in 1988 and 1989 may have been related to a decrease in fish predation (see next section) and an increase in lake productivity from nutrient upwelling events in 1986. The mechanisms associated with an increase in the density of *S. oblonga* between 1988 and 1990 were unclear. Year-to-year changes in the density of *Daphnia* were expected because such variations were reported in earlier studies of Crater Lake (Hasler & Farner, 1942; Hoffman, 1969).

Introduced Fish

Although originally barren of fish, about two million rainbow trout (*Oncorhynchus mykiss*), brown trout (*Salmo trutta*), coho salmon (*Oncorhynchus kisutch*), cutthroat trout (*Salmo clarki*), and steelhead (*Oncorhynchus mykiss*) were stocked into the lake between 1888 and 1941 (Buktenica, 1989). Wallis and Bond (1950) examined fish samples collected in 1939 and 1947 at Crater Lake and identified six kokanee salmon (*Oncorhynchus nerka*). Apparently, kokanee salmon were either inadvertently mixed with other fish that were stocked into the lake or were incorrectly identified as coho salmon. Based on samples collected between 1986 and 1994, rainbow trout and kokanee salmon are the only species of fish found in the lake.

The biomass of the kokanee salmon was relatively low between 1985 and 1987 (Fig. 21.9). The population was dominated numerically by one-year class. In January 1988, most of this year-class spawned and died. Kokanee biomass rose through 1990 and then declined in 1991 (Buktenica & Larson, 1993). Biomass of kokanee salmon continued to decline between 1992 and 1994 (M. Buktenica, personal communication).

The rainbow trout population was composed of multiple age classes and size classes between 1985 and 1994. The relative abundance of older age classes and larger fish increased from 1986 to 1991. The increase in these population characters corresponded to an increase in the relative abundance of kokanee salmon (Buktenica & Larson, 1993). These data suggest that the average length and perhaps the maximum length of trout in Crater Lake may be dependent on the availability of kokanee salmon as part of the prey base for the trout. As the abundance of kokanee salmon is cyclic in Crater Lake, length-frequency trends for rainbow trout may also be cyclic. Previous investigations at Crater Lake found

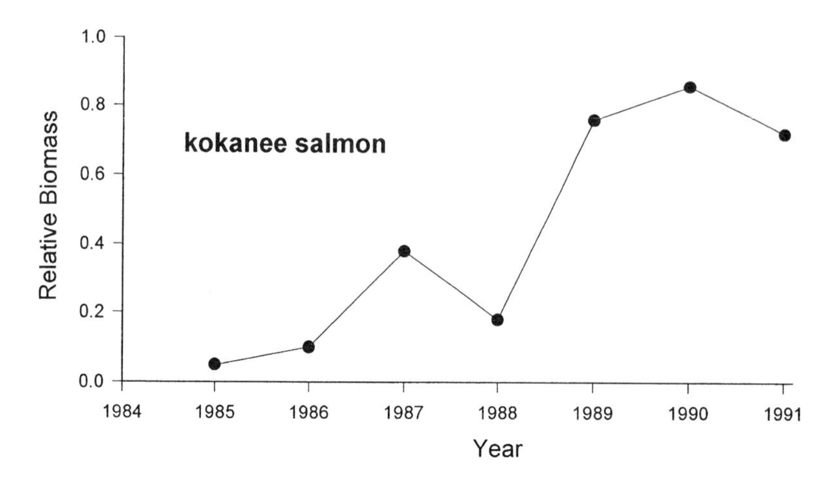

Figure 21.9 Relative biomass of kokanee salmon in Crater Lake between 1985 and 1991. (After M Buktenica, GL Larson, Ecology of Kokanee Salmon and Rainbow Trout in Crater Lake. In GL Larson, CD McIntire, RW Jacobs [eds], Crater Lake Limnological Studies Final Report. *National Park Service Technical Report* [NPS/PNROSU/NRTR-93/03] 1993:559–618.)

the growth and average size of rainbow trout and coho salmon to be greater than those found in the present study (Hasler, 1938; Hasler & Farner, 1942).

Conclusions

The results of the limnological study conducted between 1983 and 1992 when compared to data collected before 1983, clearly demonstrated that some of the components measured do not exhibit much long-term variation, while other components exhibit considerable long-term change. However, none of the components exhibited the long-term unidirectional or exponential changes that are usually associated with increased lake productivity (Goldman, 1988). With the exception of impacts caused by introduced fish, the lake was judged to be pristine within the limits of the methods of study.

The results of the ten-year limnological study also demonstrated the difficulty of describing "normal" conditions of many components of the system. Large variations in water level, Secchi disk clarity, total chlorophyll, and abundances of zooplankton species were apparent. Defining "normal" conditions of the lake system, therefore, will require studies over even longer time periods to define the range of variability of components of the system. Similar arguments have been voiced by Goldman (1988) in long-term limnological studies of Lake Tahoe. For example, during the period from 1958 to 1988, Secchi disk clarity of Lake Tahoe decreased 10 meters (0.4 meter/year) in response to human-related activities in the watershed (Goldman, 1988). Goldman (1988) demonstrated that at least five consecutive years of data were required to obtain statistically significant declines in Secchi disk readings. Goldman et al. (1989) also showed that long-term climatic cycles corresponded well to observed 30-year interannual changes in the primary productivity in Castle Lake, a subalpine lake in northern California.

These examples support the conclusion that determining "normal" conditions of lakes requires extensive data sets (Likens, 1983, 1985; Schindler et al., 1990; Catalan & Fee, 1994). Short-term studies may be important to assess particular components and processes of lake systems, but long-term data sets provide a context within which limnologists can document the dynamics of lake systems in pristine condition, such as Crater Lake and Castle Lake, or lakes increasing in productivity from human-related activities, such as Lake Tahoe. For these reasons, an ongoing limnological study of Crater Lake was initiated in 1993.

Long-term limnological investigations of pristine lakes in protected watersheds are essential to provide a data base from which to address changing conditions caused by natural phenomena and environmental impacts from distant human-related activities, such as degraded air quality and global climate change. Nonetheless, long-term limnological monitoring should be an adaptive process. The growing base of knowledge about a lake should be regularly reviewed, models that describe the structure and functioning of the lake should be revised based on this knowledge, and new hypotheses should be generated for future studies relative to this knowledge and relative to management priorities.

In general, a dynamic, adaptive monitoring program ensures that long-term monitoring progresses in an orderly fashion, in accordance with human understanding of the lake system, and of the value of the lake to society. Furthermore, developing an understanding of those components and processes that are common to lake systems will provide a basis for comparing and assessing the status and trends in lakes elsewhere, especially those in watersheds already impacted by human-related activities.

Acknowledgments

I express my thanks to the staffs of Crater Lake National Park and National Park Service Pacific Northwest Regional Office for their support of the limnological studies of Crater Lake. In particular, I thank the following for their many contributions to this paper—M. Buktenica, R. Collier, J. Dymond, C. R. Goldman, R. W. Jacobs, D. W. Larson, S. Loeb, C. D. McIntire, P. O. Nelson, K. Redmond, and R. Truitt.

Literature Cited

Bacon CR, Lanphere MA. The Geological Setting of Crater Lake, Oregon. In ET Drake, GL Larson, J Dymond, R Collier (eds), *Crater Lake: An Ecosystem Study.* San Francisco, CA: American Association for the Advancement of Science, Pacific Division, 1990:19–27.

Brode JS. The denizens of Crater Lake. *Northwest Sci* 1938;12:50–57.

Buktenica M. Ecology of Kokanee Salmon and Rainbow Trout in Crater Lake, A Deep Ultraoligotrophic Caldera Lake (Oregon) (M.S. thesis). Corvallis, OR: Oregon State University, Department of Fisheries and Wildlife, 1989.

Buktenica M, Larson GL. Ecology of Kokanee Salmon and Rainbow Trout in Crater Lake. In GL Larson, CD McIntire, RW Jacobs (eds), Crater Lake Limnological Studies Final Report. *National Park Service Technical Report* (NPS/PNROSU/NRTR-93/03) 1993:559–618.

Byrne JV. Morphology of Crater Lake. *Limnol Oceanogr* 1965;19:462–465.

Catalan J, Fee EJ. Interannual Variability in Limnic Ecosystems: Origins, Patterns, and Predictability. In R Margalef (ed), *Limnology Now: A Paradigm of Planetary Problems.* The Netherlands: Elsevier Science, 1994:81–97.

Dymond J, Collier R, Larson GL. Particle Flux Measurements. In GL Larson, CD McIntire, RW Jacobs (eds), Crater Lake Limnological Studies Final Report. *National Park Service Technical Report* (NPS/PNROSU/NRTR-93/03) 1993:215–268.

Evermann BW. United States Fish Commission investigations at Crater Lake. *Mazama* 1897;1:230–238.

Farner DS. *Annual Report, Crater Lake National Park.* Center Lake, OR: National Park Service, 1939.

Goldman CR. Primary productivity, nutrients, and transparency during the early onset of eutrophication in ultra-oligotrophic Lake Tahoe, California-Nevada. *Limnol Oceanog* 1988;33(6, pt 1):1321–1333.

Goldman CR, Jassby AZ, Powell T. Interannual fluctuations in primary production: Meteorological forcing at two subalpine lakes. *Limnol Oceanog* 1989;34:310–323.

Hasler AD. Fish biology and limnology of Crater Lake. *J Wildl Manage* 1938;2:94–103.

Hasler AD, Farner DS. Fisheries investigations in Crater Lake. *J Wildl Manage* 1942;6:319–327.

Hoffman FO. The Horizontal Distribution and Vertical Migrations of the Limnetic Zooplankton in Crater Lake, Oregon (M.S. thesis). Corvallis, OR: Oregon State University, Department of Fisheries and Wildlife, 1969.

Karnaugh EN. Structure, Abundance, and Distribution of Pelagic Zooplankton in Crater Lake, Oregon (M.S. thesis). Corvallis, OR: Oregon State University, Department of Fisheries and Wildlife, 1988.

Kemmerer GJ, Bovard JF, Boorman WR. Northwest lakes of the United States: Biological and chemical studies with reference to possibilities in production of fish. *Bull US Bur Fish* 1924;39:51–140.

Larson GL, Hurley M. Clarity of Crater Lake Measured with a Secchi Disk. In GL Larson, CD McIntire, RW Jacobs (eds), Crater Lake Limnological Studies Final Report. *National Park Service Technical Report* (NPS/PNROSU/NRTR-93/03) 1993:301–316.

Larson GL, McIntire CD. Temperature and Water Quality. In GL Larson, CD McIntire, RW Jacobs (eds), Crater Lake Limnological Studies Final Report. *National Park Service Technical Report* (NPS/PNROSU/NRTR-93/03) 1993:99–129.

Larson GL, McIntire CD, Buktenica M. Introduction to the 10-Year Study. In GL Larson, CD McIntire, RW Jacobs (eds), Crater Lake Limnological Studies Final Report. *National Park Service Technical Report* (NPS/PNROSU/NRTR-93/03) 1993a:11–16.

Larson GL, McIntire CD, Jacobs RW (eds). Crater Lake Limnological Studies Final Report. *National Park Service Technical Report* (NPS/PNROSU/NRTR-93/03) 1993b.

Larson GL, McIntire CD, Truitt R, Buktenica M, Karnaugh-Thomas E. Zooplankton. In GL Larson, CD McIntire, RW Jacobs (eds), Crater Lake Limnological Studies Final Report. *National Park Service Technical Report* (NPS/PNROSU/NRTR-93/03) 1993c:457–558.

Likens GE. A priority for ecological research. *Bull Ecol Soc Am* 1983;64:234–243.

Likens GE (ed). *An Ecosystem Approach to Aquatic Ecology: Mirror Lake and Its Environment.* New York: Springer-Verlag, 1985.

McIntire CD, Larson GL, Truitt R, Buktenica M, Debacon M. Phytoplankton. In GL Larson, CD McIntire, RW Jacobs (eds), Crater Lake Limnological Studies Final Report. *National Park Service Technical Report* (NPS/PNROSU/NRTR-93/03) 1993:359–456.

McManus J. On the Chemical and Physical Limnology of Crater Lake, Oregon (Ph.D. thesis). Corvallis, OR: Oregon State University, College of Oceanic and Atmospheric Sciences, 1992.

Nelson PO, Reilly JF, Larson GL. Chemical Solute Mass Balance of Crater Lake, Oregon. In GL Larson, CD McIntire, RW Jacobs (eds), Crater Lake Limnological Studies Final Report. *National Park Service Technical Report* (NPS/PNROSU/NRTR-93/03) 1993:269–300.

Phillips KN, Van Denburg AS. Hydrology of Crater, East and Davis Lakes, Oregon. *US Geological Survey Water Supply Paper* (1859-E) 1968.

Redmond K. Climate Variability at Crater Lake National Park and Its Effect on Water Level. In GL Larson, CD McIntire, RW Jacobs (eds), Crater Lake Limnological Studies Final Report. *National Park Service Technical Report* (NPS/PNROSU/NRTR-93/03) 1993:39–61.

Schindler DW, et al. Effects of climatic warming on lakes of the central boreal forest. *Science* 1990;250:967–970.

Wallis OL, Bond CE. Establishment of kokanee in Crater Lake, Oregon. *J Wildl Manage* 1950;14:190–193.

Index